PHYSICAL CHEMISTRY
OF FAST REACTIONS

Volume 1

Gas Phase Reactions
of Small Molecules

PHYSICAL CHEMISTRY OF FAST REACTIONS

Edited by

B. P. Levitt

Department of Chemistry
Imperial College of Science and Technology
University of London
London SW7 2AY

Volume 1
Gas Phase Reactions
of Small Molecules

PLENUM PRESS • LONDON AND NEW YORK • 1973

Plenum Publishing Company Ltd
Davis House
8 Scrubs Lane
London NW10 6SE
Telephone 01-969 4727

U.S. Edition published by
Plenum Publishing Corporation
227 West 17th Street
New York, New York 10011

ISBN-13: 978-1-4684-2687-8   e-ISBN-13: 978-1-4684-2685-4
DOI: 10.1007/978-1-4684-2685-4
Library of Congress Catalog Card Number 74-161304

Contributors

M. A. A. Clyne

Department of Chemistry, Queen Mary College, University of London, Mile End Road, London E1 4NS, England.

R. W. Getzinger

Los Alamos Scientific Laboratory, University of California, Los Alamos, New Mexico 87544, U.S.A.

F. M. Page

Department of Chemistry, University of Aston in Birmingham, Gosta Green, Birmingham 4, England.

G. L. Schott

Los Alamos Scientific Laboratory, University of California, Los Alamos New Mexico 87544, U.S.A.

J. Troe

Institut de Chemie-Physique, Ecole Polytechnique de Lausanne, Avenue des Bains 31, 1007 Lausanne, Switzerland.

H. Gg. Wagner

Institüt für Physikalische Chemie der Universität Göttingen, Burgerstrasse 50, 34 Göttingen, West Germany.

v

Introduction

The chapters in this book are devoted to the elementary reactions of small molecules in the gas phase, with some emphasis on reactions important in combustion. The first three chapters cover experimental measurements made at high temperatures, mainly using shock waves and flames; the final chapter describes discharge flow methods near room temperature. The authors—all active in the fields they describe—were asked to aim at a level intermediate between a textbook and a review, designed for readers not already familiar with this branch of chemical kinetics. We hope the book will prove especially useful to research workers in related subjects, to research students, and perhaps as source material for the preparation of lectures. The examples have been chosen to illustrate the theoretical basis of the topics rather than attempt a complete coverage.

Professors Wagner and Troe describe the remarkable progress made in recent years in measuring dissociation rates for small molecules. Tests of unimolecular reaction theories are usually made in the 'fall-off' region of pressure: the kinetics change from first order to second order as the pressure is reduced. For large molecules this region lies below atmospheric pressure and is relatively easily accessible. For molecules with four or less atoms, however, the fall-off region lies well above atmospheric pressure: it has been explored using the high pressure shock tube techniques developed by the authors. They describe current theoretical treatments of unimolecular rates: these are difficult to apply in detail even for diatomic molecules. It seems that it may not be possible to predict the rate from the molecular structure until the rates of the energy transfer processes which control the populations of vibrational and rotational levels are also fully understood.

The deceptively simple stoichiometry and the importance of the elements concerned led to much early investigation of the chain reaction between hydrogen and oxygen, mostly in static systems at moderate temperatures. Unfortunately the main kinetic features of the reaction under these conditions are controlled by heterogenous steps occurring at the walls of the reaction vessel: it is not possible to observe the separation of the kinetics into zones of initiation, exponential growth of chain centres, and heat release by recombination. However this reaction has been extensively investigated in the shock tube by Dr

Getzinger and Dr Schott: here the reaction is strictly homogenous, and the zones can be studied in detail, mainly by following the OH radical concentration. The authors describe the results of these experiments and relate them to those obtained by other techniques, e.g. ignition limit measurements. The reduction in the number of molecules as reaction proceeds is reflected in kinetics atypical of oxidation reactions in general.

In flames, the bulk of the oxidation occurs in a narrow zone where the rate of reaction is controlled not only by the kinetics but by the flow and by heat transfer. Immediately above this, the concentration of atoms and free radicals is greatly in excess of its equilibrium value. The region of burnt gas, at nearly constant temperature and pressure, is long enough in space and time to enable measurements of recombination rates to be made, and also studies of the reactions of free radicals with various metals added to the flame, and the resultant ionization. In the third chapter, Professor Page describes both the structure of flames and the kinetics of these processes which can be studied in the reaction zone.

When an electric discharge is passed through a cold diatomic gas at low pressure it is partially dissociated into atoms: in this way reasonable concentrations of O, H, D, N, halogen or other atoms can be produced in a chemically inert diluent. The recombination of these atoms, and their reaction with other molecules can be observed as the gas flows down a long tube. Many of the reactions produce molecules in excited electronic states: the resulting chemiluminescence can be used to measure the concentration of atomic species as a function of distance, and hence time, down the tube. Dr Clyne describes this important technique, which has produced direct measurements of the rates of many exothermic reactions of atoms and free radicals at room temperature and below. The reverse of the recombination steps are, of course, the dissociation reactions whose kinetics at high temperatures were described in the first chapter: if the ratio of forward and reverse rate constants is equal to the equilibrium constant, the temperature dependence of these rates can be deduced over very wide ranges of temperature.

September 1972   *B. J. Levitt*

Contents

Contents

Chapter 1

Unimolecular Dissociation
of Small Molecules

J. Troe and H. Gg. Wagner

Institut de Chimie-Physique de l'Ecole Polytechnique
Fédérale de Lausanne, Switzerland, and
Institut für Physikalische Chemie der Universität Göttingen,
Germany

1.1. INTRODUCTION

The dissociation of molecules is one of the basic processes in chemistry; the study of the kinetics of these reactions is therefore of considerable theoretical and practical interest. A simple method of obtaining information about dissociation reactions is to heat the gas to a sufficiently high temperature and then look for thermal decomposition. However for rich mixtures bimolecular reactions may well contribute to the reaction: their influence must be separated out so that the unimolecular dissociation can be isolated. The rate of the primary dissociation is determined by elementary physical processes including both energy transfer between particles and internal energy flow. Dissociation reactions, isomerisation processes, photolytic reactions, dissociation of ions (e.g. in a mass spectrometer) and chemical activation experiments are closely related processes.

Special experimental techniques are required to investigate thermal unimolecular reactions under well defined conditions. This is briefly illustrated by the following example. In order to obtain 10% dissociation in pure hydrogen at equilibrium and 1 atm, one has to heat the gas to 3080 K. At this temperature the half life of H_2 is about 2 ms. 50% dissociation is obtained at 3830 K, where the half life is only 70 μs. Under these conditions the dissociation can well be termed a "fast reaction". To study it we must use experimental techniques which permit fast and well defined heating of the gas and provide for rapid and sensitive detection of the decomposing molecules and dissociation products. For hydrogen and other diatomic molecules these requirements can only be met at present by using shock waves. Any description of the experimental methods used to study thermal dissociation reactions, particularly of small molecules, will be largely concerned with this technique.

Other methods, e.g. static systems, flow systems and adiabatic compression, are restricted to observations at lower reaction rates; these methods are useful only for the dissociation reactions of relatively unstable small molecules, but become more important as the complexity of the molecule increases.

For details of the shock tube technique we refer the reader to the extensive literature.[1] New aspects of instrumentation are included in the article of Getzinger and Schott in this book.[2] A recent discussion of real shock wave behaviour and its consequences for dissociation rate studies is particularly important.[3]

In this article we restrict ourselves mainly to the chemical and theoretical aspects of dissociation rates and to the experimental results obtained for small molecules. In section 1.2 we discuss decomposition mechanisms in general. These are illustrated by some experimental examples in sections 1.3–1.5: because of the close relation between dissociation and recombination, results for the corresponding recombination reactions are also given. Individual physical processes contributing to dissociation rates are then analysed; a brief description of the simplest dissociation models is given in section 1.2. Quantitative understanding of dissociation and recombination rates, however, requires detailed analysis of "microscopic" physical processes and statistical laws. These are considered in sections 1.6–1.8. Here the relationship between dissociation and the other systems mentioned above, all of which involve similar microscopic processes, can be seen.

Experimental results are systematically presented only for molecules with 2–4 atoms. For these systems an explicit application of the theories appears to be most promising. However, the results are also immediately applicable to more complex molecules.

1.2. GENERAL PROPERTIES OF DECOMPOSITION REACTIONS

1.2.1. Decomposition Mechanism

We assume that a number of dissociation experiments have been performed at temperature T, pressure p and carrier gas concentration [M]. The rates of disappearance of the decomposing substance A and/ or the rates of appearance of decomposition products or intermediates have been measured at different [A] to [M] ratios. The first problem of interpretation is to understand the decomposition mechanism. At high enough temperature and sufficiently large excess of inert carrier gas M, the rate determining step of the dissociation is the unimolecular dissociation of the molecule A:

$$A + M \longrightarrow B + C + M \qquad (1.1)$$

In principle, with decreasing [A] at constant [M], every decomposition reaction must become unimolecular. In practice, problems of detection sensitivity and of the impurity level of the carrier gas M soon arise. Therefore, subsequent reactions must often be taken into account, e.g.

$$B + A \longrightarrow \text{products} \qquad (1.2)$$

The rates of these atom or radical reactions (1.2) must be known to interpret the decay of [A].

Information on reactions such as (1.2) is often now available from studies of isolated fast atom or radical reactions, for example from discharge-flow experiments. An extensive review of recent work is given in ref. 4. In many cases the required information on subsequent reactions can also be derived directly from the dissociation studies. A number of examples are given in sections 1.4 and 1.5. In some systems chain reactions initiated by B or C as chain centres are observed. Sometimes the unimolecular dissociation of A cannot be separated from its bimolecular reaction. Finally, some systems are extraordinarily sensitive to impurities like O_2 or H_2O. Examples of these complications will be considered in sections 1.4 and 1.5.

1.2.2. Primary Dissociation Step

After analysing the decomposition mechanism, the kinetics of the primary unimolecular dissociation step (1.1) can be investigated, to find how its rate depends on T, p and the concentration and nature of M. The quantity of primary interest to the experimentalist is the first-order rate constant k, defined by

$$k \equiv - \frac{1}{[A]} \frac{d[A]}{dt} \qquad (1.3)$$

The general nature of the elementary physical processes which contribute to the overall chemical elementary reaction step (1.1) is well known. Collisional activation processes are described symbolically by the equation

$$A + M \xrightarrow{k_1} A^* + M \qquad (1.4)$$

and collisional deactivation processes by

$$A^* + M \xrightarrow{k_2} A + M \qquad (1.5)$$

These steps couple the reacting system A with the heat reservoir M by intermolecular energy transfer. Molecules A, which have gained sufficient energy through collisional energy transfer, are denoted by "A^*". These can decompose by intramolecular processes:

$$A^* \xrightarrow{k_3} B + C \qquad (1.6)$$

Before going into more detail and before specifying the individual molecular energy levels, we discuss how k depends on k_1, k_2 and k_3. With quasistationary [A*] so that $d[A^*]/dt \approx 0$, the overall rate constant of the unimolecular dissociation (1.1) is

$$k = k_1[M] \cdot \left(\frac{k_3}{k_3 + k_2[M]} \right) \qquad (1.7)$$

The first factor $k_1[M]$ gives the rate constant for the collisional activation which initiates unimolecular reaction. The second factor in brackets is the fraction of A* which then reacts by step (1.6). As processes (1.5) and (1.6) compete, this fraction may be less than unity. At low pressures collisional deactivation is much rarer than reaction, $k_2[M] \ll k_3$, the fraction of dissociated A* becomes 1 and

$$k([M] \longrightarrow 0) \equiv k_0 = k_1[M] \tag{1.8}$$

At high pressures collisional deactivation decreases the fraction of A* which dissociates, $k_2[M] \gg k_3$, and k becomes

$$k([M] \longrightarrow \infty) \equiv k_\infty = \frac{k_1}{k_2} \cdot k_3 \tag{1.9}$$

At $[M]_{1/2}$, defined by $k_2[M]_{1/2} = k_3$, the fraction of A* which dissociates is $0 \cdot 5$, and $k = 0 \cdot 5 k_0$ for

$$[M] = [M]_{1/2} = \frac{k_3}{k_2} \tag{1.10}$$

Equation (1.7) describes the change in reaction order with pressure at constant temperature observed for unimolecular reactions. At $k_2[M] \ll k_3$, k is proportional to $[M]$, so in the low pressure limit the unimolecular reaction is second-order. At $[M] \approx [M]_{1/2}$ the reaction order falls, and at the high pressure limit k becomes independent of $[M]$; the reaction order then is unity. This behaviour corresponds to the transition from collisional activation to intramolecular dissociation as the rate determining step. The "fall-off" of k at low pressures, according to Eq. (1.7), is illustrated in Fig. 1.1. The temperature dependence

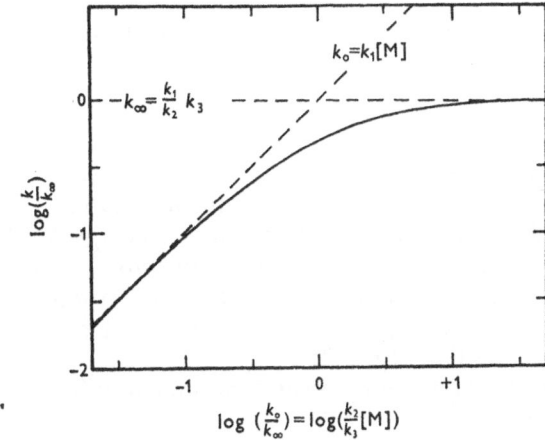

FIG. 1.1. Fall-off curve for unimolecular dissociation: simple model, equation (1.7).

of k at constant pressure or constant $[M]$ will be different for the different pressure regions. At low pressures, the temperature dependence will be that of k_1; at the high pressure limit that of $(k_1/k_2)k_3$. A more detailed discussion of the temperature dependence will be given after specifying the individual energy levels of A and A*.

In the simple mechanism (steps (1.4), (1.5) and (1.6)), the rate constant k should be dependent on the nature of the collision partner M only in the low pressure region and to a smaller extent in the transition region between low and high pressures. Different collision partners have different energy transfer efficiencies and, therefore, different values for k_1. At the high pressure limit the influence of different collisional partners cancels out on this simple mechanism, because k_1/k_2 is independent of M. This follows from the principle of detailed balancing, by which

$$k_1[A]_{equ} = k_2[A^*]_{equ} \qquad (1.11)$$

where $[A^*]_{equ}/[A]_{equ}$ gives the relative populations of A* and A at equilibrium. (k_1/k_2 is independent of the departure from equilibrium under most experimental conditions (see sections 1.6 and 1.7).)

If dissociation is investigated without dilution by inert carrier gas, the rate "constant" may change with time at low pressures during reaction. As A dissociates the composition of collisional partners will change and so hence the effective value of k_1 will change with time.

1.2.3. Relation between Dissociation and Recombination

In the reverse of a dissociation process, i.e. recombination, the same elementary processes are involved. The reaction starts with the association of the atoms and/or radicals formed by reaction (1.6):

$$B + C \rightleftarrows A^* \qquad (1.12)$$

This step is followed by sequences of collisional processes which deactivate the highly excited molecules:

$$A^* + M \longrightarrow A + M \qquad (1.13)$$

Activation and subsequent redissociation are also possible. Owing to the different starting points (molecules A in dissociation, B + C in recombination) different populations of internal energy levels occur during dissociation and recombination. It is, therefore, an interesting question whether the overall rate constants of dissociation and recombination analogous to equation (1.11) are related by the equilibrium constants (see sections 1.6–1.8).

As in dissociation, at different pressures the second-order recombination rate constant

$$k_{rec} \equiv + \frac{1}{[B][C]} \frac{d[A]}{dt} \qquad (1.14)$$

contains information on different elementary processes. On the simple model using the reverse reactions of (1.4), (1.5) and (1.6),

$$k_{rec} = k_{-3}\left(\frac{k_2[M]}{k_3 + k_2[M]}\right) \tag{1.15}$$

The first factor k_{-3} is the second-order rate constant of association of B and C forming A*, and the second factor in brackets is the fraction of A* which is deactivated at different pressures. At low pressures, $k_2[M] \ll k_3$, and the reaction becomes third-order, with deactivation rate determining:

$$k_{rec_0} = \frac{k_{-3}}{k_3} k_2[M] \tag{1.16}$$

At high pressures, $k_2[M] \gg k_3$, and the reaction is second-order:

$$k_{rec_\infty} = k_{-3} \tag{1.17}$$

Here association is the rate determining step.

Up to now only the simple "energy-transfer mechanism" for dissociation or recombination has been considered. However, the mechanism may well be different. If radiationless transitions between different electronic states are involved, reaction may proceed in several stages with rate determining steps which are different under different conditions:

$$A \underset{\xleftarrow{}}{\xrightarrow{+M}} A^* \underset{\xleftarrow{}}{\xrightarrow{}} A_{exc} \underset{\xleftarrow{}}{\xrightarrow{+M}} A_{exc}^* \underset{\xleftarrow{}}{\xrightarrow{}} B + C \tag{1.18}$$

A_{exc} indicates an electronically excited state.

Recombination reactions are most conveniently studied at low temperatures where mechanisms involving unstable intermediates may be important. In the "complex mechanism"

$$B + M(+M) \underset{\xleftarrow{}}{\xrightarrow{}} BM(+M)$$
$$BM + C \longrightarrow BC + M \tag{1.19}$$

an unstable BM is formed, which goes over to BC in a bimolecular exchange reaction. A similar mechanism has been proposed for H_2 dissociation with M = H (see section 1.3):

$$H_aH_b + H_c \longrightarrow H_a + H_bH_c^*$$
$$H_bH_c^* \longrightarrow H_b + H_c \tag{1.20}$$

The activation step is an exchange reaction which may behave quite differently from normal collisional energy transfer.

The overall behaviour of experimentally determined dissociation and recombination rates, e.g. their pressure and temperature dependence, may be presented in terms of the simple formulae derived in this section. However a quantitative understanding requires the specification

TABLE 1.1

Comparison of dissociation and recombination rate constants for O₂, N₂ and H₂ obtained from both dissociation and atom recombination experiments using equation (1.26) (see text).

	Study	D_0^0 [kcal mol⁻¹]	T [K]	$k/[\mathrm{Ar}][\mathrm{cm^3\,mol^{-1}\,s^{-1}}]$, E_a in cal mol⁻¹	$k_{rec}/[\mathrm{Ar}][\mathrm{cm^6\,mol^{-2}\,s^{-1}}]$	Ref.
O₂	Diss.	118·0	5000–	$1·7 \times 10^{14} \exp(-110\,000/RT)$		
			18 000	$2·9 \times 10^{14}\left(\dfrac{T}{7800}\right)^{-0·5} \exp(-D_0^0/RT)$	$2·0 \times 10^{13}\left(\dfrac{T}{7800}\right)^{+0·2}$	9a
	Rec.		2000 (1500–3000)	$5·4 \times 10^{14} \exp(-D_0^0/RT)$	$2·5 \times 10^{13}$	9a
	Rec.		196–327	$5·1 \times 10^{15}\left(\dfrac{T}{298}\right)^{-1·7} \exp(-D_0^0/RT)$	$6·0 \times 10^{14}\left(\dfrac{T}{298}\right)^{-2·9}$	18a
N₂	Diss.	225	8000–	$1·1 \times 10^{14} \exp(-192\,000/RT)$		
			15 000	$5·2 \times 10^{14}\left(\dfrac{T}{10\,400}\right)^{-1·6} \exp(-D_0^0/RT)$	$3·2 \times 10^{13}\left(\dfrac{T}{10\,400}\right)^{-1·6}$	10a
	Rec.		196–327	$1·5 \times 10^{16}\left(\dfrac{T}{298}\right)^{-1·2} \exp(-D_0^0/RT)$	$1·4 \times 10^{15}\left(\dfrac{T}{298}\right)^{-2·1}$	19a
H₂	Diss.	103·3	2300–3800	$1·9 \times 10^{14} \exp(-95\,500/RT)$		
				$7·8 \times 10^{14}\left(\dfrac{T}{2870}\right)^{-1·4} \exp(-D_0^0/RT)$	$2·3 \times 10^{14}\left(\dfrac{T}{2870}\right)^{-1·2}$	7a
	Rec.		190–350	$3·3 \times 10^{15}\left(\dfrac{T}{298}\right)^{+0·5} \exp(-D_0^0/RT)$	$2·3 \times 10^{15}\left(\dfrac{T}{298}\right)^{-0·5}$	17a

of individual quantum states and of transitions between these states. This is discussed in sections 1.6–1.8.

1.3. Experimental Studies of Dissociation–Recombination Kinetics for Diatomic Molecules

1.3.1. Homonuclear Diatomic Molecules

(a) *Dissociation of* H_2, D_2, O_2 *and* N_2

Dissociation of these molecules in the second-order region has been studied by several authors using shock waves. Temperatures of several thousand degrees K have been used, e.g. for H_2 2300–5300 K, for O_2 2500–18 000 K, and for N_2 6000–15 000 K.

A useful measure of temperature is the ratio D_0°/RT, where D_0° is the bond energy at 0 K. If $D_0^{\circ}/RT < 10$, coupling of vibrational relaxation of lowest levels and dissociation become important during the main course of reaction; if $D_0^{\circ}/RT \geq 10$, vibrational relaxation is nearly complete before appreciable dissociation occurs. $D_0^{\circ}/RT = 3 \cdot 3 - 23 \cdot 5$ for O_2 and $7 \cdot 5 - 18 \cdot 8$ for N_2, so in these cases this coupling must be taken into account: it was detected experimentally by a time dependent "rate constant" k with an incubation time in the early period of reaction.[5] In most other studies this effect could be neglected.

Dissociation experiments have generally been performed with the rare gases (usually Ar), with the dissociating molecules (X_2) or with the atoms produced (X) as collision partners (M). The rate constants normally have been found to increase with changing M in the order $k(\text{Ar}) < k(X_2) < k(X)$. Numerical values are of the order $1:10:30$ for O_2 dissociation, $1:4:30$ for H_2 dissociation and $1:3:20$ for N_2 dissociation. The agreement between different studies is best for M = Ar, now often better than a factor of 2. Accuracy appears to be somewhat worse for M = X_2 and X. A systematic comparison of literature data up to 1967 is given in ref. 6; later references up to the end of 1969 are included in this bibliography for H_2,[7] D_2,[8] O_2[9] and N_2.[10] Only some representative data chosen somewhat arbitrarily are discussed below.

In order to compare different aspects of rate constants, different forms of the temperature dependence have been used: (i) the simple Arrhenius equation

$$k = A \exp\left(-E_a/RT\right) \tag{1.21}$$

and (ii) an expression of the form

$$k = A'(T/\bar{T})^n \exp\left(-D_0^{\circ}/RT\right) \tag{1.22}$$

The first expression describes the observed temperature dependence adequately for most experiments. However, as small errors in E_a introduce a large uncertainty in the pre-exponential factor A, values of A and absolute values of k are better compared by the second expression. Here the spectroscopic value of D_0° is used. The temperature coefficients of the expressions (1.21) and (1.22) are related by

$$E_a = D_0^\circ + nR\bar{T} \tag{1.23}$$

$\bar{T}$ is the mean temperature with respect to a $\log k - 1/T$ plot: if the extreme experimental temperatures are T_a, T_b,

$$\bar{T} = [\tfrac{1}{2}(T_a^{-1} + T_b^{-1})]^{-1} \tag{1.24}$$

Expressions (1.21) and (1.22) are used in the comparison of rate data given in Tables 1.1 and 1.2 for different homonuclear diatomic molecules. In all cases the E_a values were found to be less than D_0°. Values of n are in the order of -0.5 to -2; n is unlikely to remain constant over large ranges of temperature. This becomes even more evident for the

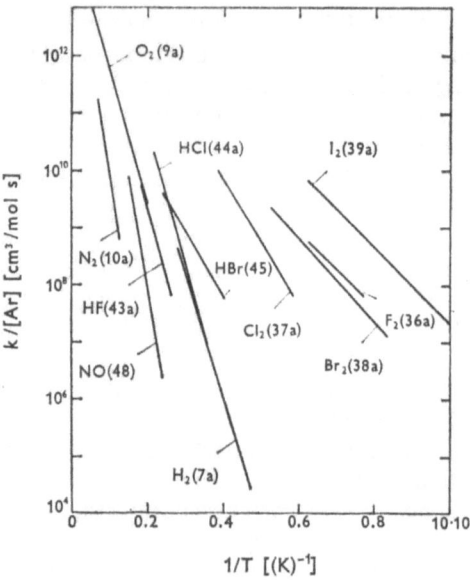

FIG. 1.2. Dissociation of diatomic molecules: references in brackets.

halogen recombination data (section 1.3.1.c). A particularly unusual temperature dependence of n has been found in H_2 dissociation for $M = H$.[11] This has been interpreted in terms of the mechanism (1.20) mentioned earlier.[12] Dissociation rate constants for several diatomic molecules are shown in Fig. 1.2.

(b) *Recombination of* H, O *and* N *Atoms*

The reverse reactions to the dissociations discussed in section 1.3.1(a), i.e. the recombinations of atoms, have been studied in the third-order region in discharge-flow experiments near room temperature (sometimes over a large temperature range, e.g. 90–600 K for $N + N + N_2 \rightarrow 2N_2$),[13] at temperatures above 1000 K in flames or at still higher temperatures in shock waves. A summary of recent values is given in ref. 4. In general it is much easier to study different efficiencies of collision partners M in recombination reactions than in shock wave dissociation experiments. For the recombination of N atoms at 298 K rate constants k_{rec} relative to argon were:[14]

$$:M \qquad He: H_2: N_2: CO_2: N_2O: H_2O$$
$$k_{rec}/k_{rec}(Ar) \quad 1{\cdot}9:1{\cdot}8:1{\cdot}0: 1{\cdot}0 : 0{\cdot}9 : 2{\cdot}7$$

A much higher efficiency of H_2O as collision partner is observed in H atom recombination, where $k_{rec}(H_2O)/k_{rec}(Ar)$ is 20 at room temperature[15] and at flame temperatures of 1000–2000 K.[16] The temperature dependence of recombination rate constants is expressed in Table 1.1 in the form

$$k_{rec} = A''(T/T_0)^m \qquad (1.25)$$

where T_0 is usually 298 K. It was found however, especially in ref. 13, that m changes with temperature and under some circumstances an Arrhenius expression with $E_a < 0$ is a better description of the experimental results. Only a few representative studies are listed in Table 1.1; other references included in the bibliography are for the recombination of H,[17] O [18] and N.[19]

Measured dissociation and recombination results are compared in

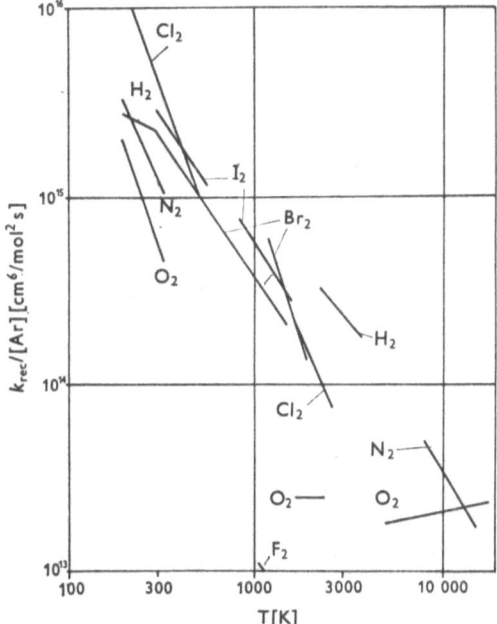

FIG. 1.3. Recombination rate constants for atoms, data of Tables 1.1 and 1.2: low temperature values measured directly, high temperature values calculated from measured rates of dissociation.

Table 1.1 using the relation

$$k_{\text{diss}}/k_{\text{rec}} = K_{\text{equ}} \tag{1.26}$$

where K_{equ} is the equilibrium constant. Dissociation rates have been converted to recombination rates and vice versa, using (1.26) and taking K_{equ} from ref. 20. There is still some dispute whether equation (1.26) holds. Most experimental evidence is in its favour[4] and theoretical arguments also strongly support it (sections 1.6–1.8). However, in the halogen system there appears to be some experimental evidence against equation (1.26); this, however, needs further verification (see below).

Values of K_{rec} from dissociation rates and values from direct measurement are compared in Table 1.1 and Fig. 1.3. m appears to decrease with temperature for H, O and N. m is between -2 and -3 at $T \approx 298$ K and $|m| < 2$ for $T > 2000$ K. Recombination of halogens (Table 1.2) seems to behave somewhat differently. As individual values are uncertain within a factor of 2 and as temperature coefficients have been obtained only over small ranges, we cannot be certain of the exact variation of m. Although low and high temperature data can be reconciled quite well with a mean value between -1 and -2, local values may be different. In Table 1.1 and Fig. 1.4 pre-exponential

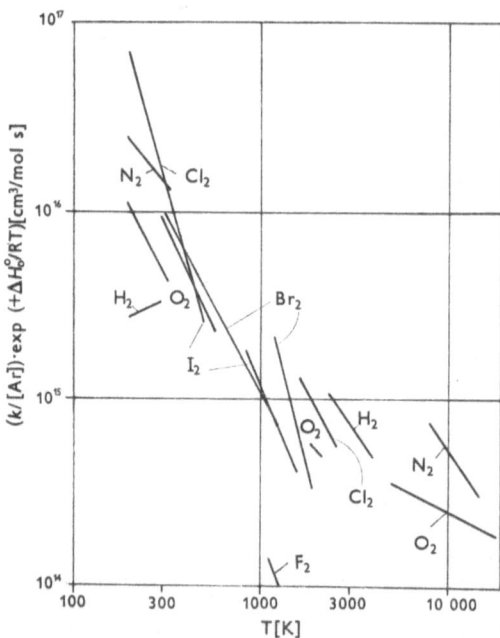

FIG. 1.4. Pre-exponential factors for dissociation rate constants of diatomic molecules: data of Tables 1.1 and 1.2; high temperature values measured directly, low temperature values calculated from measured rates of recombination.

factors $k \exp (+D_0^{\circ}/RT)$, corresponding to equation (1.22), are given for dissociation experiments and converted recombination data. The temperature coefficient n, defined by equation (1.22), again seems to change somewhat with temperature. Apparently $|n|$ decreases with temperature. Low and high temperature data can be fitted together best with $n \approx -2$. Again the uncertainty in k (by a factor of 2) and in local n values is too large for the exact variation of n over large ranges of temperature to be determined.

(c) Dissociation and Recombination of Halogens

The dissociation of F_2, Cl_2, Br_2 and I_2 has been extensively studied with the shock-wave technique at temperatures between about 800 and 3000 K. Recombinations of halogen atoms have been investigated using discharge-flow and flash photolysis between 200 and 600 K, for Br_2 up to 1300 K. In the case of iodine there is only a small gap (600–850 K) between dissociation and recombination. In this case dissociation and recombination data converted using equation (1.26) fit together rather well. For Br_2, flash photolysis studies of recombination have been extended up to temperatures where dissociation has also been studied.[21] Dissociation was investigated by following the disappearance of Br_2 molecules and also the appearance of Br atoms.[22] The observation of a discrepancy between the results for dissociation and recombination of a factor of 2 to 3 and also between the two methods used to follow dissociation has initiated much discussion. This point is worth independent verification.[3,4] As can be seen from Figs. 1.3 and 1.4 and from Table 1.2, Cl_2 high temperature dissociation and low temperature recombination results do not fit together very well, unlike I_2 and Br_2. The local values of $|n|$ and $|m|$ observed experimentally are too high for a single representation of low and high temperature data. Observations of this kind were originally taken as evidence against the validity of equation (1.26), but were discarded later on. Additional work on halogen dissociation and recombination is needed to identify the cause of these discrepancies.

Figures 1.3 and 1.4 indicate that the values of $k_{\text{reo}}/[\text{Ar}]$ for most diatomic molecules agree within a factor of about 5. The same applies for the values of $k \cdot \exp (+D_0^{\circ}/RT)/[\text{Ar}]$. The average values of m and n are around -2. A remarkable exception from this general picture seems to be the dissociation of F_2. The rate constants are apparently lower by a factor of 10 than expected by comparison with other data. This deviation has been discussed in ref. 23 in terms of large anharmonicity of vibrational levels and extensive depopulation of upper levels. However, as is shown in sections 1.6–1.7, this can lead to the opposite result, and in any case there are other possible explanations.

With the halogens it is possible to study not only the direct recombination of atoms into the electronic ground state molecules but also recombination via excited electronic states. For instance, radiative recombination of Cl atoms has been followed in shock waves[25] and in discharge-flow experiments[26] by the Cl_2 ($^3\Pi_{0^+u} \rightarrow {}^1\Sigma_u^+$) afterglow emission spectrum. Recombination of N atoms gives rise to the nitrogen afterglow emission: with this as an indicator it has been possible to study recombination of N atoms into highly excited vibrational levels of bonding excited electronic states.[27] Absolute measurements of recombination rates into excited electronic states as well as into the ground state are important for the theoretical interpretation of recombination rates.

A very detailed investigation of collision partner efficiencies has been performed for the recombination of I atoms. Rate constants with rare gases are:[28]

$$k(M)/k(Ar) = 0{\cdot}37{:}0{\cdot}53{:}1{\cdot}00{:}1{\cdot}25{:}1{\cdot}71$$

for He:Ne:Ar:Kr:Xe. A remarkable correlation between $k(M)$ and the boiling point of M has been found ref. 29; this is shown in Fig. 1.5. Obviously those intermolecular forces for M which are important in condensation are also responsible for differences in recombination

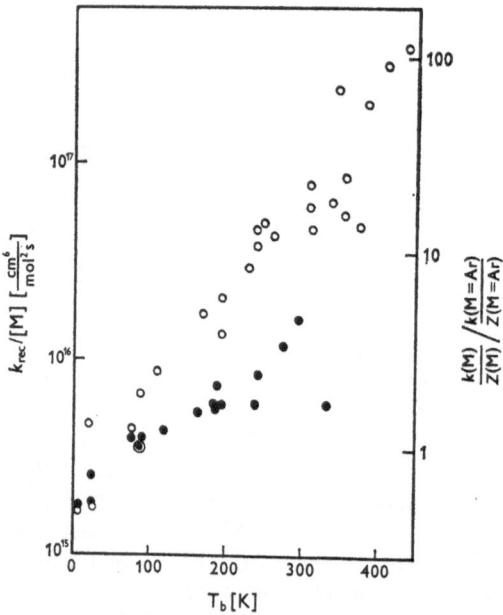

FIG. 1.5. Dependence of the recombination rate of I atoms and of the dissociation rate of nitryl chloride on the boiling point of the carrier gas (collision per collision basis). (○: Data for I + I from ref. 29, l.h. scale, ●: for NO_2Cl from ref. 75b, r.h. scale).

TABLE 1.2

Comparison of dissociation and recombination rate constants for the halogens (as Table 1.1).

	Study	D_0° [kcal mol^{-1}]	T [K]	$k/[Ar][cm^3\,mol^{-1}\,s^{-1}]$ E_a in cal mol^{-1}	$k_{rec}/[Ar][cm^6\,mol^{-2}\,s^{-1}]$	Ref.
F_2	Diss.	36·7	1100–1600	$7·1 \times 10^{12} \exp(-30\,000/RT)$ $9·8 \times 10^{13}\left(\dfrac{T}{1300}\right)^{-2·6} \exp(-D_0^\circ/RT)$	$6·4 \times 10^{12}\left(\dfrac{T}{1300}\right)^{-1·3}$	36a
Cl_2	Diss.	57·0	1700–2600	$1·1 \times 10^{14} \exp(-48\,300/RT)$ $9·6 \times 10^{14}\left(\dfrac{T}{2000}\right)^{-2·2} \exp(-D_0^\circ/RT)$	$1·2 \times 10^{14}\left(\dfrac{T}{2000}\right)^{-0·4}$	37a
	Rec.		195–500	$1·6 \times 10^{16}\left(\dfrac{T}{298}\right)^{-3·5} \exp(-D_0^\circ/RT)$	$4 \times 10^{15}\left(\dfrac{T}{298}\right)^{-3·0}$	40a
Br_2	Diss.	45·5	1200–1900	$1·4 \times 10^{13} \exp(-33\,200/RT)$ $9·0 \times 10^{14}\left(\dfrac{T}{1500}\right)^{-4·1} \exp(-D_0^\circ/RT)$	$2·9 \times 10^{14}\left(\dfrac{T}{1500}\right)^{-3·3}$	38a
	Rec.		300–1275	$1·0 \times 10^{16}\left(\dfrac{T}{298}\right)^{-1·1} \exp(-D_0^\circ/RT)$	$2·4 \times 10^{15}\left(\dfrac{T}{298}\right)^{-1·5}$	41a
I_2	Diss.	35·6	850–1600	$9·8 \times 10^{13} \exp(-30\,400/RT)$ $1·0 \times 10^{15}\left(\dfrac{T}{1100}\right)^{-2·4} \exp(-D_0^\circ/RT)$	$5·1 \times 10^{14}\left(\dfrac{T}{1100}\right)^{-1·6}$	39a
	Rec.		302–548	$8·9 \times 10^{15}\left(\dfrac{T}{298}\right)^{-2·0} \exp(-D_0^\circ/RT)$	$2·9 \times 10^{15}\left(\dfrac{T}{298}\right)^{-1·6}$	42a

rates, either via different energy transfer properties or different populations of intermediate complexes.

The most striking cases are $k(I_2)/k(Ar) = 650$ and $k(NO)/k(Ar) = 10^4$;[28,30] these values indicate chemical attraction in the intermediate complexes; with NO, INO was detected spectroscopically during the reaction,[30] and for the corresponding chlorine system Cl_3 was detected in an Ar matrix.[30a]

For iodine recombination with NO as third body the complex mechanism can be written

$$I + NO \rightleftharpoons INO$$

$$I + INO \longrightarrow I_2 + NO$$

As long as [NO] is not too large, the reaction order is three, I and NO being in equilibrium with INO. However, at large excess NO the reaction order changes to two. This has actually been observed.[30]

All the dissociation reactions of diatomic molecules mentioned above have been found to follow second-order rate laws (low pressure limit of the unimolecular dissociation). However with high pressure shock wave experiments[31] a transition to first-order region was observed for I_2 dissociation at $p \geq 100$ atm. At 1060 K the rate constant k was found to be about $\frac{1}{2}$ of the extrapolated low pressure rate constant for $[Ar] = 10^{-3}$ mol cm^{-3} and about $\frac{1}{10}$ of the extrapolated low pressure rate constant at $[Ar] = 6 \times 10^{-3}$ mol cm^{-3}. At $[Ar] \approx 5 . 10^{-3}$ mol cm^{-3}, $\partial k/\partial [Ar]$ was found to be nearly 0. This unusual behaviour for the dissociation of a diatomic molecule can be explained in terms of the long vibrational period in highly excited vibrational levels of the electronic ground state of I_2 relative to the short interval between collisions at the high pressures employed. Also the contribution from electronically excited states leading to unexcited atoms may be important both here and for other dissociation reactions. By extrapolation of the dissociation data to room temperature and conversion to recombination rates, a change in recombination reaction order is predicted at pressures in the order of 1 atm. In ref. 31 it was argued that the beginning of this transition was observed in room temperature flash-photolysis studies of I recombination near $p(Ar) = 1$ atm. This interpretation of the very unusual decay of k_{rec} at very low $[I_2]/[Ar]$ ratios, which was observed earlier,[32] is also supported by recent studies of the $[I_2]/[Ar]$ dependence of k at $p(M) \leq 1$ atm.[33] In studies with higher pressures of M, after a broad transition range from third to second order, the transition into the liquid range[34] (with $k_{rec} = 7 \times 10^{12}$ cm^3 mol^{-1} s^{-1} = $1 \cdot 2 . 10^{-11}$ cm^3 molec^{-1} s^{-1}, M = CCl_4, $T = 296$ K, and $k_{rec} = 1 \cdot 8 \times 10^{13}$ cm^3 mol^{-1} s^{-1} = $3 \cdot 0 . 10^{-11}$ cm^3 molec^{-1} s^{-1}, M = n-C_6H_{14}, $T = 323$ K) has been observed.[35] Whether the quasi-second order range of k_{rec} in dense gases is influenced by the nature of M or not is a specially interesting question.

Further details of the dissociation of halogens are given in refs. 36 (F_2), 37 (Cl_2), 38 (Br_2) and 39 (I_2), and of their recombination in ref. 40 (Cl), 41 (Br) and 42 (I).

1.3.2. Dissociation of Heteronuclear Diatomic Molecules

Dissociation of the hydrogen halides HF,[43] HCl,[44] DCl[45] and HBr[46] has been studied in shock waves. In contrast to homonuclear diatomic molecules, subsequent bimolecular reactions must be taken into account unless the ratio of reactant concentration to carrier gas concentration is kept sufficiently low. For instance, in HCl dissociation at [HCl]/[Ar] around 10^{-2} the reactions $H + HCl \rightleftharpoons H_2 + Cl$ and $Cl + HCl \rightleftharpoons Cl_2 + H$ follow the unimolecular dissociation $HCl + Ar \rightarrow H + Cl + Ar$. In addition the dissociation of Cl_2 and H_2 must be considered. The interpretation of these dissociation studies may therefore be very complex. Representative results are given in Table 1.3. These fit quite reasonably into the picture found for homonuclear diatomic molecules. No direct recombination studies appear to be available.

CO and CN dissociation, like that of the hydrogen halides, is complicated by subsequent reactions. These couple the reactions to C_2 and O_2, and the C_2 and N_2 dissociations for CO and CN respectively. By analysing the mechanisms reasonable estimates of rate constants were obtained.[47]

NO dissociation, which is important in the kinetics of high temperature air, behaves similarly. $O + NO \rightleftharpoons O_2 + N$ and $N + NO \rightleftharpoons N_2 + O$ and dissociation of O_2 contribute to the mechanism. For the rate constant of dissociation

$$k = [Ar]\, 1 \cdot 1 \times 10^{15} (T/2500)^{-1 \cdot 5} \exp\left(-150\,000/RT\right) cm^3\, mol^{-1}\, s^{-1}$$

(E_a in cal mol^{-1}) or $1 \cdot 8 \times 10^{-9}(T/2500)^{-15}$. $\exp\left(-628\,000/RT\right)$ cm^3 molecules^{-1} s^{-1} (E_a in J mol^{-1}) was obtained.[48] The rate of recombination $O + N + Ar \rightarrow NO + Ar$ at low temperatures was reported to be

$$k_{rec} = [Ar]\, 6 \times 10^{15}\, cm^6\, mol^{-2}\, s^{-1} = [Ar]\, 1 \cdot 7 \times 10^{-32}\, cm^6\, molecule^{-1}\, s^{-1}$$

at 273 K.[14] It shows a remarkably small temperature coefficient, $m = -0 \cdot 5$.

Alkali halide dissociations represent a very interesting group of reactions; these can now be studied using shock waves. Smokes of small halide crystals (diameter $\leq 1\ \mu m$) are prepared, introduced into shock tubes, heated by shock waves, vapourized and then dissociated. The dissociating molecules as well as the reaction products (atoms and

TABLE 1.3
Dissociation rates of hydrogen halides.

Study	D_0° [kcal mol⁻¹]	T (K)	$k/[\text{Ar}][\text{cm}^3\,\text{mol}^{-1}\,\text{s}^{-1}]$ E_a in cal mol⁻¹	$k_{\text{rec}}/[\text{Ar}][\text{cm}^6\,\text{mol}^{-2}\,\text{s}^{-1}]$	Ref.
HF Diss.	134.1	3800–5300	$3.4 \times 10^{14}\exp(-116\,200/RT)$ $2.5 \times 10^{15}\left(\dfrac{T}{4500}\right)^{-2}\exp(-D_0^\circ/RT)$	$5 \times 10^{14}\left(\dfrac{T}{4500}\right)^{-2}$	43a
HCl Diss.	102·17	2800–4600	$6.6 \times 10^{12}\exp(-70\,000/RT)$ $5.5 \times 10^{14}\left(\dfrac{T}{3500}\right)^{-2}\exp(-D_0^\circ/RT)$	$2.7 \times 10^{14}\left(\dfrac{T}{3500}\right)^{-2}$	44a
DCl Diss.	103·2	2800–4600	$6.6 \times 10^{12}\exp(-70\,000/RT)$ $6.5 \times 10^{14}\left(\dfrac{T}{3500}\right)^{-2}\exp(-D_0^\circ/RT)$	$2.8 \times 10^{14}\left(\dfrac{T}{3500}\right)^{-2}$	44a
HBr Diss.	88	2100–4200	$1.5 \times 10^{12}\exp(-50\,000/RT)$ $1.3 \times 10^{15}\left(\dfrac{T}{3000}\right)^{-2}\exp(-D_0^\circ/RT)$	$5.8 \times 10^{14}\left(\dfrac{T}{3000}\right)^{-2}$	46

ions) have been detected. Two alternative dissociation paths exist:

$$AlHal + M \longrightarrow Al + Hal + M \qquad (1.27a)$$

$$AlHal + M \longrightarrow Al^+ + Hal^- + M \qquad (1.27b)$$

Threshold energies of both reactions differ by the differences of halogen electron affinities and alkali ionization potentials. In all cases reaction (1.27a) is energetically favoured. However, it involves a transition from the ionic electronic ground state into the covalent excited electronic states which lead to atoms. Under some circumstances reaction (1.27b) may be favoured, since the electronic transition in step (1.27a) is forbidden.

Investigations[49] of the nature of primary products lead to the conclusion that CsBr, RbCl, RbBr, KBr and CsI primarily dissociate into ions, LiBr, NaBr, NaI and LiCl dissociate into atoms, and RbI and KI produce both ions and atoms. Investigations[50] of the temperature coefficients and absolute values of the rate constants lead to the conclusion that at pressures near 1 atm NaCl, NaI and LiF dissociate into atoms, CsCl and CsBr dissociate into ions and KCl, CsI and KI represent intermediate cases. However if the electronic transition is not collision induced, the nature of reaction products may depend on pressure. If $(AlHal)_{vib}$ denotes the electronic ground state, vibrationally excited up to threshold of step (1.27a), $(AlHal)_{el}$ is the excited homopolar electronic state and $(AlHal)^*_{vib}$ is the electronic ground state, vibrationally excited up to threshold of step (1.27b), then the mechanism can be written:

$$AlHal + M \longrightarrow (AlHal)_{vib} + M \qquad (1.28a)$$

$$(AlHal)_{vib} \longrightarrow (AlHal)_{el} \qquad (1.28b)$$

$$(AlHal)_{el} \longrightarrow Al + Hal \qquad (1.28c)$$

and

$$AlHal + M \longrightarrow (AlHal)_{vib} + M \qquad (1.29a)$$

$$(AlHal)_{vib} + M \longrightarrow (AlHal)^*_{vib} + M \qquad (1.29b)$$

$$(AlHal)^*_{vib} \longrightarrow Al^+ + Hal^- \qquad (1.29c)$$

At pressures which are low enough for the time between collisions to be sufficiently long, atoms will be formed; step (1.28a) is rate determining and the reaction order is 2. At intermediate pressures atoms will be formed; step (1.28b) is rate determining and the reaction order is 1. At high pressures ions are formed, step (1.29b) is rate determining and the reaction order again becomes 2. From the first order region the transition probability of process (1.28b) can be derived; consequently the pressure dependence of dissociation rate was investigated.[50]

The rate of NaCl dissociation around 1 atm and 2300–4800 K may be taken as representative of the absolute values found for the rate constants:

$$k = [Ar] . 5 \times 10^{14} \exp \left(-80\,000/RT\right) \text{ cm}^3 \text{ mol}^{-1} \text{ s}^{-1} \; (E_a \text{ in cal mol}^{-1})$$
$$\text{or} \quad 8 \times 10^{-10} \exp \left(-335\,000/RT\right) \text{ cm}^3 \text{ s}^{-1} \; (E_a \text{ in J mol}^{-1}).$$

With D_0° (atoms) $= 97 \cdot 4$ k cal mol^{-1} (408 kJ mol^{-1}),

$$k = [Ar] . 8 \cdot 0 \times 10^{15} \exp \left(-D_0^\circ/R\,3000\right) \text{ cm}^3 \text{ mol}^{-1} \text{ s}^{-1}$$
$$= [Ar] . 13 \times 10^{-8} \exp \left(-D_0^\circ/R\,3000\right) \text{ cm}^3 \text{ s}^{-1}$$

at 3000 K. Rate constants for other cases are comparable. The possibility of varying molecular parameters through the homologous series of alkali halides, e.g. densities of levels, should be of importance for theories of dissociation rates. At present however, experimental rate constants are not sufficiently accurate. For mercury halides, see[50a].

1.4. DISSOCIATION–RECOMBINATION KINETICS FOR TRIATOMIC MOLECULES

Dissociation of triatomic molecules can be classified according to whether the reaction is spin-allowed or spin-forbidden. In the first group the reaction takes place without change of electronic state at the singlet ground state potential surface of the dissociating molecule. As in dissociation reactions of diatomic molecules, these reactions at $[M] < 10^{-4}$ mol cm^{-3} are almost always found to be in the low pressure region of dissociation and recombination. The transition into the high pressure region has been observed only under extreme conditions with $[M] > 10^{-3}$ mol cm^{-3}, for NO_2 and O_3 (see below).

In the second group, spin-forbidden reactions, the correlation rules[51] show that reaction to ground state products is not possible without change of electronic state. For the cases studied (CO_2, CS_2, COS, N_2O) this is a singlet–triplet transition, which is partly forbidden. In spite of this the singlet–triplet reaction is always observed, owing to the favourable heat of reaction. The low singlet–triplet transition probability does not influence the rate constant in the low-pressure range. According to equation (1.8), only the rate of activation ($k_1[M]$) determines the observed reaction rate. However, the singlet–triplet transition probability is contained in k_3, equation (1.9), and therefore determines the magnitude of the high pressure rate constant. As $[M]_{1/2}$ (equation (1.10)) is decreased for transition probabilities < 1, the transition between the low and high pressure dissociation regions is shifted to lower pressures than would be expected for spin-allowed dissociation. This has been observed; both low- and high-pressure regions can be reached for all spin-forbidden dissociations of triatomic molecules.

For spin-allowed reactions the D_0° values obtained by thermochemical methods adequately represent the energy threshold of reaction. For spin-forbidden reactions this is not so. If the lowest possible transition between the potential surfaces involved occurs at energies E_0 higher than the heat of reaction D_0°, the value of E_0 acts as energy threshold. In N_2O dissociation where this happens, at present the corresponding value of E_0 can only be determined from the dissociation rate results. With other spin-forbidden reactions the lowest possible crossing of potential surfaces may take place below D_0°. Then a complicated mechanism like (1.18) with collisional processes in both electronic states can operate. Unusual temperature and pressure dependences of dissociation rate can be expected (see below).

1.4.1. Spin-allowed Reactions

Some examples of spin-allowed dissociations of triatomic molecules will now be discussed and compared with the corresponding recombination and isotope exchange reactions. Further data are given for NO_2,[52] O_3,[53] H_2O,[54] SO_2,[55] F_2O,[56] $NOCl$,[57] HO_2,[58] CF_2,[59] NF_2,[60] $BrCN$,[61] $ClCN$,[62] $ClCO$[63] and HNO.[64] A systematic comparison of literature values up to 1967 is contained in ref. 6.

(a) The Reaction $NO_2 \rightleftharpoons NO + O$

The $NO_2 \rightleftharpoons O + NO$ system appears to have been studied more extensively than any other at present: it will therefore be discussed in some detail. Dissociation was investigated with shock waves at temperatures and $[NO_2]/[M]$ ratios where bimolecular reaction between NO_2 molecules was suppressed. The subsequent reaction of O atoms with NO_2 always doubled the rate of unimolecular dissociation. As this reaction is well known from other studies, its effect on the kinetics is easily determined.

At normal pressures, the dissociation is found to be in the low pressure region. The rate constant at $1450 < T < 2000$ K being $k_0 = [Ar] \cdot 10^{16 \cdot 05} \exp(-65\,000/RT) \text{ cm}^3 \text{ mol}^{-1} \text{ s}^{-1}$ [52a] (E_a in cal mol^{-1}) or $[Ar] \cdot 1 \cdot 9 \cdot 10^{-8} \exp(-272\,000/RT) \text{ cm}^3 \text{ molecule}^{-1} \text{ s}^{-1}$ (E_a in J mol^{-1}). With $D_0^\circ = 71 \cdot 86$ kcal (300·8 kJ) mol^{-1} a high pre-exponential factor is found using equation (1.22) (see Table 1.4). The pressure dependence of the rate constant was studied with high pressure shock waves at temperatures near 1500 K.[52a] A change of reaction order was found only above 100 atm. Extrapolation to the high pressure range was just possible, leading to a value of $k_\infty \approx 10^{14 \cdot 3} \exp(-D_0^\circ/RT) \text{ s}^{-1}$. Here the temperature coefficient was chosen arbitrarily. A comparison with low pressure recombination data is given in Fig. 1.6. Low, medium and high temperature values fit well and can be expressed together by a $T^{-1 \cdot 9}$ temperature dependence. For $m = -1 \cdot 9$ the same comments apply as in diatomic molecules, i.e. local values of m may well be different. Recombination and dissociation rates are high compared to those for diatomic molecules.

Low pressure recombination data have been obtained with different experimental techniques (see summary in ref. 52b). By investigation of

TABLE 1.4

Comparison of dissociation and recombination in the low and high pressure regions and of isotopic exchange for the NO_2 system. ($D_0^\circ = 71.9 \pm 0.2$ kcal mol⁻¹).

(a) Low pressure range

Study	T [K]	$k_0/[\text{Ar}]$ [cm³ mol⁻¹ s⁻¹] E_a in cal mol⁻¹	$k_{rec_0}/[\text{Ar}]$ [cm⁶ mol⁻² s⁻¹]	Ref.
Diss.	1450–2000	$1.1 \times 10^{16} \exp(-65\,000/RT)$; $7.6 \times 10^{16}\left(\dfrac{T}{1700}\right)^{-2} \exp(-D_0^\circ/RT)$	$7.8 \times 10^{14}\left(\dfrac{T}{2000}\right)^{-1.8}$	52a
Rec.	1030	$2.4 \times 10^{17} \exp(-D_0^\circ/RT)$	2.9×10^{15}	52b, h
Rec.	300–500	$1.2 \times 10^{18}\left(\dfrac{T}{300}\right)^{-1.5} \exp(-D_0^\circ/RT)$	$3.7 \times 10^{16}\left(\dfrac{T}{300}\right)^{-2.6}$	52d
Rec.	300	$9.7 \times 10^{17} \exp(-D_0^\circ/RT)$	3.0×10^{16}	52b
Rec.	200–300	$8.5 \times 10^{17}\left(\dfrac{T}{300}\right)^{-1.8} \exp(-D_0^\circ/RT)$	$2.7 \times 10^{16}\left(\dfrac{T}{300}\right)^{-3.5}$	52g

(b) High pressure range

Study	T [K]	k_∞ [s⁻¹]	$k_{rec\,\infty}$ [cm³ mol⁻¹ s⁻¹]	Ref.
Diss.	≈1500	$2 \times 10^{14} \exp(-D_0^\circ/RT)$	2.4×10^{12}	52a
Rec.	298	$2.5 \times 10^{14} \exp(-D_0^\circ/RT)$	8×10^{12}	52i

(c) Isotope exchange

T [K]	k_{iso} [cm³ mol⁻¹ s⁻¹]	Ref.
310	1.1×10^{12}	52d

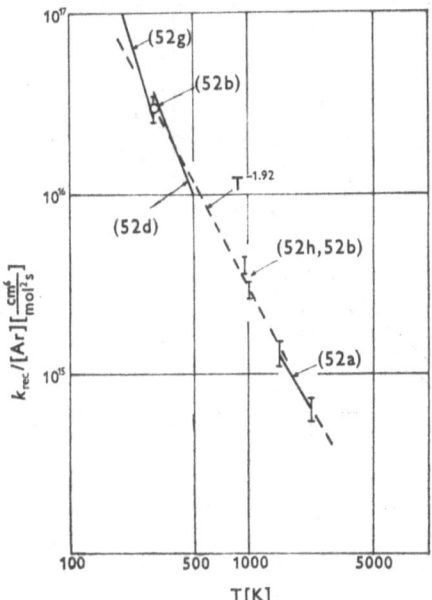

FIG. 1.6. Recombination $O + NO + Ar \rightarrow NO_2 + Ar$; references in brackets; dotted line: $k_{rec}/[Ar] \propto T^{-1.9}$.

the [NO] and [N_2] dependence of quantum yields in NO_2 photolysis the recombination $O + NO \rightarrow NO_2$ at room temperature and high pressures has been investigated: the second-order high pressure region of recombination was reached,[52c,1]

$$k_{rec_\infty} \approx 8 \cdot 10^{12} \text{ cm}^3 \text{ mol}^{-1} \text{ s}^{-1} = 1 \cdot 4 \times 10^{-11} \text{ cm}^3 \text{ molecule}^{-1} \text{ s}^{-1}$$

Conversion to dissociation rates via equation (1.26) gives $k_\infty = 10^{14.4} \exp(D_0^\circ/RT) \text{ s}^{-1}$ at 300 K, which almost agrees with the direct value at 1500 K.

Low and high pressure dissociation- and recombination rate constants are compared in Table 1.4 for both expressions. The fall-off curves of dissociation and recombination are shown in Fig. 1.7. One can see the slight shift of the curvature of the fall-off curve to higher densities as the temperature is decreased. This becomes more pronounced with unimolecular reactions of larger molecules.

For high pressures, equation (1.6) indicates that the rate determining step for dissociation in the dissociation of highly excited NO_2^*; for recombination it is the association of $O + NO$ to give NO_2^*. The question arises, whether (i) association in high pressure recombination and (ii) association as the first step of O isotope exchange with NO

2

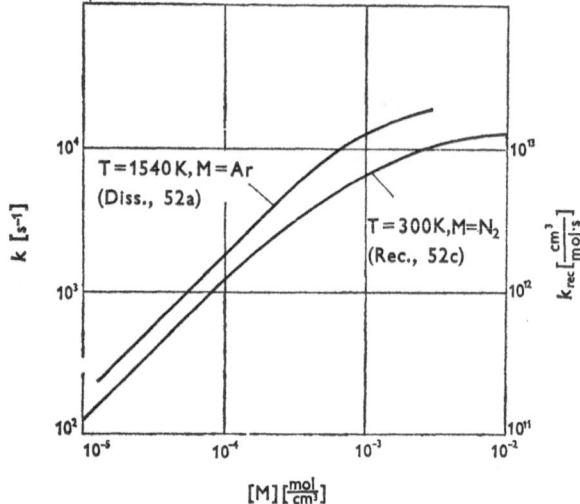

$[M] \left[\frac{mol}{cm^3}\right]$

FIG. 1.7. Fall-off curves for the dissociation $NO_2 + M \rightarrow NO + O + M$ and the recombination $O + NO + M \rightarrow NO_2 + M$.

are identical processes with similar rates or not, according to the mechanism

$$O_a + NO_b \rightleftharpoons O_aNO_b^* \longrightarrow O_aN + O_b$$
$$\downarrow {\scriptstyle +M}$$
$$O_aNO_b$$

$$(1.30)$$

This question can now be answered by comparing the high pressure recombination data with isotope exchange experiments.[52b] The rate of isotope exchange seems to be about a factor of 7 smaller than k_{rec_∞}, indicating effective redissociation of the intermediate complex (in contrast to the O_3 system, see below).

The results discussed so far correspond mainly to reactions in the electronic ground state. Some information is also available for the contribution from excited electronic states. Observation of thermal emission of electronically excited NO_2 during dissociation provided information about the population of vibrational levels of electronically excited NO_2.[52e] Marked deviations from a Boltzmann distribution were found at levels near the dissociation limit of the excited state. This observation is in accord with arguments presented in section 1.7 on the population of vibrational states in the low pressure limit of dissociation reactions. Dissociation out of the electronically excited state contributes only to a small extent to the overall rate of dissociation (a few % at 2000 K).

In low pressure recombination studies at room temperature the influence of different collision partners has also been investigated.

Relative to M = Ar the following rate constants were measured:[52f] $k_{rec}(M) = 0.8:1.5:1.0:2.2:2.2:2.3:2.7:6.3$ for M = He:N_2:O_2:CO_2: N_2O:CF_4:SF_6:H_2O. A high collision efficiency for H_2O is obviously normal, while the small effect observed in the recombination of N atoms (see section 1.3.1(b)) appears to be an exception.

(b) The Reaction $O_3 \rightleftarrows O_2 + O$

Owing to the instability of ozone it was possible to study thermal decomposition of this molecule with shock waves[53a] at temperatures around 800 K and also with flow systems and static systems[53b] at temperatures near 360 K. Reaction of O atoms with O_3 tends to double the rate of unimolecular dissociation. At shock wave temperatures the influence of this reaction decreases, leading to a transition from twice to once the unimolecular dissociation rate. As the value of D_0° is still slightly uncertain, interpretation of low temperature results is also somewhat uncertain. Assuming $D_0^\circ = 24.2$ kcal (101.3 kJ) mol^{-1}, the dissociation rates are compared with recombination rates measured at room temperature in Table 1.5. A very pronounced difference of magnitude in k and k_{rec} values is evident compared to the NO_2 data. While the NO_2 rates appear to be high compared with most other spin-allowed dissociations and recombinations of triatomic molecules, the O_3 rates are low. However, the high value of NO_2 may be due to an exceptionally high efficiency of energy transfer. This conclusion is supported by NO_2 fluorescence studies.[65] Other triatomic molecules appear to behave differently.[6] This is shown in Table 1.6 which gives some examples of low pressure data. Figure 1.8 compares dissociation rates for several spin-allowed cases.

With ozone decomposition in static systems it was possible to observe collision efficiencies[53b] $k(M)/k(Ar) = 1.3:1.5:1.6:3.9:3.7$ for He:N_2: O_2:CO_2:O_3. These values correspond to collision efficiencies obtained from recombination studies[53c] $k_{rec}(M)/k_{rec}(Ar) = 1.0:1.4:1.6:3.7:3.7:$ $8.5:15$ for He:N_2:O_2:CO_2:N_2O:SF_6:H_2O. No significant difference is observed.

Recently the high pressure region has been reached for the recombination $O + O_2 \rightarrow O_3$ at room temperature, both in the high pressure photolysis of NO_2–O_2–N_2 mixtures[53d] and from pulse radiolysis studies.[53e] As for NO_2, these high pressure recombination results agree reasonably with data on the isotope exchange of O atoms with O_2.[53f] Compared to NO_2, the high pressure recombination results are also low (see Table 1.5). This may be due to differences in the potential surfaces, e.g. the ratio of force constants to bond energies.

(c) Other Systems

Dissociation of water in shock waves at [H_2O]/[Ar] ratios $\leq 10^{-2}$ has been explained[54a] by unimolecular dissociation $H_2O + Ar \rightarrow$ $H + OH + Ar$, followed by $H + H_2O \rightarrow H_2 + OH$. At 3000 K this

TABLE 1.5

Comparison of dissociation and recombination in the low and high pressure range and of isotopic exchange for O_3 system. ($D_0^\circ = 24\cdot25 \pm 0\cdot4$ kcal mol^{-1}).

(a) Low pressure range

Study	T [K]	$k_0/$[Ar][cm^3 mol^{-1} s^{-1}] E_a in cal mol^{-1}	$k_{\mathrm{rec}_0}/$[Ar][cm^6 mol^{-2} s^{-1}]	Ref.
Diss.	769–910	$3\cdot8 \times 10^{14} \exp(-23\,150/RT)$ $7\cdot4 \times 10^{14}\left(\dfrac{T}{830}\right)^{-0\cdot3} \exp(-D_0^\circ/RT)$	$1\cdot3 \times 10^{13}\left(\dfrac{T}{1000}\right)^{-1\cdot7}$	53a
Diss.	343–383	$1\cdot3 \times 10^{15} \exp(-24\,000/RT)$ $2\cdot0 \times 10^{15}\left(\dfrac{T}{360}\right)^{-0\cdot3} \exp(-D_0^\circ/RT)$	$6\cdot8 \times 10^{13}\left(\dfrac{T}{360}\right)^{-1\cdot2}$	53b
Rec.	188–373	$3\cdot0 \times 10^{15}\left(\dfrac{T}{300}\right)^{-2\cdot3} \exp(-D_0^\circ/RT)$	$2\cdot8 \times 10^{14}\left(\dfrac{T}{273}\right)^{-3\cdot4}$	53g
Rec.	300	$2\cdot2 \times 10^{15} \exp(-D_0^\circ/RT)$	$1\cdot5 \times 10^{14}$	53c
Rec.	296	$1\cdot5 \times 10^{15} \exp(-D_0^\circ/RT)$	$1\cdot0 \times 10^{14}$	53e

TABLE 1.5 (*cont*).

(b) High pressure range

Study	T [K]	k_∞ [s⁻¹]	$k_{\text{rec}\infty}$ [cm³ mol⁻¹ s⁻¹]	Ref.
Rec.	296	$1\cdot3 \times 10^{13} \exp(-D_0^0/RT)$	5×10^{11}	53e
Rec.	298	$1\cdot6 \times 10^{13} \exp(-D_0^0/RT)$	6×10^{11}	53d

(c) Isotope exchange

T [K]	k_{iso} [cm³ mol⁻¹ s⁻¹]	Ref.
310	6×10^{11}	53f1
298–402	$7\cdot7 \times 10^{11} \left(\dfrac{T}{340}\right)^{+1\cdot6}$	53f3

TABLE 1.6

Spin-allowed dissociations of triatomic molecules in the low-pressure region.

Reaction	D_0° [kcal mol^{-1}]	T [K]	$k_0/[\text{Ar}]$[cm^3 mol^{-1} s^{-1}] E_a in cal mol^{-1}	Ref.
$NO_2 \rightarrow NO + O$	71·86	1450–1000	$1 \cdot 1 \times 10^{16} \exp(-65\,000/RT)$	52a
$O_3 \rightarrow O_2 + O$	24·2	769–910	$3 \cdot 8 \times 10^{14} \exp(-23\,150/RT)$	53a
		343–383	$1 \cdot 3 \times 10^{15} \exp(-24\,000/RT)$	53b
$H_2O \rightarrow H + OH$	117·6	3000–7000	$5 \times 10^{14} \exp(-105\,000/RT)$	54a
		2570–3250	$9 \times 10^{15} \left(\dfrac{T}{3000}\right)^{-2 \cdot 2} \exp(-D_0^\circ/RT)$	54b
$SO_2 \rightarrow SO + O$	131·3	4500–7500	$2 \cdot 5 \times 10^{14} \exp(-110\,000/RT)$	55a
$F_2O \rightarrow FO + O$	~37	820–1240	$1 \cdot 3 \times 10^{15} \exp(-34\,200/RT)$	56a
$NOCl \rightarrow NO + Cl$	36·9	473–684	$6 \cdot 9 \times 10^{15} \exp(-35\,400/RT)$	57a

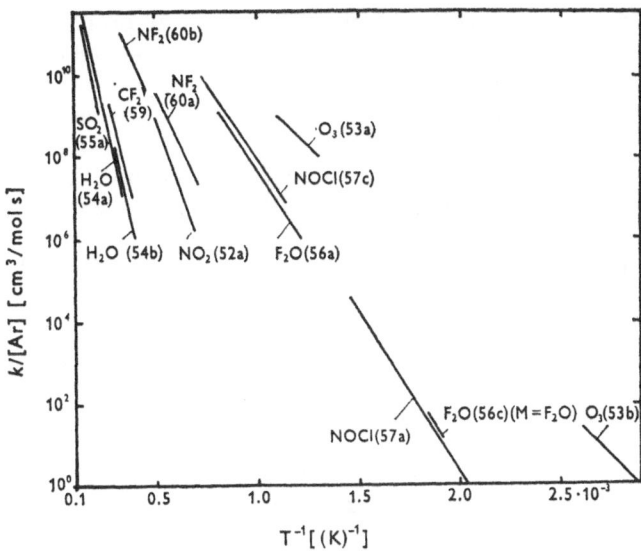

FIG. 1.8. Spin-allowed dissociations of triatomic molecules in the low pressure region; references in brackets.

second reaction doubles the rate of H_2O disappearance; at 6000 K it can be neglected. At $[H_2O]/[Ar] > 10^{-2}$ and $T \leq 3000$ K a number of bimolecular atom-transfer reactions of H and O with H_2, H_2O, OH and O_2 rapidly establish a partial equilibrium, which must be included in the analysis. This is of particular importance if [OH] is taken as a measure of reaction.[54b]

The dissociation rates can now be reconciled fairly well with recombination rates measured in flames at 1000–2000 K.[54c] The high collision efficiency of H_2O itself as collision partner M, $k(H_2O)/k(Ar) \approx 20$, must be allowed for. Numerical values are given in Table 1.6.

Only at very high temperatures ($T \geq 4500$ K) and $[SO_2]/[Ar] \leq 0.5 . 10^{-2}$ could the unimolecular dissociation $SO_2 \rightarrow SO + O$ be observed in shock waves.[55a] At lower temperatures and higher $[SO_2]/[Ar]$ secondary bimolecular reactions were obviously important; a complex mechanism is evident, but cannot yet be uniquely resolved.[55b] An exceptionally high rate is reported for the reverse reaction $O + SO \rightarrow SO_2$: $k_{reo} = [Ar] 3.2 \times 10^{17}$ cm^6 mol^{-2} s^{-1} = [Ar] . 8.7 $\times$ 10^{-31} cm^6 molecule^{-2} s^{-1} at 300 K.[55c]

The dissociation $F_2O \rightarrow OF + F$ has long been considered a good example of a unimolecular dissociation in the low pressure range. Results from early low temperature experiments and recent shock wave studies show a consistent picture for the variation of k over several orders of magnitude.[56a] Recently some evidence has been obtained against a simple mechanism,[56b] though the influence of secondary reactions is not yet completely clear. The central problem in this system is probably the magnitude of D_0°: values of 37 and 43 kcal mol^{-1} (155 and 178 kJ mol^{-1}) have been proposed.[56]

NOCl dissociation data for both static systems and shock waves are available.[57a] The recombination NO + Cl → NOCl has been studied.[57b] The CF$_2$[59] and NF$_2$[60] dissociations were followed in shock waves after dissociating C$_2$F$_4$ and N$_2$F$_4$. BrCN[61] and ClCN[62] experiments have also been used to estimate the heat of formation of the CN radical.

Recombination data alone are available for Cl + CO → ClCO,[63] H + NO → HNO[64] and H + O$_2$ → HO$_2$.[58] The latter reaction has been studied extensively during the investigation of the H$_2$–O$_2$ system in shock waves at 1000–1800 K. H$_2$O is a very effective collision partner in this recombination.

Table 1.6 compares dissociation results of spin-allowed reactions of triatomic molecules.

1.4.2. Spin-forbidden Reactions

According to the discussion at the beginning of section 1.4 the carrier gas pressures at which a transition occurs between the low and high pressure regions of spin-forbidden dissociations of triatomic molecules are lower than those in spin-allowed dissociations. It is thus easier experimentally to reach the high pressure limit, and in fact it has been possible to obtain the full fall-off curves for all examples of molecules in this group.

(a) *Dissociation of* N$_2$O

The dissociation N$_2$O → N$_2$ + O(^{3}P) is a classical example of a unimolecular reaction. It was studied relatively early in static and flow systems at comparatively low temperatures.[66a–c] Secondary reactions are relatively easy to take into account at low temperatures. The reaction was observed mainly in the low pressure regime, but first indications of a transition to the high pressure range have been found.[66b] A great variety of different carrier gases was studied. As examples the relative efficiencies for N$_2$O:He:Ne:Ar:Kr:Xe:O$_2$:N$_2$:CO$_2$:H$_2$O are 1:1:0·44:0·1–0·18:0·25:0·15:0·21:0·26:1·2:1·6. An extensive analysis of these early low temperature data is given in ref. 66d. With the availability of shock wave studies, the dissociation again was studied in detail using both optical detection of the reactants[66e,f] and detection in time-of-flight mass spectrometers coupled to shock tubes.[66g,h] Because the reactions of O atoms with N$_2$O have a large activation energy, they only become important at high temperatures; O atoms are consumed by recombination at low temperatures. There is still discussion[66i] on the absolute value of the rate of reactions O + N$_2$O → 2NO and O + N$_2$O → N$_2$ + O$_2$; these influence the accuracy of the high temperature results. A comparison of different sets of these experiments is given in refs. 6, 66h. At temperatures between the shock wave and the static system regions, additional investigations have been made using adiabatic compression.[66k] The different studies in the different ranges agree fairly well, as can be seen from Fig. 1.9.

At low temperatures ($840 \leq T \leq 1050$ K) with Ar as carrier gas[66d]

$$k = [\text{Ar}] \, . \, 10^{14.7} \exp\left(-59\,200/RT\right) \text{cm}^3 \, \text{mol}^{-1} \, \text{s}^{-1}$$

(E_a in cal mol^{-1}) or $[\text{Ar}] \, . \, 8{\cdot}3 \times 10^{-10} \exp\left(-248\,000/RT\right)$ cm^3 molecule^{-1} s^{-1} (E_a in J mol^{-1}); at high temperatures ($1500 \leq T \leq 2500$ K)[66f]

$$k = [\text{Ar}] \, . \, 10^{14.7} \exp\left(-58\,000/RT\right) \text{cm}^3 \, \text{mol}^{-1} \, \text{s}^{-1}$$

(E_a in cal mol^{-1}) or $[\text{Ar}] \, . \, 8{\cdot}3 \times 10^{-10} \exp\left(-243\,000/RT\right)$ cm^3 molecule^{-1} s^{-1} (E_a in J mol^{-1}). Part of the difference in activation energy at

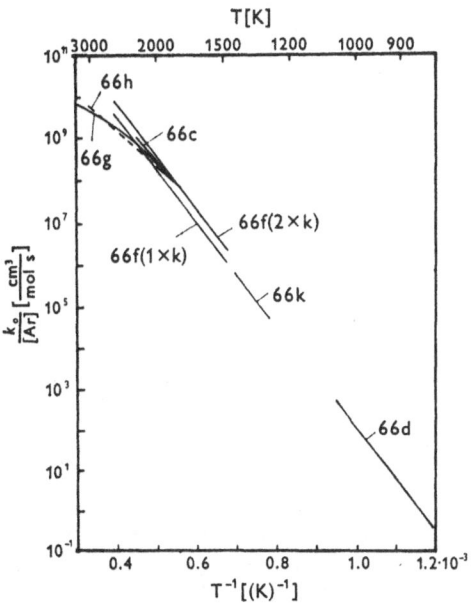

FIG. 1.9. Dissociation of N_2O: low pressure region.

highest temperature may be due to differing contributions from the reactions of O atoms. From the comparison between low and high temperature values, the apparent activation energies obtained with mass-spectrometric analysing techniques appear to be somewhat low.

By high pressure shock wave experiments it was possible to obtain the full transition to the high pressure range and to study the high pressure limiting rate in the range $1400 \leq T \leq 2000$ K. A high pressure limiting value of

$$k_\infty = 10^{11.1} \exp\left(-59\,500/RT\right) \text{s}^{-1} \quad \text{or} \quad 10^{11.1} \exp\left(-249\,000/RT\right) \text{s}^{-1}$$

2A

for E_a in cal or J respectively was found.[66f] The low value of the pre-exponential factor shows that the reaction occurs via the spin-forbidden reaction path. The same follows immediately from the apparent activation energies in the low and high pressure range. Both values may be reconciled with[66f] an E_0 value around 63 kcal (263 kJ) mol^{-1}. This is considerably higher than the thermochemical ΔH_0° value of 39 kcal (161 kJ) mol^{-1} for the reaction $N_2O \rightarrow N_2 + O(^3P)$, but lower than the ΔH_0° value of 84 kcal (351 kJ) mol^{-1} for the spin-allowed reaction $N_2O \rightarrow N_2 + O(^1D)$. Thus a section through the potential surfaces of the two contributing electronic states must look similar to the picture given in Fig. 1.10. Whether the upper electronic state, mainly

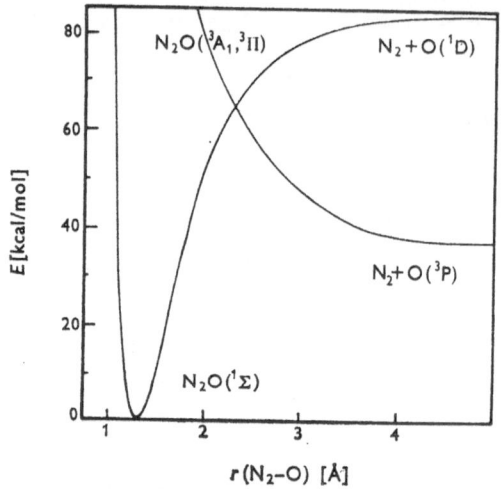

FIG. 1.10. Electronic states involved in N_2O dissociation.

responsible for the reaction, is linear ($^3\Pi$) or bent ($^3A_2$) cannot be decided yet. The correlation between the low pressure apparent activation energies and E_0 may be obtained in the same way as in spin-allowed reactions. (This will be discussed in detail in section 1.7, see e.g. equation (1.80).) However, in the high pressure region E_a and E_0 are connected in a complicated way, which can only be understood from the properties of the surface of intersection between the electronic states (see ref. 66f).

In Fig. 1.11 a fall-off curve of the N_2O dissociation[66f] at 2000 K is compared to the fall-off curve of the spin-allowed dissociation of NO_2 at 1540 K.[52a] The fall-off of the first-order rate constants occurs at higher pressures for NO_2. The low pressure part of a N_2O fall-off curve obtained in a static system[66b] at 888 K is also shown: compared to larger polyatomic molecules only a slight shift of the transition range to lower pressures with decreasing temperature is observed.

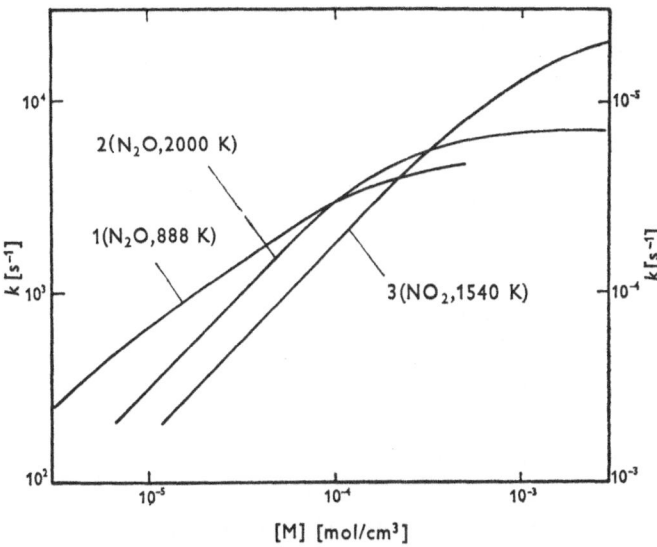

FIG. 1.11. Fall-off curves of dissociation of N_2O (curves 1 and 2) and NO_2 (curve 3). Curve 1: $T = 880$ K, M = N_2O, after ref. 66b, r.h. scale; curve 2: $T = 2000$ K, M = Ar, after ref. 66f, l.h. scale; curve 3: $T = 1540$ K, M = Ar, after ref. 52a, l.h. scale.

An important relation exists between the thermal dissociation of N_2O and the electronic deactivation of $O(^1D)$ atoms in collisions with N_2.[66l] In both cases highly excited vibrational levels of the electronic ground state N_2O molecules are involved as intermediates. In addition, crossing from the singlet to the triplet potential surface takes place in both reactions. However, this crossing occurs at different excitation energies. In the reaction of $O(^1D)$ with N_2 at very high pressure the intermediate N_2O is also stabilized.[66m] Similar relations between dissociation and deactivation of $O(^1D)$ and $S(^1D)$ exist for the CO_2, CS_2 and COS system and probably other examples (see e.g. ref. 66n).

(b) *Dissociation of* CO_2, COS, CS_2

Whereas a pronounced potential barrier must be overcome in the N_2O dissociation, as illustrated in Fig. 1.10, a different situation arises for CO_2, CS_2 and COS. From the temperature dependence of low and high pressure dissociations, and also from the fact that recombination and radiative combination of O with CO can be observed, one can conclude that the potential surfaces involved have quite different shapes. This is illustrated in Fig. 1.12. The triplet states cross the ground states at energies which are either lower or which are only slightly higher than the ΔH_0° values of the reaction. Therefore the final step of dissociation, i.e. the separation of the reaction products in the triplet state, will be

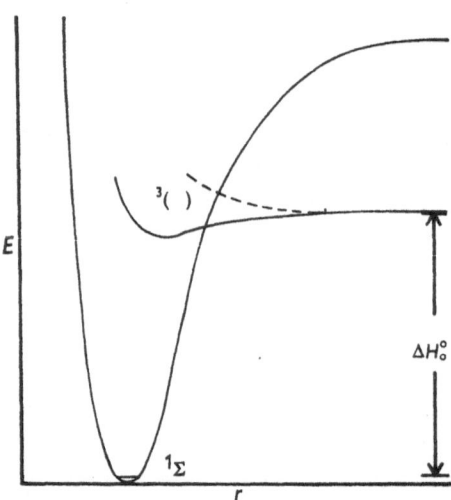

FIG. 1.12. Electronic states involved in the dissociation of CO_2, CS_2 and COS (Schematically).

much slower than for N_2O, where it is a simple downhill motion. Instead, if crossing occurs at $E \leq \Delta H_0^\circ$ additional energy may be required for dissociation. This can be provided by further collisions or may possibly be taken from the initial rotational energy, which remains practically unchanged at the moment of potential crossing. A complex mechanism may operate, involving interference from initial activation in the singlet state, singlet–triplet transitions and final dissociation out of the triplet state. This may lead to unexpected properties of both dissociation rate constants and recombination rate constants.

The dissociations of CO_2,[67] CS_2[68] and COS[69] in the low pressure region have been studied by several authors in shock waves: they are not accessible by other techniques. Agreement for the absolute values of rate constants is normally fairly good. However, remarkable uncertainties have arisen in the activation energies of the CO_2 dissociation. This may be due partly to small influences of H_2O impurities.[67a] The apparent activation energies appear to decrease with increasing temperature; this is much too pronounced to be explained by the arguments given in section 1.7 (see e.g. equation (1.80)). We think these low activation energies should be verified. A comparison of the low pressure and high pressure limiting rates for spin-forbidden dissociations is given in Table 1.7.

With high pressure shock waves the high pressure regions for CO_2,[67b] CS_2[68a] and COS[69a] dissociation could also be observed. As shown in Table 1.7 the low pre-exponential factors of the rate constants again indicate that the reactions are spin-forbidden. An analysis of these values, similar to ref. 66f, leads to the conclusion that the electronic transition-matrix elements are of the order $0 \cdot 1$ kcal ($0 \cdot 4$ kJ) mol^{-1} for separating O atoms but are larger for S atoms. This is in agreement with the magnitudes of spin-orbit coupling measured for O and S atoms. Also the thermally averaged singlet–triplet transition probabilities derived from Table 1.7 are different: 10^{-3}–10^{-2} for O atoms and 10^{-2}–10^{-1} for S atoms.

The reverse reaction of CO_2 dissociation, i.e. the recombination of O atoms with CO, has been studied in considerable detail giving a clear picture only recently. High temperature shock wave studies,[70a] intermediate temperature shock wave studies[70b] and low temperature flow system experiments[70c] appear to indicate changing temperature coefficients from slightly positive apparent activation energies at low T to slightly negative activation energies at high T. The reaction order at 300 K was found to be 3 below 1 atm and became 2 above 1 atm.[70e] Water impurities apparently have disturbed many measurements. A somewhat complex reaction mechanism may result from the character of the potential surfaces. Experiments on the O + CO luminescence[70d] must also be explained from the complex potential surface. Important additional information is provided by studies of the O + CO isotope exchange.[70c,53f3]

Thermal emission during the decomposition of CS_2, like NO_2, shows peculiar behaviour.[71] There must be highly nonequilibrium populations of excited electronic states. As these states are rather strongly coupled to vibrationally excited states in the ground state,[65] remarkable deviations from equilibrium populations must also be produced in the ground state.

1.5. TETRATOMIC MOLECULES

In experimental investigations of the unimolecular dissociation of tetratomic molecules the problems of suppressing secondary reactions and the influence of impurities become more and more important. A summary of some dissociation results of tetratomic systems is given in Table 1.8.

Dissociation of H_2O_2, leading in the first step to two OH radicals, has been measured under flow system conditions[72a] and in shock waves.[72b] Both sets of measurements consistently fit together: the rate constant has been studied over 7 orders of magnitude. Secondary reactions of OH with H_2O_2 could be suppressed only for $T \geq 1400$ K. At $T \leq 1000$ K quasistationary concentrations of OH and HO_2 simplify the evaluation of k. At $1000 \leq T \leq 1400$ K the condition of quasistationarity of HO_2 and OH is not fulfilled, requiring a detailed analysis of the complete mechanism.[72b] In flow system studies the influence of different carrier gases was investigated. At carrier gas

TABLE 1.7

Spin-forbidden dissociation of triatomic molecules.

Reaction	E_0 [kcal mol^{-1}]	T [K]	$k_0/[Ar]$ [cm^3 mol^{-1} s^{-1}] E_a in cal mol^{-1}	k_∞ [s^{-1}]	Ref.
$N_2O(^1\Sigma) \to N_2 + O(^3P)$	~63	840–1050	$5 \cdot 2 \times 10^{14} \exp(-59\,200/RT)$		66cd
		1500–2500	$5 \times 10^{14} \exp(-58\,000/RT)$		66f
		1700–2200	$5 \times 10^{13} \exp(-49\,500/RT)$		66e
		1400–2000		$1 \cdot 3 \times 10^{11} \exp(-59\,000/RT)$	66f
$CO_2(^1\Sigma^+) \to CO + O(^3P)$	~128	2800–4400	$5 \times 10^{14} \exp(-99\,000/RT)$		67c
		6000–11\,000	$1 \cdot 8 \times 10^{13} \exp(-77\,000/RT)$		67d
		3500–6000	$3 \cdot 6 \times 10^{13} \exp(-80\,000/RT)$		67d
		2800–3700		$2 \times 10^{11} \exp(-110\,000/RT)$	67b
$CS_2(^1\Sigma^+) \to CS + S(^3P)$	~100	2250–3350	$4 \times 10^{15} \exp(-81\,800/RT)$		68b
		1950–2800		$8 \times 10^{12} \exp(-89\,000/RT)$	68a
$COS(^1\Sigma) \to CO + S(^3P)$	~71	1600–3100	$1 \cdot 5 \times 10^{14} \exp(-60\,700/RT)$		69a
				$3 \cdot 7 \times 10^{11} \exp(-68\,300/RT)$	69a

TABLE 1.8

Dissociation of tetratomic molecules.

Reaction	D_0° [kcal mol^{-1}]	T [K]	$k_0/[\text{Ar}] [\text{cm}^3\text{ mol}^{-1}\text{ s}^{-1}]$ E_a in cal mol^{-1}	Ref.
$H_2O_2 \rightarrow 2OH$	49·6	720–950	$5\cdot7 \times 10^{16} \exp(-46\,300/RT)$	72a
		950–1450	$2 \times 10^{16} \exp(-43\,000/RT)$	72b
$NH_3 \rightarrow NH_2 + H$	102(108?)	3000	$8\cdot5 \times 10^9$	74a
$NO_2Cl \rightarrow NO_2 + Cl$	29·5	453–523	$6\cdot4 \times 10^{15} \exp(-27\,500/RT)$	75b
		400–1000	$1\cdot3 \times 10^{15} \exp(-26\,700/RT)$	75c
$CH_2O \rightarrow CHO + H$	~82	1400–2200	$5 \times 10^{16} \exp(-72\,000/RT)$	73a

pressures near 1 atm the reaction has always been found to be in the low pressure region. The transition to the high pressure range was observed with shock waves at about 20 atm.[72b]

Extrapolation of the low pressure high temperature data to room temperature and conversion to recombination rates gives good agreement with direct measurements of recombination of OH radicals.[72c]

Dissociation of formaldehyde, CH_2O, at comparably low temperatures is obviously determined by a complex decomposition mechanism. Conclusions on the unimolecular dissociation can only be drawn from measurements at high temperatures under shock wave conditions.[73a] In this system the primary dissociation leading to formyl radicals is followed by decomposition of CHO and subsequent reactions of H atoms with CH_2O and CHO. By analysing the chain mechanism the rate constant of the unimolecular reaction was derived.[73a]

As with CH_2O, the ammonia dissociation[74] has been shown to be sensitive to both secondary reactions and impurities. By lowering the reactant concentration as far as possible and by suppressing the water content of reaction mixtures in the shock tube, the unimolecular reaction was probably observed over a broad temperature range.[74a] In earlier shock wave investigations probably only the high temperature end gave the true rate of the unimolecular dissociation. In addition to the low pressure limit, the transition to the high pressure region was observed at carrier gas pressures between 10 and 300 atm. Comparison with the reverse recombination, studied at room temperature[74b] is very interesting. As expected for polyatomic molecules the change in reaction order corresponding to the transition between the low and high pressure regions is shifted to considerably lower pressures at lower temperatures.

Dissociation of NO_2Cl has been studied both in the low pressure region and in the transition region up to the high pressure limit,[75] as well as in the liquid phase.[75a] The results have been compared with theory in Slater's book (see sections 1.7 and 1.8). A great variety of collision partners was studied.[75b] As with iodine, there is a correlation between boiling points of collision partners and their collision efficiencies for dissociation. This is shown in Fig. 1.5. Limiting high pressure rates and liquid phase rates disagree by about a factor of 10. This may be understood by cage effects in the liquid.

Dissociation of $(CN)_2$ was investigated[76a] in connection with the question of NC—CN bond energy. Recombination of CN radicals was studied.[76b] Studies of the dissociation of ClF_3,[77] F_2O_2[78] and $FClO_2$ have been described.[79] Data on recombination rates are available only in the following systems. Recombination $O + NO_2 \rightarrow NO_3$ was studied in the low pressure third-order region[80a] as well as in the full transition range up to the high pressure limit.[80b] As in the NO_2 system (section 1.4.1(a)) isotope exchange data are also available.[80c] Data on NO_3 formation at high pressures and on isotope exchange between O and NO_2 may be compared using arguments similar to those for the NO_2 and O_3 systems (see section 1.4). Comparison of the high pressure data with the rate of the bimolecular reaction $O + NO_2 \rightarrow NO + O_2$ and its dependence on carrier gas pressure[80b] shows that the latter reaction involves intermediates with much shorter lifetimes than those in the former reaction.

Investigations have been made of the recombinations $O + SO_2 \rightarrow SO_3$,[81] $H + SO_2 \rightarrow HSO_2$,[82] and $NO + CN \rightarrow NOCN$.[83]

1.6. ELEMENTARY PHYSICAL PROCESSES IN DISSOCIATION AND RECOMBINATION

1.6.1. Individual Reaction Paths

The simple three-step mechanism of dissociation and recombination, given in sections 1.2.2 and 1.2.3, must be generalized to obtain a quantitative description of dissociation and recombination rates. Collisional activation and deactivation steps, (1.4) and (1.5), and the final dissociation of highly excited molecules (1.6), never proceed in a unique way. Instead, many different individual reaction paths exist which contribute to the overall reaction. This will be illustrated briefly in the following by looking at the fate of one particular molecule.

The energy required for dissociation is transferred to the molecule in a large number of collisions. In these collisions either (i) internal energy, i.e. vibrational and rotational energy, is exchanged between the collision partners, or (ii) translational energy of the relative motion is converted to internal energy or vice versa. The molecule will not gain large amounts of internal energy in one collision very often. Normally the internal energy of the molecule will change many times before the energy threshold is obtained. This is illustrated in Fig. 1.13. Here, the internal energy E of one molecule is plotted against time t. Vertical steps correspond to energy transfer in collisions, horizontal parts of the curve correspond to times between collisions. The situation shown will be observed at fairly high pressures where deactivation of molecules with $E > E_0$ becomes important. The energy E may thus exceed the threshold E_0 without dissociation occurring simply because subsequent collisions deactivate the molecule. The time lag between the last collision and final occurrence of dissociation is represented by the interval between t'' and t'. It is assumed that at $t > t'$ further collisions can no longer deactivate the molecule.

At low pressures the horizontal parts of the curve become so long that final dissociation occurs at t''': every molecule which reaches E_0 decomposes before possible deactivation by subsequent collisions. Therefore at low pressures the "mean first passage time" of molecules from $E = 0$ to $E = E_0$ is a measure of the dissociation rate (see refs. 84, 85).

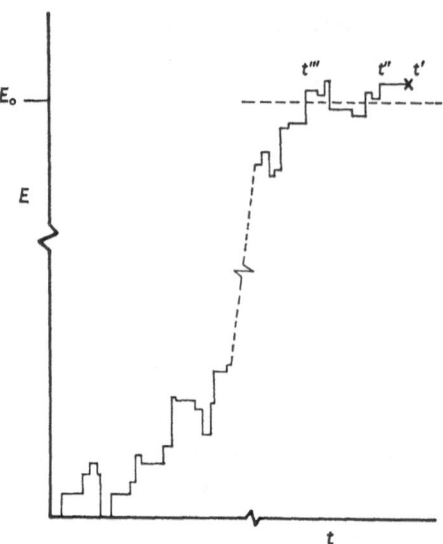

FIG. 1.13. Internal energy in a dissociating molecule.

At high pressures the internal dynamics in the molecule during the time interval t'' to t' is of further interest. A schematic description is given by Fig. 1.14. Here, for a polyatomic molecule the stretching coordinate q of the bond which finally breaks is plotted against time. At time t'' the molecule is perturbed by its last collision before dissociation: the finite duration is indicated on the figure. At $t > t''$ the stretching of the bond is determined only by the coupling to the other parts of the molecule. As the bond will not be a normal vibration of the molecule, even without anharmonicity a complex time behaviour of the stretching is to be expected. This will become still more pronounced if anharmonicity at large bond extensions is taken into account. At time t' the extension becomes so large that neither coupling of the particular bond to other bonds nor a subsequent collision can prevent dissociation. However, no definite critical bond extension can be defined from this picture. Healing of the bond may take place between the times t'' and t' shown in Fig. 1.14.

As the fate of individual molecules cannot be investigated experimentally yet, in practice many different reaction paths of the kinds illustrated in Figs. 1.13 and 1.14 occur simultaneously. Statistical laws which determine the probabilities of different paths play an important role. Before these laws are discussed in section 1.6.3, a presentation of the "bottle-neck" picture of a dissociation in terms of a reaction path with only few stages during reaction is instructive: this is discussed in section 1.6.2.

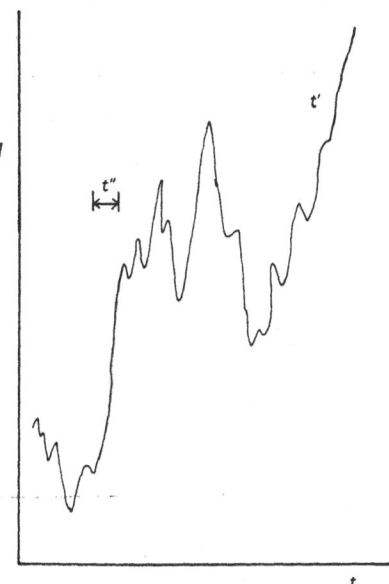

FIG. 1.14. Breaking of a bond in a dissociating molecule.

1.6.2. Bottle-neck Representation of Dissociation

In the following a very simple specification of the quantum states is used; only three levels of internal energy and three regions of extension of the dissociating bond are considered. Possible transitions between these states are shown in Fig. 1.15: only transitions to neighbouring states are considered. Often one of these transitions will be the slowest and therefore rate determining step. If this step is known, the description becomes particularly simple. For example, in the low pressure region of dissociation the transitions

$$A''' \longrightarrow A'' \longrightarrow A' \longrightarrow P$$

will be fast compared to collisional processes: only the rates of the transitions

$$A_0 \rightleftarrows A_1 \longrightarrow A_2$$

will be contained in the dissociation rate. The number of transitions per unit time from A_0 to A_1 is denoted by

$$k_{10}[A_0] \tag{1.31}$$

Corresponding definitions are used for the other rate constants, k_{01} and k_{21}. Two cases can be distinguished: If $k_{21} > k_{01} > k_{10}$, the "bottle-neck" of the reaction will be the transition $A_0 \rightarrow A_1$, and the overall rate constant of dissociation k will be equal to k_{10}. If instead $k_{21} < k_{01}$ and $k_{01} > k_{10}$, the "bottle-neck" will be the transition $A_1 \rightarrow A_2$ and k will be equal to $k_{10}k_{21}/k_{01}$. If it can be assumed that

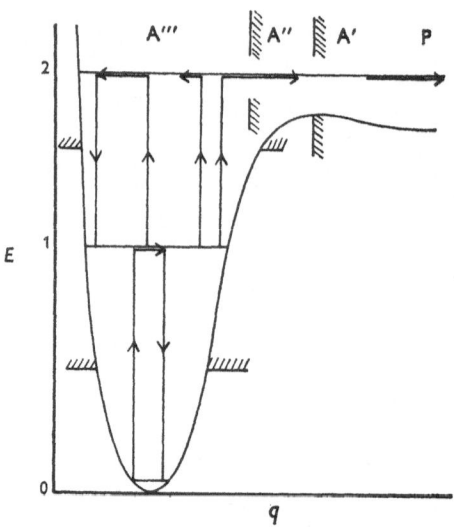

FIG. 1.15. Bottle-neck representation of dissociation.

k_{10}/k_{01} is equal to $([A_1]/[A_0])_{equ}$ independent of the occurrence of reaction, one obtains

$$k = ([A_1]/[A_0])_{equ} \cdot k_{21} \qquad (1.32)$$

Details of the fast processes before the bottle-neck are lost. These processes only ensure that an equilibrium population is maintained up to the bottle-neck. Also details of the fast processes behind the bottle-neck do not enter into the overall rate.

In general, the rate is given by the equilibrium population before the bottle-neck times the flow rate through the bottle-neck. At high pressures, the transition $A''' \rightarrow A''$ can be the rate determining step, if intramolecular coupling is rate determining (one example of this case would be a radiationless transition between electronic states). Often the bottle-neck will be the crossing ($A'' \rightarrow A'$) of a "critical surface", which may be located at potential barriers or, if these do not exist, at places with minima of the local entropy,[117] as shown in Fig. 1.15. At very high densities the separation of products $A' \rightarrow P$ may become rate determining.

The expressions for the rate constant given above apply to the quasistationary period of reaction. During this period the concentrations of all states of A, i.e. of A_0, A_1, A_2, A''', and A'', decay with the same rate constant k. In the initial period of reaction, where the quasistationary concentrations are built up, this is not true. During this time a time-dependent "rate constant" will normally be observed. This is easily seen from the complete kinetic description of a three-stage mechanism, which is given, e.g. in ref. 86.

Comparison of the experimental results with the different possible rate constants obtained from the simple model immediately allows some fundamental conclusions. The quasistationary stage of reaction is normally approached in times which are shorter than the time resolution of the experimental set up. The pre-quasistationary stage has been observed only at extremely high temperatures for a few diatomic molecules. In all other cases time independent rate constants are observed: these are not to be confused with the final approach to chemical equilibrium, when recombination starts to compete with dissociation. In the low pressure range in general it is not transitions between low vibrational levels but transitions between states near the energy threshold which are rate determining. Otherwise, quite different temperature coefficients from those actually observed experimentally would be expected. The effective bottle-neck well may be smeared out over several levels near the energy threshold, but must be located at energies near E_0. In the high pressure range in some cases slow intramolecular coupling, e.g. through forbidden electronic transitions, is found to be rate determining. In many cases, however, the crossing of the activated complex at large bond extensions is rate determining. This is defined more precisely in section 1.8.

1.6.3. General Transport Equations at Low Pressures

In the following we express the overall rates of dissociation and recombination in terms of the "microscopic" physical processes considered in section 1.6.1. We restrict ourselves at first only to the low pressure limit. Here occurrence of reaction may be defined by the passage of molecules at the energy threshold E_0, as shown in sections 1.6.1 and 1.6.2.

The vibrational energy levels E_i of the dissociating molecules with $E_i < E_0$ are denoted by the indices i and j, see Fig. 1.16. The number of transitions per unit time from level i to level j produced by energy transfer in collisions between molecules A and collision partners M are denoted by

$$k_{ji}[A_i] \tag{1.33}$$

The population of state i, $[A_i]$, is thus determined by the result of gains and losses due to transitions to all other states j:

$$\frac{\partial [A_i]}{\partial t} = + \sum_{j}^{*} k_{ij}[A_j] - \sum_{j} k_{ji}[A_i] \tag{1.34}$$
$$\scriptstyle E_j < E_0$$

The first sum $\sum^{*}$ describing the gain is restricted to the discrete levels below E_0. E_0 is defined by the fact that all states with $E_i > E_0$ are negligibly populated in the low pressure limit, because the dissociative lifetimes are shorter than the time between collisions. The second sum also includes transitions from discrete states $E_i < E_0$ to states above the energy threshold.

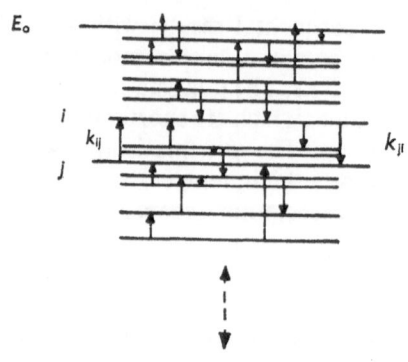

FIG. 1.16. Transitions between discrete states of a dissociating molecule.

The collisional transition rates k_{ij} and k_{ji} can be expressed in terms of cross-sections for the individual collision processes. Of course, k_{ij} and k_{ji} are quantities averaged over many parameters, e.g. the relative velocities of the collision partners and their orientations, the phases and amplitudes of the vibrations in A and M, the internal states of M etc. However, for individual cases these averages are well defined if it is assumed that the distribution of translational states is Maxwellian and also that the distribution of the phases of the oscillators etc. before collision are in equilibrium. These assumptions appear to be valid in many experimental situations. Consequently if the dissociation of a diatomic molecule A in large excess of a monatomic carrier gas M is considered, k_{ji} is given by

$$k_{ji} = [\text{M}] \left\langle \iint \sigma_E(j \mid i; v) f_\text{A}(v_\text{A}) f_\text{M}(v_\text{M}) \left| v_\text{A} - v_\text{M} \right| dv_\text{A} \, dv_\text{M} \right\rangle \text{ph., or.}$$

$$(1.35)$$

Here σ_E is the total cross section for energy transfer from vibrational state i to j by collision at a relative collision velocity v; f_A and f_M are the equilibrium distributions of velocities of A and M. The right hand side is averaged over all initial phases of the vibrations (ph.) and orientations of the collision partners (or.) before collision. The symbols used in equation (1.35) are similar to those given in ref. 87. If collision partners M are polyatomic, the averaging must also be performed over the phases and distributions of internal states of M; these are assumed to be in equilibrium.

Up to now rotation of the molecules A has been neglected: it is assumed that dissociation is brought about by vibrational energies larger than E_0. Rotational energy may promote dissociation by centrifugal forces and must be taken into account. A complete description including the combined transfer of vibrational and rotational energy will be very complicated: the effective energy threshold will depend on the particular rotational states. We shall correct for rotational contributions in a simplified manner below.

Following the discussion of microscopic reversibility for cross-sections and detailed balancing for rate coefficients given in ref. 87, one can derive an essential property of the thermal averaged k_{ji}:

$$k_{ji}[A_i]_{\mathrm{equ}} = k_{ij}[A_j]_{\mathrm{equ}} \qquad (1.36)$$

By the thermal averaging (equation (1.35)) and the principle of detailed balancing (equation (1.36)), the reacting molecules A are coupled to the heat reservoir M; these molecules therefore have "a temperature" in spite of the occurrence of reaction.

The logical way to proceed would be to calculate the averaged energy transfer rates k_{ji} and, after this, to solve the system of coupled differential equations (1.34). However, the theory of energy transfer processes between non- or slightly-excited oscillators (see e.g. ref. 88) used to describe vibrational relaxation phenomena is probably not applicable to the collisions which are rate determining in dissociation. Firstly for these collisions pre-excitation of oscillators must be taken into account; secondly the adiabatic approximation may well break down. Even in the case of the relatively slight excitation produced in the fluorescence studies of I_2, the "normal" theory of energy transfer did not apply.[89] New attempts are being made to describe energy transfer in highly excited oscillators.[90] Also the number of states i which are important for dissociation increases enormously with increasing complexity of the molecules: a complete calculation of all k_{ji} would be both extremely tedious and, because of the final averaging, not very useful. Thus a formal solution of equation (1.34) with later specification of k_{ji} appears to be advisable. This solution together with the corresponding results for the recombination at the low pressure limit is developed in section 1.7.

1.6.4. General Transport Equations at High Pressures

As was mentioned in sections 1.6.1 and 1.6.2, dissociation and recombination in the high pressure range can no longer be described in terms of the energy content of the molecule alone. The dynamics of internal vibrations, and in particular the occurrence of some critical extension of one bond, determines the rate determining step.

At low pressures to a good approximation the mean first passage time of a molecule at E_0 depends only on collision properties. However at high pressures collisional processes also determine where the bottle-neck of reaction is located and must be considered in addition to intramolecular processes: A complete description thus becomes much more complicated. A suitable starting point in this region seems to be given by the generalized Liouville equation including collisions:[91]

$$\frac{\partial[A(q, p, t)]}{\partial t} + \{[A(q, p, t)], H\}$$

$$= \iint (k(q, p; q', p')[A(q', p', t)] - k(q', p'; q, p)[A(q, p, t)]) \, dq' \, dp'$$

$$(1.37)$$

Here q and p determine the complete set of internal coordinates and momenta of the molecule A in classical phase space. $[A(q, p, t)]$ describes the density of molecules A at some point in phase space. The collisional integrals on the right-hand side contain the contribution of collisional gains and losses: $k(q, p; q', p')$ is the collisional transition rate from the point (q', p') to the point (q, p). The Poisson bracket $\{[A], H\}$ determines the flux due to intramolecular motion undisturbed by collisions; $H(q, p)$ is the intramolecular Hamiltonian. The collisional transition rates are again averaged over the distribution of translational velocities on collision, as in equation (1.35). However the phase average is not carried out at this stage. Equation (1.37) is not restricted to the high pressure limit but is quite general. By integrating over energy surfaces E with only mild restrictions,[92] equation (1.34) can be derived from equation (1.37).

The solution of equation (1.37) at the high pressure limit is given in section 1.8. Another transport equation which is less general is given below. This equation is very useful particularly for the fall-off region and to compare data with other dissociation systems (see section 1.1). The system is described in terms of variable energy, with a specific dissociation rate constant $k(E_i)$. With this notation one obtains at general pressures

$$\frac{\partial[A_i]}{\partial t} = \sum_j k_{ij}[A_j] - \sum_j k_{ji}[A_i] - k(E_i)[A_i] \qquad (1.38)$$

Of course if this form of transport equation is used the boundary which divides undissociated molecules and dissociation products must be specified. It determines the rates of dissociation $k(E_i)$. A boundary of this sort represents the bottle-neck of reaction, and can only ultimately be derived from the more general equation (1.37). The use of equation (1.38) requires additional assumptions, which may be difficult to justify. In particular, if the rate constant $k(E_i)$ is used to compare dissociation systems with differing modes of activation, different values of $k(E_i)$ may be found owing to the different populations in phase space.

1.7. RATE COEFFICIENTS AT LOW PRESSURES

1.7.1. Populations during the Pre-Quasistationary Period

With an assumed set of collisional transition rates, k_{ji}, equation (1.34) has been solved numerically[93] for the 14 vibrational levels of H_2. The populations obtained at different times after the sudden temperature rise from "0 K" to 2000 K is shown in Fig. 1.17. As under the assumed conditions the half life is $> 10^{-5}$ s, one can see how rapidly an almost equilibrium population is established. Similarly the overall rate constant for dissociation increases from 0 to the quasistationary value in a time much shorter than the half life of H_2. The length of the pre-quasistationary period may be assumed to be of the order of the vibrational relaxation time. We shall consider only the quasistationary

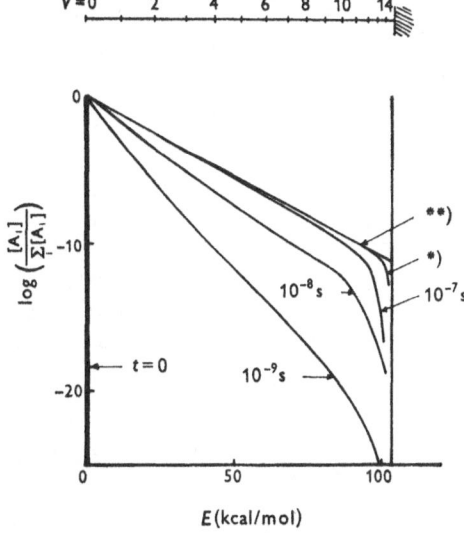

FIG. 1.17. Relative populations during the initial phase of H_2 dissociation (after ref. 93),

$$*g(E_i) = f(E_i)\left(1 - \exp\left(-\frac{E_0 - E_i}{kT}\right)\right)$$

$$**g(E_i) = f(E_i), \quad T = 2000 \text{ K}$$

period; this corresponds to the portion of the reaction observed at all except the very highest temperatures.[5]

1.7.2. Populations and Rate Coefficients during the Quasistationary Period

During the main period of dissociation, i.e. the quasistationary period, only the smallest nontrivial eigenvalue of equation (1.34) determines the change in the system with time. Therefore in this period all populations of states with $E_i < E_0$ decay with the same rate constant and the relative concentrations of $g(E_i)$ become nearly stationary:

$$\frac{\partial g(E_i)}{\partial t} \longrightarrow 0 \quad \text{with} \quad g(E_i) = \frac{[A_i]}{\sum_i [A_i]} \tag{1.39}$$

The relative populations $g(E_i)$ during reaction will in general be non-equilibrium. As the overall rate constant k is intimately related to $g(E_i)$, one may determine the shape of $g(E_i)$ first and then derive k from $g(E_i)$.

Introducing equation (1.39) into equation (1.34) leads to

$$g(E_i) \simeq \frac{1}{Z(E_i)} \sum_{\substack{j \\ E_j < E_0}}^* k_{ij} g(E_j) \tag{1.40}$$

with $Z(E_i)$ defined as

$$Z(E_i) = \sum_j k_{ji} \tag{1.41}$$

As will be shown in section 1.7.4, the vibrational states at energies of the order of E_0 form a quasicontinuum. This is due to the large number of ways of distributing large amounts of vibrational energy amongst the oscillators of polyatomic molecules. To some extent this also holds for levels near the dissociation limit in diatomic molecules. Thus, to a first approximation one may replace the set of linear equations in (1.40) by a single integral equation:

$$g(E_i) = \frac{1}{Z(E_i)} \int_0^{E_0} k(E_i \mid E_j) g(E_j) \, \mathrm{d}E_j \tag{1.42}$$

with

$$Z(E_i) = \int_0^\infty k(E_j \mid E_i) \, \mathrm{d}E_j \tag{1.43}$$

The principle of detailed balancing for k_{ij} then takes the form

$$k(E_i \mid E_j) f(E_j) = k(E_j \mid E_i) f(E_i) \tag{1.44}$$

where $f(E_i)$ is the equilibrium value of the relative population $g(E_i)$ of vibrational states.

The solution of the integral equation (1.42) was derived in ref. 6 for special types of collisional transition rates $k(E_j \mid E_i)$; these must of course obey equation 1.44. If, for example, $k(E_j \mid E_i)$ has the exponential cusp-like form shown in Fig. 1.18, with (i) a range which is not too large compared to the changes in the densities of vibrational levels and (ii) with a shape which does not vary with energy between $E_i \approx E_0 - kT$

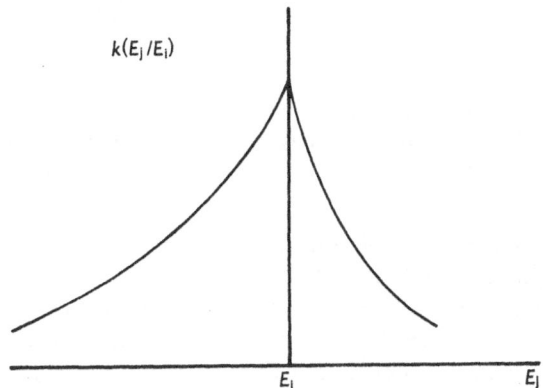

FIG. 1.18. Exponential model of collisional transition rates.

and E_0, the solution of equation (1.42) is

$$g(E_i) \approx f(E_i)\left[1 - \frac{\exp\left(-\dfrac{E_0 - E_i}{kT}\right)}{1 + \sqrt{\langle \Delta E^2(E_0)\rangle}/\sqrt{2}\,kT}\right] \qquad (1.45)$$

with

$$\langle \Delta E^2(E_0)\rangle = \frac{1}{Z(E_0)}\int_0^\infty (E_j - E_0)^2 k(E_j \mid E_0)\,dE_j \qquad (1.46)$$

and $Z(E_0)$ given by equation (1.43). Then

$$g(E_i) \approx f(E_i) \quad \text{at} \quad \sqrt{\langle \Delta E^2(E_0)\rangle} \gg kT \qquad (1.47)$$

This is the "strong collision limit". At $\sqrt{\langle \Delta E^2(E_0)\rangle} \ll kT$, one obtains

$$g(E_i) \approx f(E_i)\left[1 - \exp\left(-\frac{E_0 - E_i}{kT}\right)\right] \qquad (1.48)$$

This situation is illustrated in Fig. 1.19. In the "weak collision limit" the bottle-neck of reaction, used in the sense of section 1.6.2, becomes

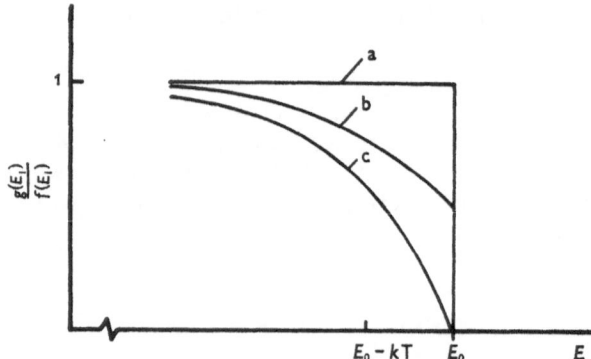

FIG. 1.19. Populations during low pressure dissociation, model for $k(E_j | E_i)$ described in Fig. 1.18. a = strong collision limit, b = intermediate case, c = weak collision limit.

smeared out over the levels around $E_0 - kT$; in the strong collision limit, it is exactly localized at E_0.

Having determined the population distribution during reaction one can immediately derive the overall rate constant k by simple integration. k represents the collisional flux from levels below E_0 to levels above E_0,

$$k = \int_{E_0}^{\infty} \left\{ \int_0^{E_0} k(E_j \mid E_i) g(E_i)\, dE_i \right\} dE_j \qquad (1.49)$$

The equilibrium distribution $f(E_i)$ is represented by

$$f(E_i) = \frac{\rho(E_i) \exp(-E_i/kT)}{Q_{\mathrm{vib}}(T)} \qquad (1.50)$$

where $Q_{\mathrm{vib}}(T)$ is the vibrational partition function. $\rho(E_i)$ is the density of vibrational levels, i.e. the number of vibrational states per unit energy interval. If $\rho(E_i)$ changes slowly compared to the exponential function,

$$k \approx Z_0 P_\sigma P_{\Delta E^2} \frac{\rho(E_0)kT}{Q_{\mathrm{vib}}(T)} \exp\left(-\frac{E_0}{kT}\right) \qquad (1.51)$$

Here the hard-sphere collision number Z_0 is introduced as a reference collision number,

$$Z_0 \equiv [\mathrm{M}]\sigma_0 N_{\mathrm{L}} \sqrt{\frac{8RT(\mathrm{M_A} + \mathrm{M_M})}{\pi \mathrm{M_A M_M}}} \qquad (1.52)$$

where σ_0 = hard-sphere cross section, N_L = Loschmidt's number, $\mathrm{M_A}$ and $\mathrm{M_M}$ = molar masses of A and M. The efficiency per collision,

$P_\sigma P_{\Delta E^2}$, is divided into two factors. P_σ is given by

$$P_\sigma = \frac{\sigma_{tot}(E_0)}{\sigma_0} \tag{1.53}$$

with a "total cross section for energy transfer"

$$\sigma_{tot}(E_0) \equiv \frac{1}{Z_0} \int_0^\infty k(E_j \mid E_0)\, dE_j \tag{1.54}$$

which can be derived immediately from equations 1.43 and 1.35 by averaging the cross section $\sigma_E(j \mid i; v)$ for individual transitions between vibrational states. $P_{\Delta E^2}$ is given by

$$P_{\Delta E^2} = \left(1 + \frac{\sqrt{2}\, kT}{\sqrt{\langle \Delta E^2(E_0)\rangle}}\right)^{-2} \tag{1.55}$$

with $\langle \Delta E^2(E_0)\rangle$ defined in equation (1.46). Whereas the first part of the collision efficiency, P_σ, is a measure only of the total number of energy transfer processes, the second part, $P_{\Delta E^2}$, describes the influence of the energy range of the function k_{ji} shown in Fig. 1.18. The second part in the strong collision limit approaches 1, whereas in the weak collision limit it becomes

$$P_{\Delta E^2} \longrightarrow \langle \Delta E^2(E_0)\rangle / 2(kT)^2 \tag{1.56}$$

The expression (1.51) derived for the overall rate constant k shows that most details of individual collision processes are lost by thermal averaging. However, two averaged quantities of collisional transitions at the bottle-neck near E_0, i.e. $\sigma_{tot}(E_0)$ and $\langle \Delta E^2(E_0)\rangle$, enter into the rate constant. From low pressure dissociation studies one can thus derive information on the product $\sigma_{tot}(E_0)\langle \Delta E^2(E_0)\rangle$ at the weak collision limit or $\sigma_{tot}(E_0)$ at the strong collision limit. Little use has yet been made of this source of information on collision processes.

The exponential model used so far for the averaged collisional transition probabilities, appears to be a very reasonable starting point. It obeys the principle of detailed balancing and also the probability of transition decreases with increasing separation of the levels. If other functional forms are chosen, for instance a stepladder model, one can look for a solution of equation (1.42), e.g. by the method of Fourier transforms.[6] Another solution of the step-ladder model with discrete levels is given in section 7.4. Instead of discussing other models, we shall restrict ourselves to the weak and strong collision limits; here a solution is possible which is independent of the functional form. The solution (1.45) can then be considered as one way of correlating the expressions obtained at the limits.

Since equation (1.47) is valid at the strong collision limit for all possible values of k_{ji}, we can perform a simple expansion around this limit. The integral equation (1.42) can be solved by iteration:[96,97] it converges rapidly near the strong collision limit. $g^{(0)}(E_i) = f(E_i)$ is used as zeroth approximation. From equation 1.42 one obtains the first approximation

$$g^{(1)}(E_i) = f(E_i)\left(1 - \frac{\displaystyle\int_{E_0}^{\infty} k(E_j \mid E_i)\,\mathrm{d}E_j}{Z(E_i)}\right) \tag{1.57}$$

Calculation of k from equation (1.57) shows that for general $k(E_j \mid E_i)$ the first approximation results agree with equation 1.55 up to terms of first order in $\sqrt{2}\,kT/\sqrt{\langle \Delta E^2(E_0)\rangle}$. The smaller $\sqrt{\langle \Delta E^2(E_0)\rangle}/\sqrt{2}\,kT$, the slower the iteration converges. Therefore near the weak collision limit a different procedure must be used. This is described in the next section.

1.7.3. Diffusion Theory of Dissociation at Low Pressures

The important weak collision limit, $\sqrt{\langle \Delta E^2(E_0)\rangle} \ll kT$, can be treated in a rather general way[6,94,98–100] without assuming a particular functional dependence of $k(E_j \mid E_i)$ on E_j and E_i. In this case we can convert the classical equivalent of equation (1.34) into a simpler equation in the form of a diffusion equation. Therefore, the weak collision limit may also be described as the "diffusion limit of dissociations".

It has been shown in ref. 98, that by expanding the collision integrals corresponding to the right-hand side in equation (1.34) about $E_j = E_i$ one can derive from the integral equivalent of equation (1.34) the equation

$$\frac{\partial [A(E_i)]}{\partial t} \approx \frac{\partial}{\partial E_i}\left\{\tfrac{1}{2} D(E_i) \frac{\partial [A(E_i)]}{\partial E_i}\right\} \tag{1.58}$$

with

$$D(E_i) = \int_{-\infty}^{\infty} k([E_i + \tfrac{3}{2}\Delta E] \mid [E_i + \tfrac{1}{2}\Delta E])\,\Delta E^2\,\mathrm{d}\,\Delta E \tag{1.59}$$

or, if the change of D with E_i is not too large, with

$$D(E_i) \approx \langle \Delta E^2(E_i)\rangle Z(E_i) \tag{1.60}$$

where

$$\langle \Delta E^2(E_i)\rangle \approx \frac{1}{Z(E_i)}\int_{-\infty}^{\infty} k(E_j \mid E_i)(E_j - E_i)^2\,\mathrm{d}E_j \tag{1.61}$$

and

$$Z(E_i) = \int_{-\infty}^{\infty} k(E_j \mid E_i)\,\mathrm{d}E_j \tag{1.62}$$

Expressions (1.59) and (1.60) may be understood as the "diffusion coefficient" for the "diffusion process" at the energy ladder, illustrated in Fig. 1.13. The "diffusion coefficients" in general depend on E_i. Further ways of converting the master equation (1.34) into simpler forms are given in ref. 98 for $k(E_j \mid E_i) = F_1(E_j)F_2(E_i)$ and in ref. 100 by conversion into a Fokker–Planck equation.

A general solution[98] of the diffusion equation (1.58) during the quasistationary period of reaction is given by

$$g(E_i) \approx f(E_i) \left[1 - \frac{\displaystyle\int_0^{E_i} \frac{dE_j}{f(E_j)D(E_j)}}{\displaystyle\int_0^{E_0} \frac{dE_j}{f(E_j)D(E_j)} + \frac{1}{2f(E_0)C(E_0)}} \right] \qquad (1.63)$$

with

$$C(E_0) = \int_{-\infty}^{\infty} k([E_0 + \tfrac{3}{2}\Delta E] \mid [E_0 + \tfrac{1}{2}\Delta E]) \, |\Delta E| \, d\,\Delta E \qquad (1.64)$$

$$\approx \langle |\Delta E(E_0)| \rangle Z(E_0)$$

and

$$\langle |\Delta E(E_0)| \rangle = \frac{1}{Z(E_0)} \int_{-\infty}^{\infty} k(E_j \mid E_0) \, |E_j - E_0| \, dE_j \qquad (1.65)$$

An absorbing barrier at E_0, corresponding to the low pressure limit of dissociation, is used as boundary.

From the population equation (1.63) the overall rate constant k is obtained using equation (1.49):

$$k \approx \left\{ 2 \int_0^{E_0} \frac{dE_i}{f(E_i)D(E_i)} + \frac{1}{f(E_0)C(E_0)} \right\}^{-1} \qquad (1.66)$$

From this general equation one can see immediately the connection with equation (1.51). As long as $D(E_i) \approx Z(E_i)\langle \Delta E^2(E_i) \rangle$ near $E_i \approx E_0$ varies much slower with E_i than $f(E_i)$ and as long as corresponding to the assumptions in a diffusion theory $\langle \Delta E^2(E_0) \rangle \ll \langle |\Delta E(E_0)| \rangle 2kT$ is fulfilled, one obtains from equation (1.66):

$$k \approx Z(E_0) \frac{\langle \Delta E^2(E_0) \rangle}{2(kT)^2} \frac{\rho(E_0)kT}{Q_{\mathrm{vib}}(T)} \exp\left(-\frac{E_0}{kT} \right) \qquad (1.67)$$

in agreement with the diffusional, weak collision limit of equation (1.51).

1.7.4. Step-ladder Models

As long as the integral description with a continuous energy scale in equation (1.34) is adequate, the weak collision limit is described by equation (1.66) and the strong collision limit by equation (1.51) with $P_{\Delta E^2} = 1$. For a special functional shape of $k(E_j \mid E_i)$ i.e. the exponential model, a complete formula connecting both limiting cases is given by equations (1.51) and (1.55). Therefore, most cases of $k(E_j \mid E_i)$ can be treated reasonably well. However, situations where only next neighbour transitions between discrete levels or pronounced discrete contributions superimposed on a quasi-continuum of $k(E_j \mid E_i)$ are important, must be treated separately. In this case the solutions of equation (1.34) for step-ladder models with only nearest neighbour transitions from ref. 94, 98, 101, 102 are applicable:

$$k \approx \left\{ \sum_{i=1}^{P^*} (k_{i(i-1)} f_i)^{-1} \right\}^{-1} \tag{1.68}$$

$P^* = $ total number of states below E_0.

1.7.5. Densities of Vibrational States

A major point of investigation has been the density of vibrational levels $\rho(E)$; this is important in determining the equilibrium population of states. The value of ρ at energies near E_0 is needed for dissociation reactions, see e.g. equation (1.51). The first approach appears to be that of ref. 103, where the number of vibrational states in the range $[E + dE, E]$ was derived for molecules consisting of classical oscillators. Models for quantized oscillators with identical or commensurable frequencies have been considered in ref. 104. The following expressions appear to be the most suitable for practical applications.

The density of states of one oscillator of the energy eigenvalue ε_i can be adequately represented by

$$\rho_i(E_i) = \frac{1}{\varepsilon_i} \tag{1.69}$$

Correspondingly, the number of states $n(E_i)$ in the energy interval $[E_i, -\varepsilon_i/2]$ is approximately

$$n_i(E_i) \approx \int_{-\varepsilon_i/2}^{E_i} \rho_i(E) \, dE = \left(E_i + \frac{\varepsilon_i}{2} \right) \Big/ \varepsilon_i \tag{1.70}$$

Here the lowest vibrational level is chosen as $E_i = 0$. The choice of $-\varepsilon_i/2$ as lower limit of integration in equation (1.70) corresponds to the representation of the step function for $n_i(E_i)$ in a quantized oscillator by a continuous function. This function distributes the increase of $n_i(E_i)$ at $m\varepsilon_i$ smoothly over the energy interval $[m\varepsilon_i + \varepsilon_i/2, m\varepsilon_i - \varepsilon_i/2]$, see Fig. 1.20 ($m$ integer).

Densities $\rho(E)$ and numbers of levels $n(E)$ for systems of s weakly coupled harmonic oscillators are obtained from equation (1.69) by integration:

$$\rho(E) \approx \frac{d}{dE} \int \cdots \int_{\Sigma E_i \leq E} \rho_1(E_1) \cdots \rho_s(E_s) \, dE_1 \cdots dE_s \tag{1.71}$$

Again we must ask which lower limit of integration in equation (1.71) represents the best value in the smoothing procedure. An upper limit of $\rho(E)$ is given by the integration limits $-\sum_{i=1}^{s} \varepsilon_i/2 \leq \sum_{i=1}^{s} E_i \leq E$. After

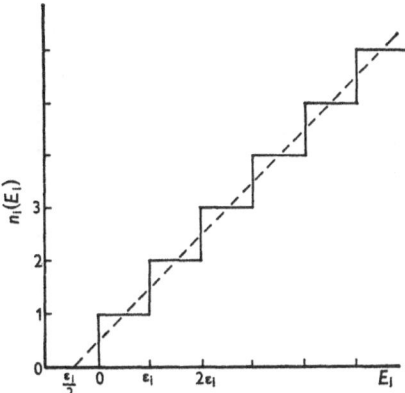

FIG. 1.20. Number of vibrational levels in one oscillator.

integration by equation (1.71), together with equation (1.69) for the different $\rho_i(E_i)$ this immediately leads to:

$$\rho(E) \approx \frac{(E + a(E)E_z)^{s-1}}{(s-1)! \prod\limits_{i=1}^{s} \varepsilon_i} \qquad (1.72)$$

$E_z = \frac{1}{2} \sum\limits_{i=1}^{s} \varepsilon_i =$ zero point energy of vibration and $a(E) = 1$. The correct expression should really use the best smoothed function of $\rho_i(E_i)$ as shown in Fig. 1.20: for every oscillator $-\varepsilon_i/2$ must be chosen as lower limit of integration in equation (1.71). By performing the integration of equation (1.71) in this way, a correction factor $a(E) < 1$ is obtained for equation (1.72). This factor has been derived, e.g. in ref. 106, 107. It is important at $E/E_z < 1$, but gives corrections smaller than 1% in $\rho(E)$ at $E/E_z \geq 1$. Clearly $a(E)$ must be known for large polyatomic molecules, where E_0 may become $\leq E_z$ because of the large number of oscillators. However, for small molecules at energies which are not too low, $a(E) = 1$ is a good approximation.

A comparison of the smoothed number of states $n(E)$ with directly counted values for discrete states is shown in Fig. 1.21. One can see that the step function valid for discrete states approaches the smoothed curve at high energies. The assumption of a quasicontinuum of vibrational states in molecules with several oscillators thus is an excellent approximation at high energies.

Smoothed densities of levels $\rho(E)$ for Morse oscillators coupled with harmonic oscillators etc. are given in refs. 6, 108. From these expressions the anharmonicity contributions to $\rho(E_0)$ can be obtained easily. A further important expression which allows us to estimate the maximum contribution from rotation is derived as follows. If one assumes

3

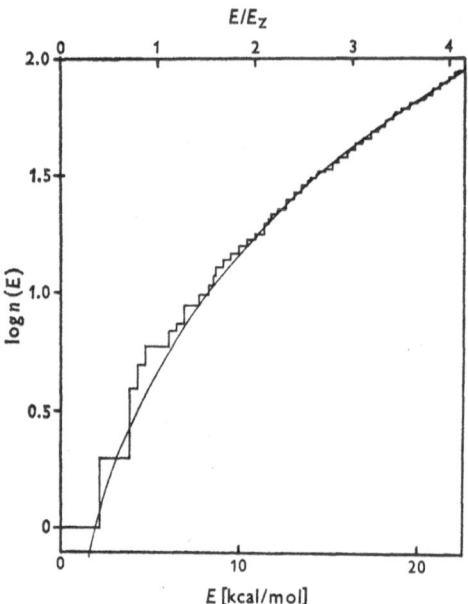

FIG. 1.21. Number of vibrational levels $n(E)$ below energy E for NO_2. Step function exactly counted; smoothed function from equation (1.72) with $a(E) = 1$.

(i) that no centrifugal barriers (see section 1.8) limit the availability of rotational energy for dissociation, so that all rotational degrees of freedom may be "active", and (ii) that collisional energy transfer can be represented by one function $k(E_j \mid E_i)$ independent of the rotational and vibrational contributions, then $\rho(E_0)$ in equation (1.51) may include rotational degrees of freedom. Then, S in $S - 1$ and $(S - 1)!$ of equation (1.72) will be increased by $r/2$, where r equals the number of rotational degrees of freedom, see ref. 106. For diatomic molecules a Morse oscillator coupled to a rigid rotator appears to be an important model.[6] Considerable contributions in determining the properties of $\rho(E)$ have been made by the statistical theory of mass spectra.[109]

1.7.6. Collision Models

While the densities of vibrational states at energies near E_0 are usually of the order 1–10 kcal^{-1} (0·24–2·4 kJ^{-1}) mol for diatomic molecules, $\rho(E_0)$ is much larger for polyatomics: Table 1.9. Even with triatomic molecules, $\rho(E_0)$ is already of the order 10^3 kcal^{-1} (240 kJ^{-1}) mol. Thus in general there are very many different ways of transferring energy which are important for dissociation. In particular, in highly excited polyatomic molecules "complex collisions", i.e. collisions in which several oscillators change their state, will play a central role

TABLE 1.9

Densities $\rho(E_0)$ of vibrational levels at E_0. (For H_2 and I_2: number of vibrational levels in the interval $[(E_0 - 1 \text{ kcal mol}^{-1}), E_0]$)

	H_2	I_2	O_3	H_2O	NO_2	N_2O	NO_2Cl	N_2H_4
kcal^{-1} mol ~ 1	28	13	16	79	880	3×10^3	$1\cdot2 \times 10^6$
kJ^{-1} mol			3·1	3·8	19	210	720	$2\cdot9 \times 10^5$

because of the decreasing separation of the vibrational states of the total molecule. In these collisions the amounts of energy exchanged can be smaller than the different ε_i by combined excitation and de-excitation of oscillators in one collision. The corresponding "continuous" collision model was used in the continuous description given in sections 1.7.2 and 1.7.3.

In sections 1.7.2 and 1.7.3 it was shown that of all the details of individual collisional processes only the total cross section for energy transfer, $\sigma_{\text{tot}}(E_0)$, and the mean squared energy transferred per collision, $\langle \Delta E^2(E_0) \rangle$, enter into the dissociation rate constant k. Consequently we are mainly interested in determining these quantities rather than trying to describe the individual processes.

The total cross section $\sigma_{\text{tot}}(E_0)$ will be sensitive to the intermolecular potential, which operates at the different collisional orientations. From the order of magnitude of experimentally determined dissociation rate constants, one can conclude that $\sigma_{\text{tot}}(E_0)$ is of the order of the gas kinetic cross section. However, pronounced differences in collision efficiencies, e.g. of water or some atoms as collision partners, may be ascribed to long range forces which increase $\sigma_{\text{tot}}(E_0)$. Information on the range of intermolecular forces obtained from vibrational relaxation studies[88] can probably also be used for dissociation. However, the influence of these forces on vibrational relaxation times and on dissociation rates is completely different, owing to the difference between complex collisions on the one hand and simple transitions between levels separated by large energy intervals on the other.

The mean squared energies $\langle \Delta E^2(E_0) \rangle$ are of course also determined by the intermolecular potentials. The duration of the collision or the lifetime of the collision complex will be of primary importance. The statistical collision model assumes a statistical distribution of the energies of all oscillators in A and M during collision. If before collision A is highly excited but M is not excited, this results in very effective energy transfer. With the statistical theory of reaction rates as discussed in section 1.8 one can easily calculate for this model values of $\langle \Delta E^2(E_0) \rangle$, see e.g. ref. 97. One finds in general $\sqrt{\langle \Delta E^2(E_0) \rangle} > kT$, and so $P_{\Delta E^2} = 1$ in equation (1.55). Details of $\langle \Delta E^2(E_0) \rangle$ for this model are of no interest for dissociation. However, this model appears to be very important in chemical activation and photolysis studies.

The statistical model of energy exchange may be applicable for cases with long lived collision complexes of polyatomic molecules. In small molecules with short collision times, two limiting cases have been considered, the adiabatic limit at low collision velocities and the nonadiabatic, "impulsive" limit at high collision velocities. For collisions between a diatomic homonuclear molecule A_2 and an atom M, the impulsive limit has been investigated in detail in refs. 92, 100. Here, different cases have to be distinguished. If the mass ratio $\gamma^* = m(M)/m(A)$ is much smaller or larger than 1, energy transfer is not very efficient, leading to $P_{\Delta E^2} \ll 1$. Maximum values of $P_{\Delta E^2}$ are obtained at $\gamma^* \approx 1$. For the impulsive collision model $P_{\Delta E^2}$ may be calculated—for $\gamma \ll 1$ it is approximately $\gamma^* E_0/2kT$. In any case, in the high velocity limit nonequilibrium distributions at $E < E_0$ must be taken into account except for $\gamma^* \approx 1$. These result in $P_{\Delta E^2} < 1$.

In the adiabatic limit, the results normally obtained from theories of energy transfer are not immediately applicable, because pre-excitation of the oscillators must be taken into account. The general formulation may be derived from ref. 88. Also in polyatomic molecules the coupling of the driven oscillator with other parts of the molecule has been shown to influence energy transfer.[90] Much work remains to be done. If the intermediate case between the adiabatic and nonadiabatic limit is considered, one might estimate[94] $\langle \Delta E^2 \rangle$ by

$$\sqrt{\langle \Delta E^2 \rangle} \approx \frac{\hbar \langle v \rangle}{a^*} \tag{1.73}$$

where $\langle v \rangle$ is an averaged collision velocity and a^* the range of the interaction potential. Indeed, this estimate sometimes reproduces the experimental orders of magnitude of k.[6,94] A temperature dependence $\sqrt{\langle \Delta E^2 \rangle} \propto T^{1/2}$ which corresponds to equation (1.73) was found to be in reasonable agreement with measurements of apparent activation energies of k (at least, if rotational contributions may be neglected).[6] Classical trajectory calculations of molecular collisions[98,110] appear to be most promising at the moment. In particular with rotating Morse oscillators near the dissociation limit an empirical representation of computer results was obtained in ref. 98, which enables $k(E_j | E_i)$ to be formulated. One important result is that $k(E_j | E_i)$ near $E_i \approx E_0$ depends not only on $\Delta E = E_j - E_i$, but also on E_i. According to equation (1.63), this results in a smearing out of the bottle-neck to energies lower than those shown in Fig. 1.19. $g(E_i) = \frac{1}{2} f(E_i)$ is obtained at $E_0 - E_i \approx 0.5kT$ for $\gamma = 0$, at $E_0 - E_i \approx kT$ for $\gamma = 1$, at $E_0 - E_i \approx 2kT$ for $\gamma = 2$. Here γ is a measure of the adiabaticity of collision,

$$\gamma = \left(\frac{\pi}{3} \beta a^* \right)^2 \mu_3/\mu_{12} \tag{1.74}$$

where β is the Morse parameter, $a*$ is the range of interaction potential, μ_3 is the reduced mass of the collision partners and μ_{12} is the reduced mass of the molecule A. $k(E_j \mid E_i)$ may be derived from ref. 98 and used to calculate k. Correct orders of magnitude of rate constants are found for dissociation of homonuclear diatomic molecules with rare gases as collision partners.[98]

1.7.7. Earlier Strong Collision Theories of Dissociation at Low Pressures

While nonequilibrium effects at the weak collision limit or for intermediate cases have been treated only recently, the strong collision limit was discussed much earlier. The relation with these earlier theories is reviewed briefly. At the strong collision limit k is given by

$$k \approx Z_0 P_\sigma \frac{\rho(E_0)kT}{Q(T)} \exp\left(-\frac{E_0}{kT}\right) \qquad (1.75)$$

If expression (1.72) is used for the density of vibrational levels ρ and equation (1.76) for the vibrational partition function Q,

$$Q_{\text{vib}}(T) = \prod_{i=1}^{s} (1 - \exp(-\varepsilon_i/kT))^{-1} \qquad (1.76)$$

one obtains for the strong collision limit

$$k \approx Z_0 P_\sigma \frac{(E_0 + E_z)^{s-1} kT}{(s-1)! \prod_{i=1}^{s} \varepsilon_i \prod_{i=1}^{s} (1 - \exp(-\varepsilon_i/kT))^{-1}} \exp\left(-\frac{E_0}{kT}\right) \qquad (1.77)$$

In this form k takes the value given by the HKRRM theory[103–105,111,112] of unimolecular reactions. The high temperature limit of equation (1.77) at $T \gg \max\{\varepsilon_i/k\}$

$$k \approx \frac{Z_0 P_\sigma}{(s-1)!} \left(\frac{E_0}{kT}\right)^{s-1} \exp\left(-\frac{E_0}{kT}\right) \qquad (1.78)$$

had already been derived in ref. 103. Expressions with constant values of ε_i for all oscillators were given in ref. 104. The derivation of equation (1.77) given in the previous sections shows the meaning of the empirical collision efficiencies for different M in the strong collision formulation of k. The maximum contribution of rotation to equation (1.77) is given by multiplication with the factor

$$P_{\text{rot}} = \frac{(s-1)!}{\left(s + \dfrac{r}{2} - 1\right)!} \left(\frac{E_0}{kT}\right)^{r/2} \qquad (1.79)$$

(r = number of rotational degrees of freedom).

In most cases this will be in conflict with the conservation of angular momentum. If by using the individual centrifugal barriers[113] of rotation conservation of angular momentum is taken into account, P_{rot} will lie between 1 and expression (1.79). P_{rot} may be considerably higher than 1 and in any case must not be neglected. A recent discussion is given in ref. 104a. Anharmonicity corrections to equation (1.77) will be $P_{anh} \approx (s-1)/(s-1\cdot5)$ for a molecule consisting of a Morse oscillator coupled to $s-1$ harmonic oscillators. Larger anharmonicity corrections are expected only for diatomic molecules. For polyatomic molecules the average energy per oscillator is small even if the total energy of the molecule exceeds E_0, and decreases with increasing s. The anharmonicity correction is therefore lower.

1.7.8. Comparison with Experiments

For molecules with more than two atoms equation (1.51) is a suitable starting point for comparison with experimental results on low pressure dissociations. As the statistical factors $\rho(E_0)$, $Q(T)$ and $\exp(-E_0/RT)$ are fairly accessible, one may derive $P_\sigma \cdot P_{\Delta E^2} \cdot P_{rot} \cdot P_{anh}$. For this product one obtains[6] values which in general are of the order 10^{-2}–10, and for normal inert collision partners are about $0\cdot1$–$0\cdot3$. A few collision partners exhibit large values pointing to large P_σ by "chemical interaction". "Normal" collision partners show relative values of $P_\sigma \cdot P_{\Delta E^2}$, which scatter within a factor of 10 in most cases. The differences appear to be larger than would be expected just from the differing values of P_σ. Therefore one can expect weak collision contributions of $P_{\Delta E^2} < 1$ for simple collision partners, whereas with complex partners $P_{\Delta E^2} \to 1$ corresponding to the strong collision limit. As $P_{rot} \cdot P_{anh} \geq 1$, from the relatively small absolute values of the total product $P_\sigma \cdot P_{\Delta E^2} \cdot P_{rot} \cdot P_{anh}$ one would also expect weak collision contributions $P_{\Delta E^2} < 1$. Values of P_σ smaller than 1 are, however, not impossible, according to the absolute measurements of collision numbers for vibrational energy transfer in polyatomic molecules after activation by light absorption.[114] As $P_{rot} \cdot P_{anh}$ can be determined in individual cases, low pressure dissociation rates will become important sources for the product $P_\sigma \cdot P_{\Delta E^2}$.

Whereas the absolute values of k depend significantly on the values of $P_\sigma \cdot P_{\Delta E^2}$, normally $P_\sigma \cdot P_{\Delta E^2}$ makes only a small contribution to the temperature dependence of k in polyatomic molecules. For small polyatomic molecules an empirical relation for the apparent activation energy E_a (see equation (1.21)) was derived from experiments:[6]

$$E_a \approx E_0 + k\frac{\partial \ln Q_{vib}(T)}{\partial 1/T} + \tfrac{1}{2}kT$$

$$= E_0 - (s_{eff} - \tfrac{1}{2})kT \qquad (1.80)$$

s_{eff} is defined by equation (1.80). This is in agreement with the predictions of the weak collision limit of equation (1.51) if (i) $\sqrt{\langle \Delta E^2(E_0) \rangle} \propto T^{1/2}$ according to equation (1.73) and (ii) there is no significant contribution from rotation according to equation (1.79). The uncertainty in the calculated activation energies E_a due to unknown properties of collision processes is probably of the order of kT. Thus in polyatomic molecules the contribution of the vibrational partition function determines the main part of the difference between E_a and E_0. Equation (1.80) allows values of the dissociation energies E_0 to be derived from low pressure dissociation studies.

Collisional activation in polyatomic molecules at total energies $E \approx E_0$ for the whole molecule occurs to a large extent with individual oscillators, only slightly activated. This is not so for diatomic molecules: here the collision processes for molecules with $E \approx E_0$ may change in character because the only oscillator is highly excited. This leads to a more pronounced smearing out of the bottle-neck compared to polyatomic molecules and to different contributions from collision processes to the absolute values and temperature coefficients of k, as discussed in the last part of section 1.7.6, see e.g. equation (1.74). By using equation (1.51), comparison with experimental data in ref. 6 leads to values of $P_\sigma \cdot P_{\Delta E^2} \cdot P_{\text{rot}} \cdot P_{\text{anh}}$ for diatomic molecules with Ar as collision partner in the order of 1–10. As anharmonicity and rotational contributions are relatively large, these higher values for diatomic molecules can be expected from the statistical factors alone. In addition, the shift of the bottle-neck to lower energies leads to larger absolute values of k and to smaller apparent activation energies than expected from equation (1.51). For details see ref. 98. The low apparent activation energies in halogen dissociations may be due to this effect, which should be peculiar to diatomic molecules. Finally, it should be noted that for complex dissociations which involve bimolecular exchange reactions of the type (1.20), the picture may be completely different.

1.7.9. Recombination at Low Pressures

In comparing experimental dissociation results with the corresponding reverse recombinations, sections 1.2–1.5, the relationship

$$k_{\text{rec}} = K_{\text{equ}}^{-1} k \qquad (1.81)$$

was always used. It is clear that equation (1.81) cannot be valid under all circumstances: it is, for example, meaningless in the initial pre-quasistationary period of reaction where k is time dependent. The conditions necessary for equation (1.81) to be valid must therefore be determined. In ref. 98 it was shown that (i) high dilution of the reactants by inert carrier gas and (ii) restriction to the quasistationary period of reaction with rapid relaxation to a nearly equilibrium distribution of the vibrational levels of A are sufficient conditions for equation (1.81) to be applied. Note that this is still true in spite of the nonequilibrium distribution of levels near E_0. For recombination, redissociation of complexes at $E > E_0$ must also be taken into account.

We shall not discuss further the theoretical problems of the validity of equation (1.81) under general conditions. Instead, we shall continue to use this relation to convert calculated dissociation rates into recombination rates. For this, we shall take the statistical expression for the equilibrium constant K_{equ} for the reaction

$$A \rightleftharpoons B + C \qquad (1.82)$$

given by ref. 103

$$K_{equ} \equiv \left(\frac{[B][C]}{[A]}\right)_{equ} = \frac{g_B g_C}{g_A} \frac{1}{N_L} \left(\frac{2\pi\mu_{12}kT}{h^2}\right)^{3/2} \frac{Q(B)Q(C)}{Q(A)} \exp\left(-\frac{\Delta H_0^\circ}{kT}\right) \qquad (1.83)$$

where g_B, g_C, g_A are electronic degeneracies and μ_{12} is the reduced mass of B and C. With equation (1.83) the strong collision limit of the low pressure recombination rate constant becomes

$$k_{rec} \approx Z_0 \cdot P_\sigma \cdot P_{rot} \cdot P_{anh}$$

$$\times \frac{g_A Q_{rot}(A)\rho_{vib}(E_0)kTN_L}{g_B g_C Q_{vib}(B)Q_{rot}(B)Q_{vib}(C)Q_{rot}(C)} \left(\frac{h^2}{2\pi\mu_{12}kT}\right)^{3/2} \qquad (1.84)$$

The weak collision limit by using equations (1.51) and (1.56) becomes

$$k_{rec} \approx Z_0 P_\sigma P_{rot} P_{anh}$$

$$\times \frac{g_A Q_{rot}(A)\rho_{vib}(E_0)kTN_L}{g_B g_C Q_{vib}(B)Q_{rot}(B)Q_{vib}(C)Q_{rot}(C)} \left(\frac{h^2}{2\pi\mu_{12}kT}\right)^{3/2} \frac{\langle \Delta E^2(E_0)\rangle}{2(kT)^2} \qquad (1.85)$$

A step-ladder model may use equations (1.68) and (1.84). Further step-ladder models are discussed in ref. 115. Quantitative results for the continuum model of recombination of atoms have been derived by computer calculation of $P_{\Delta E^2}$ and P_σ for a rotating Morse oscillator.[98]

1.8. RATE COEFFICIENTS AT HIGH PRESSURES

1.8.1. Different High Pressure Regions. Activated Complexes

In order to obtain the overall rate constants for dissociation at high pressures, equation (1.37) must be solved. This requires a complete set of collisional transition rates in phase space, i.e. $k(q, p; q', p')$. This set of course is extremely difficult to obtain. However, for the high velocity limit of collisions, the "impulsive limit", $k(q, p; q', p')$ can be determined.[91,92,100] With these values equation (1.37) can be simplified by expansion of the collision integrals, analogous to the conversion of the master equation (1.34) into a diffusion equation (1.58), at least for inefficient collisions with $\gamma^* = m(M)/m(A) \ll 1$. In this way instead of equation (1.37) a Fokker–Planck equation is derived

$$\frac{\partial[A(q, p, t)]}{\partial t} + \{[A(q, p, t)], H\}$$

$$= \beta\nabla(p + m(M)kT\nabla)[A(q, p, t)] \quad (1.86)$$

with the "viscosity"

$$\beta = \tfrac{8}{3}[M]\sigma_0\sqrt{\frac{2\pi\gamma^*kT}{m(M)}} \quad (1.87)$$

For the one-dimensional case a solution of equation (1.86) has been obtained using the theory of Brownian motion in a force field.[116] To apply this solution, it is assumed that a potential hill is located between dissociating molecules and dissociation products at some bond extension $q^{\ddagger}$. At very low pressures the solutions discussed in section 1.7 are valid. At increasing pressure equilibrium population is established at bond extensions $q \ll q^{\ddagger}$ for all energies including $E > E_0$; this region of equilibrium population is extended up to $q \approx q^{\ddagger}$ and finally exceeds $q^{\ddagger}$. The intermediate case where the equilibrium population is established just as far as $q \approx q^{\ddagger}$ corresponds to the normal high pressure limit of gas phase dissociations. The real high pressure case requires densities as high as are realized in liquids. Here the cage effect will be of primary importance. The bottle-neck of reaction shifts from the "activated complex" at $q^{\ddagger}$ to larger q. In the "impulsive" collision case the dissociation will become diffusion limited. With other types of collisions different kinds of interaction between dissociating molecules and carrier medium are also possible.

From the arguments just given it becomes evident, that a true high pressure limit of gas phase dissociations as used in section 1.2 does not exist. However, the transition region between low pressure dissociation and very high density dissociation is fairly broad (except probably for diatomic molecules, e.g. the experiments on I_2 mentioned in section 1.3.1(c)). Therefore, the usual "activated complex" assumption for the limiting high pressure dissociation rate normally appears to be a good approximation for gas phase dissociations. In this model it is assumed that a "critical surface" in phase space can be defined between reactants and products. As far as to this surface an equilibrium population is established; behind this surface the population is negligible. The reaction rate is given by the flux through this surface in one direction only. In ref. 116 it was shown that this model is a solution of equation (1.86) if low pressure behaviour is not taken into account. As this model is not influenced by the collision transition rates $k(q, p; q', p')$, it is generally used. In addition it appears to be reasonable to generalise this model to polyatomic molecules. Either the "activated complex" is localised at the top of some energy barrier in the reaction coordinate (lowest possible reaction path), or the point of lowest density of states is chosen as "activated complex".[117]

This activated complex model will be discussed in the following sections. The very high density rate constant derived from equation (1.86) is first discussed briefly. Instead of the result in the activated complex model, i.e.

$$k = \nu \exp\left(-\frac{E_0}{kT}\right) \tag{1.88}$$

one obtains

$$k = \frac{\nu}{\beta}\,\omega_c \exp\left(-\frac{E_0}{kT}\right) \tag{1.89}$$

Here, ν is the frequency of the unexcited oscillator in the dissociating diatomic molecule; ω_c is a parameter which depends on the shape of potential at the potential hill between reactants and products, k is proportional to M^{-1} corresponding to diffusion limited reaction. It is well known that unimolecular reactions in liquids[118] are in general not influenced by the carrier medium in the way predicted by equation (1.89). Obviously, the "impulsive" collision model is not generally applicable. Interaction potentials more realistic than the hard sphere potential used in the "impulsive" collision model are necessary. In particular, a further increase of k with [M] is possible. For solvent effects, see ref. 118.

1.8.2. High Pressure Limit in Gases

According to section 1.8.1 the assumption of an activated complex defined by the properties discussed in section 1.8.1 appears to be a reasonable approach to the normal high pressure limit in gas phase dissociation. Therefore, the limiting rate constant in general will be of the form

$$k = \int_{\ddagger} v_\sigma^{\ddagger} f(\sigma)\,d\sigma \tag{1.90}$$

where σ denotes the critical surface of reaction. $f(\sigma)\,d\sigma$ is the equilibrium population at the critical surface. This concerns only those states from which molecular motion directly crosses the critical surface with velocity $v_\sigma^{\ddagger}$. The integral extends over the whole critical surface.

It can be shown[119] that equation (1.90) is the $Z \rightarrow \infty$ limit of Slater's new approach[120] to the first order rate constant for the strong collision case,

$$k = \frac{1}{Q}\int\cdots\int \exp[-Zs(q,p)]Z\exp\left(-\frac{H}{kT}\right)\frac{dq\,dp}{h^n} \tag{1.91}$$

Here $s(q,p)$ is the time lag between activation to a state (q,p) and crossing of the critical surface; Q is the partition function. Equation (1.91) is the classical description; the corresponding quantum forms are derived in refs. 120, 121. $s(q, p)$ may be determined by trajectory calculations in solving the equations of intramolecular motion. This was first done for harmonic oscillators in refs. 120, 122. Other potentials are treated in Monte Carlo studies of dissociation rates[123] and energy transfer.[124] In particular, after adequate averaging over $s(q, p)$, the specific rate constants $k(E)$ have been calculated in this way (see section 1.8.3). If just the overall rate constant k at the high pressure limit is required, more details are determined by trajectory calculations than are finally used.

Equation (1.90) appears to be an equivalent, but simpler, starting point; more explicitly, it may be written:[119,125]

$$k = \frac{1}{\phi} \int_{E_0}^{\infty} \left[\int \cdots \int d\sigma \right] \exp\left(-\frac{E}{kT} \right) dE \qquad (1.92)$$

with

$$\phi = \int \cdots \int \exp\left[-\frac{H(q, p)}{kT} \right] dq\, dp \qquad (1.93)$$

Integration in equation (1.92) extends over the positive half of the critical surface corresponding to flux from reactants to products. If the critical surface may be characterized by some value $q_1^{\ddagger}$ in one bond q_1, equation (1.92) simplifies to

$$k = \frac{kT}{h} \exp\left(-\frac{E_0}{kT} \right) \frac{\phi^{\ddagger}/h^{n-1}}{\phi/h^n} \qquad (1.94)$$

where the "partition function of the activated complex" $\phi^{\ddagger}$ is given by

$$\phi^{\ddagger} = \int \cdots \int \exp\left(-\frac{H^{\ddagger}}{kT} \right) \prod_{i=2}^{n} dq_i\, dp_i \qquad (1.95)$$

The Hamiltonian $H^{\ddagger}$ of the activated complex equals

$$H^{\ddagger} = H(q_1 = q_1^{\ddagger}, \ldots, q_n, p_1 = p_1^{\ddagger}, \ldots, p_n) - E_0 \qquad (1.96)$$

$p_1^{\ddagger}$ is given by solving the equation

$$\partial H'/\partial p_1 = 0 \qquad (1.97)$$

with

$$H' = H(q_1 = q_1^{\ddagger}, \ldots, q_n, p, \ldots, p_n) - E_0 \qquad (1.98)$$

In order to apply equation (1.94) to experimental data one must replace the classical partition functions by the corresponding quantum partition functions. Then, one obtains

$$k = \frac{kT}{h} \frac{Q^{\ddagger}}{Q} \exp\left(-\frac{E_0}{kT} \right) \qquad (1.99)$$

With separation of the different degrees of freedom, equation (1.99) takes the form initially given by transition state theory.[126] At the high temperature limit equation (1.99) approaches (1.88). A useful way of describing activated complexes is given in the HRKRM formulation.[105,111] Systematic applications of equation (1.99) to polyatomic molecules has provided a powerful guide for the estimation of entropies of activation etc. for many compounds, mainly organic.

With many degrees of freedom the statistical contributions from a large number of oscillators to $\phi^{\ddagger}$ must be taken into account; however with small molecules the problems in choosing the correct activated complexes are much easier to investigate. In this case there is a good chance of obtaining adequate potential surfaces. In addition the assumption of separability of degrees of freedom near the activated complex may be abandoned without unduly complicating the calculations of equation (1.95). One can thus concentrate on the detailed properties of the critical surface. In particular, if centrifugal barriers are chosen as activated complexes, the dependence on the rotational quantum number can be investigated. This has been done for the NO_2 system, where experimental results on the high pressure limit are available at 1500 and 300 K.[127] It was argued that the activated complexes must have a complicated form: both centrifugal barriers and bending vibrations play an important role. Comparison of the calculated and experimental values of k provides a check on the quality of the semiempirical potential surface used.

1.8.3. Forbidden Reactions

The high pressure limiting rates of spin-forbidden dissociations (e.g. N_2O, CO_2, CS_2 and CO_2) and of other forbidden reactions (e.g. alkali halides) cannot be described by the equations given in section 1.8.4. In these cases the critical surface can be localized relatively easily at some intersection line or surface of the electronic states involved. In addition, however, the transition probabilities must be included in $\phi^{\ddagger}$ (see equation (1.90)); these are different at different points of the surface. For this, the theory of transition probabilities of Landau and Zener[128] and extended theories[129] must be used. For details see refs. 6, 66f.

1.8.4. Specific Rate Constants

At the high pressure limit of dissociation (section 1.8.5) only the thermal average of the individual dissociation rates $k(E)$ of molecules at energy E is required. However a detailed determination of this quantity is necessary for the fall-off region between the low and high pressure limits and for other dissociation systems (mass spectra, photolysis, chemical activation etc.)

In general, $k(E)$ will depend on the activation process. By activation some distribution of starting points (q, p) at the energy hypersurface

E is produced. From these points the representative points of the molecules move by intramolecular motion until a critical surface is crossed. If the initial distribution is known, $k(E)$ may be obtained by trajectory calculations and suitable averaging. As in general the initial distributions will be different for different activation systems, different $k(E)$ will result. However, a useful limiting model is treated in the statistical theory of reaction rates, where an equilibrium distribution of starting points is assumed. With this assumption, expression (1.90) may again be used restricted to molecules at the energy hypersurface E. Note also that the critical surface for a calculation of $k(E)$ may be different from the surface used for the high pressure limit. General equations for $k(E)$ and the derivation of the harmonic oscillator expression, e.g. at $E \gg E_0$

$$k(E) \approx \frac{\prod_{i=1}^{s} \nu_i}{\prod_{i=1}^{s-1} \nu_i^{\ddagger}} \cdot \left(\frac{E - E_0}{E}\right)^{s-1} \qquad (1.100)$$

are described in refs. 6, 104, 105, 111, 120, 130. Corrections to $k(E)$ at energies near $E \approx E_0$ may be obtained as shown in ref. 107. Detailed formulae are given by the HKRRM treatment[105,111] of $k(E)$; explicit calculations[107] of $k(E)$ follow this theory. For further details the reader is referred in particular to ref. 107.[127]

1.8.5. Fall-off Curves

The simplest expression for the overall dissociation rate constant in the transition range between low and high pressure limits is given by equation (1.7)

$$k = k_1[M] \left(\frac{k_3}{k_3 + k_2[M]}\right) \qquad (1.101)$$

A complete description of the transition range needs the correct treatment of collision processes and of intramolecular processes. It is therefore much more complicated than the limiting cases. In order to distinguish between the different problems, it is advisable to express k in the transition range in reduced form and to fit it to both limiting cases. This is easy using equation (1.101). One obtains immediately with $k_0 \equiv k_1[M]$ and $k_\infty \equiv (k_1/k_2)k_3$

$$\frac{k}{k_0} = \frac{1}{1 + (k_0/k_\infty)} \qquad (1.102)$$

or

$$\frac{k}{k_\infty} = \frac{1}{1 + (k_\infty/k_0)} \qquad (1.103)$$

As $k_0 = k_1[M]$ is a reasonable measure of [M], the [M]-dependence is expressed in terms of k_0. At $k_0 = k_\infty$, $k = 0.5k_0 = 0.5k_\infty$. At $k_0 = 0.1k_\infty$, $k \approx 0.9k_0$; at $k_0 = 10k_\infty$, $k \approx 0.9k_\infty$. The effects of different factors on both the limiting rate constants k_0 and k_∞ and the shape of the reduced fall-off curve must be studied. First the influence of different energy states can be discussed for the strong collision model. Here k is given by

$$k = \int_{E_0}^{\infty} \frac{Zf(E)k(E)}{k(E) + Z}\, dE \qquad (1.104)$$

Reduced fall-off curves may be calculated using $f(E)$ and $k(E)$ from equations (1.72) and (1.100). The results[120] are presented in Table 1.10

TABLE 1.10

Reduced fall-off curves of unimolecular reactions. (k_0/k_∞ is a measure of the carrier gas concentration; at $\log(k_0/k_\infty) = 0$ the extrapolations of limiting low and high pressure rates intersect, see text. s_{eff} is given by equation (1.80). The Table gives the correction factors, by which the extrapolated limiting low- (at $\log(k_0/k_\infty) \leq 0$) or high- (at $\log(k_0/k_\infty) \geq 0$) pressure rate constants must be multiplied, to obtain the rate constants in the transition range. For details see ref. 6.)

s_{eff} \ $\log\left(\dfrac{k_0}{k_\infty}\right)$	−6	−5	−4	−3	−2	−1	0	1	2	3
1					0·99	0·91	0·50	0·91	0·99	
3					0·96	0·80	0·40	0·84	0·98	
5			0·98	0·93	0·84	0·62	0·29	0·70	0·94	0·99
7	0·98	0·95	0·90	0·82	0·65	0·43	0·19	0·54	0·87	0·98

and Fig. 1.22. The statistical theory of Kassel and the dynamical theory of Slater appear to give very similar results. In general the transition region is broadened as s increases. In comparisons with experiment, classical formulae must usually be replaced by quantum formulae. It was therefore suggested in ref. 6 that s_{eff} (defined by equation (1.80)) should be used as the quantity "s" in Table 1.10. This procedure proved successful for moderately complex molecules, where both limiting cases are accessible experimentally. Corrections to this procedure are possible by using the complete HKRRM formulae for the strong collision model.

Inefficient energy transfer tends to broaden the transition curves as though s were increased by about 1–2.[131] Inefficient intramolecular

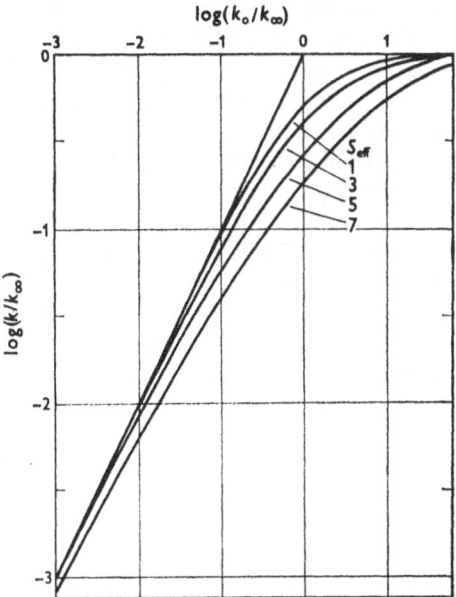

FIG. 1.22. Reduced fall-off curves for dissociation and recombination.
(See Table 1.10.)

coupling also produces a similar broadening.[132] In addition $k(E)$ may
be pressure dependent in the transition region (section 1.8.4). All these
factors are extremely difficult to treat correctly; this limits the validity
of corrections to Table 1.10. Therefore, improvement must first be
obtained in interpretations of the limiting rates; only after this can the
shape of reduced fall-off curves be refined. In spite of these uncertainties, good agreement with calculated and measured fall-off curves
has been found for many cases.

1.8.6. High Pressure Recombination-Isotope Exchange

As was discussed in section 1.7.9 one also may transform the high
pressure dissociation rate constant by equation (1.81) into a high
pressure recombination rate constant, i.e. into the recombination rate
constant for the region where the reaction order approaches two. To do
this one must combine the equilibrium constant (equation (1.83)) with
equation (1.99):

$$k_{\text{rec}} = \frac{kT}{h} \frac{g_A}{g_B g_C} \frac{Q^{\ddagger}}{Q(B)Q(C)} N_L \left(\frac{h^2}{2\pi\mu_{12}kT} \right)^{3/2} \qquad (1.105)$$

As in section 1.8.2 the main problem is the calculation of the partition
function $Q^{\ddagger}$ of the activated complex. This requires a knowledge of the

location of the activated complexes and thus of the interaction potential between B and C. As a first approximation one can use either (i) a potential, related to the equilibrium properties of A, e.g. a potential surface with a Morse potential for the B–C separation,[127] or (ii) a potential better for very large distances, e.g. an r^{-6} potential for B–C. In the case (i) the potential parameters in the first approximation are taken from the equilibrium properties of A. In the case (ii), the potential parameters are taken from the long range interaction constants, e.g. the van der Waals constants for B and C which are related to their polarizabilities. The latter approach was applied in ref. 113, using the centrifugal barriers as locations of activated complexes (see however[127]).

In high pressure recombinations the rate determining step is the passing of the activated complex; one can immediately compare this step with an isotope exchange:

recombination: $B + C \longrightarrow A^* \xrightarrow{(+M)} A$ (1.106)

isotope exchange: $B + C \longrightarrow A^* \longrightarrow B' + C'$ (1.107)

In step (1.107) the prime denotes particles with exchanged isotopes. Both processes have nearly the same first step, i.e. the formation of A^*, and this is rate determining. There may be slight differences between A^* in the two reactions because the region of phase space A^* occupies is determined by different processes. In step (1.106) this region is determined by collisions; in step (1.107) it is determined by the possibility of isotope exchange. The fraction of A^* which reacts approaches unity in the high pressure recombination (1.106). However it may be slightly smaller in step (1.107) if isotope exchange involves a small energy barrier due to different zeropoint energies of $B + C$ and $B' + C'$. Also because redissociation of A^* to $B + C$ is possible, isotope exchange is sometimes less efficient than recombination, e.g. it is slower for the O isotope exchange in NO_2.

Experimental rates of high pressure recombination and isotope exchange have been obtained for the NO_2,[52a] O_3[53d] and NO_3[52a] systems. If the molecular configurations of highly excited A, i.e. of A^*, are identical in steps (1.106) and (1.107), isotope exchange reactions should become suppressed at high pressures due to the formation of stabilized A molecules, and reaction (1.106) should change from third order to second order. In this connection it is interesting that similar rates are observed for the high pressure recombination $O + NO_2 \rightarrow NO_3$ and the isotope exchange between O and NO_2. However, the reaction $O + NO_2 \rightarrow NO + O_2$ has a different rate and even at inert gas pressures around 100 atm it could not be suppressed by deactivation of the intermediate complex.[52a] This suggests that intramolecular coupling is incomplete in this case and that different configurations of A^* exist which are only weakly coupled.

REFERENCES

1. E. F. Greene and J. P. Toennies, Chemische Reaktionen in Stoßwellen (Vol. 3 of Fortschritte der Physikalischen Chemie, W. Jost Editor, Steinkopff Verlag, Darmstadt 1959)
   E. F. Greene and J. P. Toennies, Chemical Reactions in Shock Waves (Edward Arnold, London, 1964)
   J. N. Bradley, Shock Waves in Chemistry and Physics (Methuen, London, 1962)
   A. G. Gaydon and I. R. Hurle, The Shock Tube in High Temperature Chemical Physics (Chapman & Hall, London, 1963)
   Ye. V. Stupochenko, S. A. Losev and A. I. Osipov, Relaxation in Shock Waves (Springer-Verlag, Berlin, Heidelberg and New York, 1967)
2. R. W. Getzinger and G. L. Schott, This book, page 81
3. R. L. Belford and R. A. Strehlow, *Ann. Rev. Phys. Chem.*, **20**, 247 (1969)
4. F. Kaufman, *Ann. Rev. Phys. Chem.*, **20**, 45 (1969)
5. K. L. Wray, 10th Int. Symp. Combustion (Combustion Institute, Pittsburgh, 1965) p. 523
   M. Camac and A. Vaughan, *J. Chem. Phys.*, **34**, 460 (1961)
6. J. Troe and H. Gg. Wagner, *Ber. Bunsenges. Physik. Chem.*, **71**, 937 (1967)
   J. Troe, *Ber. Bunsenges. Physik. Chem.*, **72**, 908 (1968)
7. (a) A. L. Myerson and W. S. Watt, *J. Chem. Phys.*, **49**, 425 (1968)
   E. A. Sutton, *J. Chem. Phys.*, **36**, 2923 (1962)
   R. W. Patch, *J. Chem. Phys.*, **36**, 1919 (1962)
   J. P. Rink, *J. Chem. Phys.*, **36**, 262 (1962)
   W. C. Gardiner and G. B. Kistiakowsky, *J. Chem. Phys.*, **35**, 1765 (1961)
   I. R. Hurle, A. Jones and J. L. J. Rosenfeld, *Proc. Roy. Soc.*, **A310**, 253 (1969)
   I. R. Hurle, 11th Int. Symp. Combustion (Combustion Institute, Pittsburgh, 1967), p. 827
   T. A. Jacobs, R. R. Giedt and N. Cohen, *J. Chem. Phys.*, **47**, 54 (1967)
8. E. A. Sutton, *J. Chem. Phys.*, **36**, 2923 (1962)
   J. P. Rink, *J. Chem. Phys.*, **36**, 1398 (1962)
   T. A. Jacobs, R. R. Giedt and N. Cohen, *J. Chem. Phys.*, **48**, 947 (1968)
9. (a) K. L. Wray, 10th Int. Symp. Combustion (Combustion Institute, Pittsburgh, 1965), p. 523
   W. S. Watt and A. L. Myerson, *J. Chem. Phys.*, **51**, 1638 (1969)
   V. N. Kondratjev and E. E. Nikitin, *J. Chem. Phys.*, **45**, 1078 (1966)
   K. L. Wray, *J. Chem. Phys.*, **37**, 1254 (1962); **38**, 1518 (1963)
   J. P. Rink, *J. Chem. Phys.*, **36**, 572 (1962)
   J. P. Rink, H. Knight and R. Duff, *J. Chem. Phys.*, **34**, 1942 (1961)
   M. Camac and A. Vaughn, *J. Chem. Phys.*, **34**, 460 (1961)
   S. R. Byron, *J. Chem. Phys.*, **30**, 1380 (1959)
   S. A. Losev, *Dokl. Akad. Nauk SSSR*, **141**, 894 (1961)
   N. A. Generalov and S. A. Losev, *J. Quant. Spectry Radiative Transfer*, **6**, 101 (1966)
10. (a) J. P. Appleton, M. Steinberg and D. J. Liquornik, *J. Chem. Phys.*, **48**, 599 (1968); **49**, 2468 (1968)
    S. R. Byron, *J. Chem. Phys.*, **44**, 1378 (1966)
    B. Cary, *Phys. Fluids*, **8**, 26 (1965); **9**, 1047 (1966)
    R. A. Allen, J. C. Keck and J. C. Camm, *Phys. Fluids*, **5**, 284 (1962)
    K. L. Wray and S. R. Byron, *Phys. Fluids*, **9**, 1046 (1966)
11. I. R. Hurle, 11th Int. Symp. Combustion (Combustion Institute, Pittsburgh, 1967) p. 827

12. J. L. J. Rosenfeld, *Discussions Faraday Soc.*, **44**, 89 (1967)
    I. R. Hurle, P. Mackey, J. L. J. Rosenfeld, *Ber. Bunsenges. Physik. Chem.*, **72**, 991 (1968)
13. M. A. A. Clyne and D. H. Stedman, *J. Phys. Chem.*, **71**, 3071 (1967)
14. I. M. Campbell and B. A. Thrush, *Proc. Roy. Soc.*, **A296**, 201 (1967); *Trans. Faraday Soc.*, **64**, 1275 (1968)
15. H. Eberius, K. Hoyermann and H. Gg. Wagner, *Ber. Bunsenges. Physik. Chem.*, **73**, 962 (1969)
16. W. E. Kaskan, *Combust. Flame*, **2**, 229 (1958)
    E. M. Bulewicz, C. C. James and T. M. Sugden, *Proc. Roy. Soc.*, **A235**, 89 (1965)
    E. M. Bulewicz and T. M. Sugden, *Trans. Faraday Soc.*, **54**, 1855 (1958)
    G. Dixon-Lewis, M. M. Sutton and A. Williams, *Discussions Faraday Soc.*, **33**, 205 (1962)
17. (a) F. S. Larkin, *Can. J. Chem.*, **46**, 1005 (1968)
    J. E. Bennett and D. R. Blackmore, *Proc. Roy. Soc.*, **A305**, 553 (1968)
    R. E. Roberts, R. B. Bernstein and C. F. Curtiss, *Chem. Phys. Lett.*, **2**, 366 (1968)
    H. Eberius, K. Hoyermann and H. Gg. Wagner, *Ber. Bunsenges. Physik. Chem.*, **73**, 962 (1969)
    V. V. Azatyan, *Kin. i. Kataliz*, **9**, 1188 (1968)
    F. Kaufman, *Can. J. Chem.*, **47**, 1917 (1969)
    F. S. Larkin and B. A. Thrush, 10th Int. Symp. Combustion (Combustion Institute, Pittsburgh, 1965), p. 397
    E. M. Bulewicz and T. M. Sugden, *Trans. Faraday Soc.*, **54**, 1855 (1958)
18. (a) I. M. Campbell and B. A. Thrush, *Proc. Roy. Soc.*, **A296**, 222 (1967); *Trans. Faraday Soc.*, **64**, 4275 (1968)
    H. I. Schiff, *Can. J. Chem.*, **47**, 1903 (1969)
    K. L. Wray, 10th Int. Symp. Combustion (Combustion Institute, Pittsburgh, 1965), p. 523
    J. H. Kiefer and R. W. Lutz, *J. Chem. Phys.*, **42**, 1709 (1965)
    J. E. Morgan and H. I. Schiff, *J. Chem. Phys.*, **38**, 1495 (1963)
    R. R. Reeves, G. Manella and H. I. Schiff, *J. Chem. Phys.*, **32**, 632 (1960)
19. (a) I. M. Campbell and B. A. Thrush, *Proc. Roy. Soc.*, **A296**, 201 (1967); *Trans. Faraday Soc.*, **64**, 1275 (1968)
    K. M. Evenson and D. S. Burch, *J. Chem. Phys.*, **45**, 2450 (1967)
    M. A. A. Clyne and D. H. Stedman, *J. Phys. Chem.*, **71**, 3071 (1967)
    E. S. Shane and W. Brennen, *Chem. Phys. Lett.*, **4**, 31 (1969)
    W. Groth and K. H. Becker, private communication 1969
    H. Gg. Wagner and J. Wolfrum, *Angew. Chem.*, **10**, 604 (1971)
    B. Brocklehurst and K. R. Jennings., *Progr. Reaction Kinetics* (G. Porter, Editor), **4**, 1 (1967)
    G. R. Brown and C. A. Winkler, *Angew. Chem.*, **82**, 187 (1970)
20. JANAF Thermochemical Tables 1965 (U.S. Dept. of Commerce, NBS)
21. J. K. K. Ip and G. Burns, *J. Chem. Phys.*, **51**, 3414, 3425 (1969); *Discussions Faraday Soc.*, **44**, 241 (1967)
22. R. K. Boyd, G. Burns, T. R. Lawrence and J. H. Lippiatt, *J. Chem. Phys.*, **49**, 3804 (1968)
    R. K. Boyd, J. D. Brown, G. Burns and J. H. Lippiatt, *J. Chem. Phys.*, **49**, 3822 (1968)
23. D. J. Seery, *J. Phys. Chem.*, **70**, 1684 (1966)
24. V. H. Dibeler, J. A. Walker and K. E. McCulloh, *J. Chem. Phys.*, **51**, 4230 (1969)
25. R. A. Carabetta and H. B. Palmer, *J. Chem. Phys.*, **46**, 1325, 1538 (1967); **49**, 2466 (1968)
    H. B. Palmer, *J. Chem. Phys.*, **47**, 2116 (1967)

26. M. A. A. Clyne and D. H. Stedman, *Trans. Faraday Soc.*, **64**, 1816 (1968)
27. W. Groth and K. H. Becker, private communication 1969 (for N_2^*)
    R. W. Fair and B. A. Thrush, *Trans. Faraday Soc.*, **65**, 1208 (1969) (for S_2^*)
28. M. I. Christie, *J. Am. Chem. Soc.*, **84**, 4066 (1962)
29. K. E. Russell and J. Simons, *Proc. Roy. Soc.*, **A217**, 271 (1953)
30. G. Porter, Z. G. Scabo and M. G. Townsend, *Proc. Roy. Soc.*, **A270**, 493 (1962)
30a. L. Y. Nelson and G. C. Pimentel, *J. Chem. Phys.*, **47**, 3671 (1967)
31. J. Troe and H. Gg. Wagner, *Z. Physik. Chem.*, **NF55**, 326 (1967)
32. M. I. Christie, A. G. Harrison, R. G. W. Norrish and G. Porter, *Proc. Roy. Soc.*, **A231**, 446 (1955)
    H. H. Kamer, M. H. Hanes and E. J. Bair, *J. Opt. Soc. Am.*, **51**, 775 (1961)
    R. L. Strong, J. C. W. Chien, P. E. Graf and J. E. Willard, *J. Chem. Phys.*, **26**, 1287 (1957)
33. G. Burns, R. J. LeRoy, O. J. Morriss and J. A. Blake, *Proc. Roy. Soc.*, **A316**, 81 (1970)
34. S. Aditya and J. E. Willard, *J. Am. Chem. Soc.*, **79**, 2680 (1957)
35. J. Troe, to be published
36. (a) C. D. Johnson and D. Britton, *J. Phys. Chem.*, **68**, 3032 (1964)
    D. J. Seery and D. Britton, *J. Phys. Chem.*, **70**, 4074 (1966)
    R. W. Diesen, *J. Chem. Phys.*, **44**, 3662 (1966); *J. Phys. Chem.*, **72**, 108 (1968)
    D. J. Seery, *J. Phys. Chem.*, **70**, 1684 (1966)
37. (a) R. A. Carabetta and H. B. Palmer, *J. Chem. Phys.*, **46**, 1333 (1967); **47**, 2202 (1967)
    T. A. Jacobs and R. R. Giedt, *J. Chem. Phys.*, **39**, 749 (1963)
    H. Hiraoka and R. Hardwick, *J. Chem. Phys.*, **36**, 1715 (1962)
    R. W. Diesen and W. J. Felmlee, *J. Chem. Phys.*, **39**, 2115 (1963)
    M. van Thiel, D. J. Seery and D. Britton, *J. Phys. Chem.*, **69**, 1333 (1965)
38. (a) M. Warshay, *J. Chem. Phys.*, **75**, 2700 (1971); *J. Chem. Phys.*, **54**, 4060 (1971)
    D. Britton, *J. Phys. Chem.*, **64**, 742 (1960)
    D. Britton and N. Davidson, *J. Chem. Phys.*, **25**, 810 (1956)
    H. B. Palmer and D. F. Hornig, *J. Chem. Phys.*, **26**, 98 (1957)
    R. K. Boyd, G. Burns, T. R. Lawrence and J. H. Lippiatt, *J. Chem. Phys.*, **49**, 3804, 3822 (1968)
39. (a) J. Troe and H. Gg. Wagner, *J. Physik. Chem.*, **NF55**, 326 (1961)
    D. Britton, N. Davidson, W. Gehman and G. Schott, *J. Chem. Phys.*, **25**, 804 (1956)
    D. Britton, N. Davidson and G. Schott, *Discussions Faraday Soc.*, **17**, 58 (1954)
    J. Troe, to be published
40. (a) M. A. A. Clyne and D. H. Stedman, *Trans. Faraday Soc.*, **64**, 1816, 2689 (1968)
    R. A. Carabetta and H. B. Palmer, *J. Chem. Phys.*, **46**, 1325, 1333, 1538 (1967); **49**, 2466 (1968)
    H. B. Palmer, *J. Chem. Phys.*, **47**, 2116 (1967)
    L. W. Bader and E. A. Ogryzlo, *Nature*, **201**, 491 (1964)
    E. Hutton and M. Wright, *Trans. Faraday Soc.*, **61**, 78 (1965)
41. (a) J. K. K. Ip and G. Burns, *J. Chem. Phys.*, **51**, 3414, 3425 (1969)
    R. L. Strong, J. C. W. Chien, P. E. Graf and J. E. Willard, *J. Chem. Phys.*, **26**, 1287 (1957)
    E. Rabinowitch and W. C. Wood, *Trans. Faraday Soc.*, **32**, 907 (1936)
    M. I. Christie, R. S. Roy and B. A. Thrush, *Trans. Faraday Soc.*, **55**, 1139 (1939)
    M. R. Basila and R. L. Strong, *J. Phys. Chem.*, **67**, 521 (1963)
    G. Burns and D. F. Hornig, *Can. J. Chem.*, **38**, 1702 (1960)
    G. Burns, *Can. J. Chem.*, **46**, 3229 (1968)
42. (a) D. L. Bunker and N. Davidson, *J. Am. Chem. Soc.*, **80**, 5085 (1958)
    G. Porter and J. A. Smith, *Proc. Roy. Soc.*, **A261**, 68 (1961)

G. Porter, Z. G. Szabo and M. G. Townsend, *Proc. Roy. Soc.*, **A270, 493** (1962)

M. I. Christie, A. G. Harrison, R. G. W. Norrish and G. Porter, *Proc. Roy. Soc.*, **A231**, 446 (1955)

M. I. Christie, *J. Am. Chem. Soc.*, **84**, 4066 (1962)

R. L. Strong, J. C. W. Chien, P. E. Graf and J. E. Willard, *J. Chem. Phys.*, **26**, 1287 (1957)

R. Engleman and N. Davidson, *J. Am. Chem. Soc.*, **82**, 4770 (1960)

H. H. Kramer, M. H. Hanes and E. J. Bair, *J. Opt. Soc. Am.*, **51**, 775 (1961)

E. Rabinowitch and W. C. Wood, *J. Chem. Phys.*, **4**, 497 (1936)

G. Porter, *Discussions Faraday Soc.*, **33**, 198 (1962)

43. (a) T. A. Jacobs, R. R. Giedt and N. Cohen, *J. Chem. Phys.*, **43**, 3688 (1965)
    J. A. Blauer, *J. Phys. Chem.*, **72**, 79 (1968)

44. (a) T. A. Jacobs, N. Cohen and R. R. Giedt, *J. Chem. Phys.*, **46**, 1958 (1967)
    D. J. Seery and C. T. Bowman, *J. Chem. Phys.*, **48**, 4314 (1968)
    E. S. Fishburne, *J. Chem. Phys.*, **45**, 4053 (1966)

45. T. A. Jacobs, N. Cohen and R. R. Giedt, *J. Chem. Phys.*, **46**, 1958 (1967)

46. R. R. Giedt, N. Cohen and T. A. Jacobs, *J. Chem. Phys.*, **50**, 5374 (1969)

47. A. R. Fairbairn, *J. Chem. Phys.*, **51**, 972 (1969); *Proc. Roy. Soc.*, **A312**, 207 (1969)

48. K. L. Wray and J. D. Teare, *J. Chem. Phys.*, **36**, 2582 (1962)

49. R. S. Berry, T. Cernoch, M. Coplan and J. J. Ewing, *J. Chem. Phys.*, **49**, 127 (1968)
    J. Ewing, R. Milstein and R. S. Berry, *J. Chem. Phys.*, **54**, 1752 (1971)

50. R. Hartig, H. A. Olschewski, J. Troe and H. Gg. Wagner, *Ber. Bunsenges. Physik. Chem.*, **72**, 1016 (1968)
    K. Luther and J. Troe, to be published

50a. D. G. Horne, R. Gosavi and D. P. Strausz, *J. Chem. Phys.*, **48**, 4758 (1968)

51. K. E. Shuler, *J. Chem. Phys.*, **21**, 624 (1953)

52. (a) J. Troe, *Ber. Bunsenges. Physik. Chem.*, **73**, 144 (1969)
    (b) J. Heicklen and N. Cohen, *Advan. Photochem.*, **5**, 157 (1968)
    (c) J. Troe, *Ber. Bunsenges. Physik. Chem.*, **73**, 906 (1969)
    (d) F. S. Klein and J. T. Herron, *J. Chem. Phys.*, **40**, 2731 (1964); **41**, 1285 (1963); **44**, 3645 (1966);
    (e) B. P. Levitt, *Trans. Faraday Soc.*, **59**, 59 (1963)
    (f) F. Kaufman and J. R. Kelso, ref. 253 in ref. 4
    (g) M. A. A. Clyne and B. A. Thrush, *Proc. Roy. Soc.*, **A269**, 404 (1962)
    (h) F. Kaufman, N. J. Gerri and R. E. Bowman, *J. Chem. Phys.*, **25**, 106 (1956)
    (h) F. Kaufman, *J. Chem. Phys.*, **28**, 352 (1958)
    R. E. Huffman and N. Davidson, *J. Am. Chem. Soc.*, **81**, 2311 (1959)
    E. S. Fishburne, D. M. Bergbauer and R. Edse, *J. Chem. Phys.*, **43**, 1847 (1965)
    H. Hiraoka and R. Hardwick, *J. Chem. Phys.*, **39**, 2362 (1963)
    (i) H. Gaedtke, K. Glaenzer, H. Hippler, K. Luther and J. Troe, 14th Symp. on Combustion (Pittsburgh 1972, in press)

53. (a) W. M. Jones and N. Davidson, *J. Am. Chem. Soc.*, **84**, 2868 (1962)
    (b) A. Glissmann and H. J. Schumacher, *Z. Physik. Chem.*, **B21**, 323 (1933)
    (b) S. W. Benson and A. E. Axworthy, *J. Chem. Phys.*, **26**, 1718 (1957); **42**, 2614 (1964)
    (b) E. Castellano and H. J. Schumacher, *Z. Physik. Chem.*, **NF34**, 198 (1962)
    (c) F. Kaufman and J. R. Kelso, *J. Chem. Phys.*, **46**, 4541 (1967)
    (d) H. Hippler and J. Troe, *Ber. Bunsenges. Physik. Chem.*, **74** (1970)
    (e) M. C. Sauer, *J. Phys. Chem.*, **71**, 3311 (1967)
    (e) M. C. Sauer and L. M. Dorfman, *J. Am. Chem. Soc.*, **87**, 3801 (1965)
    (f$_1$) F. S. Klein and J. T. Herron, *J. Chem. Phys.*, **44**, 3645 (1966)
    (f) W. Brennen and H. Niki, *J. Chem. Phys.*, **42**, 3725 (1965)
    (f$_2$) S. Jaffe and F. S. Klein, *Trans. Faraday Soc.*, **62**, 3135 (1966)

(f) S. H. Garnett, G. B. Kistiakowsky and B. V. O'Gradey, *J. Chem. Phys.*, **51**, 84 (1969)

(g) M. A. A. Clyne, D. J. McKenney and B. A. Thrush, *Trans. Faraday Soc.*, **61**, 2701 (1965)

J. A. Zaslowky, H. B. Urbach, F. Leighton, R. J. Wnuk and J. Wojtowicz, *J. Am. Chem. Soc.*, **82**, 2682 (1960)

N. Basco, *Proc. Roy. Soc.*, **A283**, 302 (1964)

M. F. R. Mulcahy and D. J. Williams, *Trans. Faraday Soc.*, **64**, 59 (1968)

G. M. Meaburn, D. Perner, J. LeCalvre and M. Bourene, *J. Phys. Chem.*, **72**, 3920 (1968)

54. (a) H. A. Olschewski, J. Troe and H. Gg. Wagner, *Z. Physik. Chem.*, **NF47**, 383 (1965); 11th Int. Symp. Combustion (Combustion Institute, Pittsburgh, 1966) p. 155

(b) J. B. Homer and I. R. Hurle, *Proc. Roy. Soc*, **A314**, 585 (1970)

(c) C. J. Halstead and D. R. Jenkins, 12th Int. Symp. Combustion (Combustion Institute, Pittsburgh, 1968) p. 979

(c) R. W. Getzinger and L. S. Blair, *Combust. Flame*, **13**, 271 (1968)

J. L. J. Rosenfeld and T. M. Sugden, *Combust. Flame*, **8**, 44 (1964)

R. W. Getzinger, 11th Int. Symp. on Combustion (Combustion Institute, Pittsburgh, 1966) p. 117

55. (a) H. A. Olschewski, J. Troe and H. Gg. Wagner, *Z. Physik. Chem.*, **NF44**, 173 (1965)

(b) B. P. Levitt and D. B. Sheen, *Trans. Faraday Soc.*, **63**, 2955 (1967)

(b) A. G. Gaydon, G. H. Kimbell and H. B. Palmer, *Proc. Roy. Soc.*, **A276**, 461 (1963)

(c) J. C. Halstead and B. A. Thrush, *Proc. Roy. Soc.*, **A295**, 363 (1966)

56. (a) J. Troe, H. Gg. Wagner and G. Weden, *Z. Physik. Chem.*, **NF56**, 238 (1967)

(b) J. A. Blauer and W. C. Solomon, *J. Phys. Chem.*, **72**, 2307 (1968)

(b) W. C. Solomon, J. A. Blauer and F. C. Jaye, *J. Phys. Chem.*, **72**, 2311 (1968)

(b) M. C. Lin and S. H. Bauer, *J. Am. Chem. Soc.*, **91**, 7737 (1969)

(c) W. Koblitz and H. J. Schumacher, *Z. Physik. Chem.*, **B25**, 283 (1934)

57. (a) P. G. Ashmore and M. G. Burnett, *Trans. Faraday Soc.*, **58**, 1801 (1962)

(a) P. G. Ashmore and M. S. Spencer, *Trans. Faraday Soc.*, **55**, 1868 (1959)

(b) T. C. Clark, M. A. A. Clyne and D. H. Stedman, *Trans. Faraday Soc.*, **62**, 3354 (1966)

(c) B. Deklau and H. B. Palmer, 8th Int. Symp. Combustion (Williams & Wilkins, 1962) p. 139

R. B. Timmons and B. deB. Darwent, *J. Phys. Chem.*, **73**, 2208 (1969)

58. D. Gutman, E. A. Hardwidge, F. A. Dougherty and R. W. Lutz, *J. Chem. Phys.*, **47**, 4400 (1967)

R. W. Getzinger and G. L. Schott, *J. Chem. Phys.*, **43**, 3237 (1965)

R. W. Getzinger, 11th Int. Symp. Combustion (The Combustion Institute, 1967) p. 117

R. L. Wadlinger and B. deB. Darwent, *J. Phys. Chem.*, **71**, 2057 (1967)

59. A. P. Modica, *J. Chem. Phys.*, **44**, 1585 (1966); *J. Phys. Chem.*, **72**, 4594 (1968)

A. P. Modica and S. J. Sillers, *J. Chem. Phys.*, **48**, 3283 (1968)

60. (a) A. P. Modica and D. F. Hornig, *J. Chem. Phys.*, **43**, 2739 (1965); **49**, 629 (1968)

(b) R. W. Diesen, *J. Chem. Phys.*, **41**, 3526 (1964); **44**, 3662 (1966); **45**, 759 (1966)

A. P. Modica, *J. Chem. Phys.*, **46**, 3663 (1967)

61. W. L. Patterson and E. F. Greene, *J. Chem. Phys.*, **36**, 1146 (1962)

B. P. Levitt and A. B. Parsons, *Trans. Faraday Soc.*, **65**, 1199 (1969)

62. D. Schofield, W. Tsang and S. H. Bauer, *J. Chem. Phys.*, **42**, 2132 (1965)

63. T. C. Clark, M. A. A. Clyne and D. H. Stedman, *Trans. Faraday Soc.*, **62**, 3354 (1966)
64. D. B. Hartley and B. A. Thrush, *Proc. Roy. Soc.*, **A297**, 520 (1967)
    F. C. Kohout and F. W. Lampe, *J. Chem. Phys.*, **46**, 4075 (1967)
    M. A. A. Clyne and B. A. Thrush, *Trans. Faraday Soc.*, **57**, 1305 (1961); *Discussions Faraday Soc.*, **33**, 139 (1962)
    K. H. Hoyermann, Ph.D. Thesis, Göttingen (1968)
65. S. E. Schwartz and H. S. Johnston, *J. Chem. Phys.*, **51**, 1286 (1969)
66. (a) R. M. Lewis and C. N. Hinshelwood, *Proc. Roy. Soc.*, **A168**, 441 (1938)
    (a) F. F. Musgrave and C. N. Hinshelwood, *Proc. Roy. Soc.*, **A106**, 284 (1932)
    (b) E. Hunter, *Proc. Roy. Soc.*, **A144**, 386 (1934)
    (c) M. Volmer and H. Froehlich, *Z. Physik. Chem.*, **B19**, 85, 89 (1932)
    (c) M. Volmer and M. Bogdan, *Z. Physik. Chem.*, **B21**, 257 (1933)
    (d) H. S. Johnston, *J. Chem. Phys.*, **19**, 663 (1961)
    (e) E. S. Fishburne and R. Edse, *J. Chem. Phys.*, **41**, 1297 (1964); **44**, 515 (1966)
    (f) W. Jost, K. W. Michel, J. Troe and H. Gg. Wagner, *Z. Naturforsch*, **19a**, 59 (1964)
    (f) H. A. Olschewski, J. Troe and H. Gg. Wagner, *Ber. Bunsenges. Physik. Chem.*, **70**, 450 (1966)
    (g) D. Gutman, R. L. Belford, A. J. Hay and R. Pancirov, *J. Phys. Chem.*, **70**, 1793 (1966)
    (g) A. A. Borisov, *Kin. i Kataliz*, **9**, 482 (1968)
    (g) A. P. Modica, *J. Phys. Chem.*, **69**, 2111 (1965)
    (g) J. N. Bradley and G. B. Kistiakowsky, *J. Chem. Phys.*, **35**, 256 (1961)
    (g) S. H. Garnett, G. B. Kistiakowsky and B. V. O'Grady, *J. Chem. Phys.*, **51**, 84 (1969)
    (h) S. C. Barton and J. E. Dove, *Can. J. Chem.*, **47**, 521 (1969)
    (i) H. Henrici and S. H. Bauer, *J. Chem. Phys.*, **50**, 13333 (1969)
    (k) A. Martinengo, J. Troe and H. Gg. Wagner, *Z. Physik. Chem.*, **NF51**, 104 (1966)
    (l) D. R. Snelling and E. J. Bair, *J. Chem. Phys.*, **47**, 228 (1967); **48**, 5737 (1968)
    (m) W. DeMore and O. F. Raper, *J. Chem. Phys.*, **37**, 2048 (1962)
    (n) D. Husain, *Advan. Chem. Phys.*, in press
67. (a) H. Vasatko and W. Hardy, unpublished results, Göttingen 1970
    (a) R. C. Millikan, unpublished measurements 1965
    (b) H. A. Olschewski, J. Troe and H. Gg. Wagner, *Ber. Bunsenges. Physik. Chem.*, **70**, 1060 (1966)
    (c) K. W. Michel, H. Richtering, H. A. Olschewski and H. Gg. Wagner, *Z. Physik. Chem.*, **NF39**, 129 (1963); **44**, 160 (1965)
    (d) W. O. Davies, *J. Chem. Phys.*, **41**, 1846 (1964); **43**, 2809 (1965)
    S. A. Losev, Q. A. Generalov and V. A. Maximenko, *Dokl. Akad. Nauk SSSR*, **150**, 839 (1963); *J. Quant. Spectry Radiative Transfer*, **6**, 101 (1966)
    T. A. Brabbs, F. E. Belles and S. A. Zlatarich, *J. Chem. Phys.*, **38**, 1939 (1963)
    E. S. Fishburne, K. R. Bilwakesh and R. Edse, *J. Chem. Phys.*, **45**, 160 (1966)
    T. C. Clarke, S. H. Garnett and G. B. Kistiakowsky, *J. Chem. Phys.*, **51**, 2885 (1969)
    H. A. Olschewski, J. Troe and H. Gg. Wagner, 11th Int. Symp. Combustion (The Combustion Institute, Pittsburgh, 1967) p. 155
68. (a) H. A. Olschewski, J. Troe and H. Gg. Wagner, *Ber. Bunsenges. Physik. Chem.*, **70**, 1060 (1966)
    (a) H. A. Olschewski, J. Troe and H. Gg. Wagner, *Z. Physik. Chem.*, **NF45**, 329 (1965)
    (b) A. G. Gaydon, G. H. Kimbell and H. B. Palmer, *Proc. Roy. Soc.*, **A279**, 313 (1969)

69. (a) H. G. Schecker and H. Gg. Wagner, *J. Chem. Kinetics*, 1, 54 (1969)
    A. J. Hay and R. L. Belford, *J. Chem. Phys.*, 47, 3944 (1967)
70. (a) T. A. Brabbs and F. E. Belles, 11th Int. Symp. Combustion (The Combustion Institute, Pittsburgh, 1967) p. 125
    (b) F. Zabel, Diplomarbeit, Göttingen, 1968
    (b) M. C. Lin and S. H. Bauer, *J. Chem. Phys.*, 50, 3377 (1969)
    (c) L. Avramenko and R. Kolesnikova, *Izv. Akad. Nauk SSSR Otd. Khim. Nauk*, 1562 (1959)
    (c) V. N. Kondratjev and E. I. Intezarova, *J. Chem. Kinetics*, 1, 105 (1969)
    (c, d) B. H. Mahan and R. B. Solo, *J. Chem. Phys.*, 37, 2669 (1962)
    (c, d) V. N. Kondratjev and I. I. Ptichkin, *Kin. i Kataliz*, 2, 449 (1961)
    (d) R. H. Hartunian, W. P. Thompson and E. W. Hewitt, *J. Chem. Phys.*, 44, 1765 (1966)
    (d) M. A. A. Clyne and B. A. Thrush, *Proc. Roy. Soc.*, A269, 404 (1962)
    T. G. Slanger and G. Black, *J. Chem. Phys.*, 53, 3720 (1970); R. Simonaitis and J. Heicklen, *J. Chem. Phys.*, 56, 2004 (1972)
71. F. Rosenkranz and H. Gg. Wagner, *Z. Physik. Chem.*, NF61, 302 (1968)
    S. J. Arnold, W. G. Brownlee and G. H. Kimbell, *J. Chem. Phys.*, 72, 4344 (1968); 73, 3751 (1969)
72. (a) P. A. Giguère and I. D. Liu, *Can. J. Chem.*, 35, 283 (1957)
    (a) W. Forst, *Can. J. Chem.*, 36, 1308 (1958)
    (a) C. K. McLane, *J. Chem. Phys.*, 17, 379 (1949)
    (a) C. N. Satterfield and T. W. Stein, *J. Phys. Chem.*, 61, 537 (1957)
    (a) D. E. Hoare, J. B. Prothero and A. D. Walsh, *Trans. Faraday Soc.*, 55, 548 (1959)
    (a) R. R. Baldwin and D. Brattan, 8th Int. Symp. on Combustion (Williams & Wilkins, 1962) p. 110
    (b) E. Meyer, H. A. Olschewski, J. Troe and H. Gg. Wagner, 12th Int. Symp. Combustion (The Combustion Institute, Pittsburgh, 1969) p. 345
    (b) J. Troe, *Ber. Bunsenges. Physik. Chem.*, 73, 946 (1969), H. Kijewski and J. Troe, *Helv. Chim. Acta*, 55, 205 (1972)
    (c) G. Black and G. Porter, *Proc. Roy. Soc.*, A266, 185 (1962)
    (c) J. Caldwell and A. R. Back, *Trans. Faraday Soc.*, 61, 1939 (1962)
73. (a) H. G. Schecker and W. Jost, *Ber. Bunsenges. Physik. Chem.*, 73, 521 (1969)
    I. D. Gay, G. P. Glass, G. B. Kistiakowsky and H. Niki, *J. Chem. Phys.*, 43, 4017 (1965)
74. (a) H. Henrici, Dissertation, Göttingen, 1966
    (b) M. H. Hanes and E. J. Bair, *J. Chem. Phys.*, 38, 672 (1963)
    T. A. Jacobs, *J. Phys. Chem.*, 67, 665 (1963)
    K. W. Michel and H. Gg. Wagner, 10th Int. Symp. Combustion (The Combustion Institute, Pittsburgh, 1965) p. 333
    J. N. Bradley, R. N. Butlin and D. Lewis, *Trans. Faraday Soc.*, 63, 12, 2962 (1967)
75. (a) D. Beggs, C. Block and D. J. Wilson, *J. Phys. Chem.*, 68, 1494 (1964)
    (b) M. Volpe and H. S. Johnston, *J. Am. Chem. Soc.*, 78, 3903 (1956)
    H. J. Schumacher and G. Sprenger, *Z. Physik. Chem.*, B12, 115 (1931)
    H. F. Cordes and H. S. Johnston, *J. Am. Chem. Soc.*, 76, 4264 (1954)
    P. G. Ashmore and M. G. Burnett, *Trans. Faraday Soc.*, 58, 1801 (1962)
    H. Hiraoka and R. Hardwick, *J. Chem. Phys.*, 36, 2164 (1962)
    G. Casaletto and H. S. Johnston cited in N. B. Slater (Theory of Unimolecular Reactions, Cornell Univ. Press, Ithaca, 1959)
    H. D. Knauth and H. Martin, *Ber. Bunsenges. Physik. Chem.*, 73, 922 (1969)
76. D. Schofield, W. Tsang and S. H. Bauer, *J. Chem. Phys.*, 42, 2132 (1965)
    M. Cowperthwaite, W. Tsang and S. H. Bauer, *J. Chem. Phys.*, 36, 1768 (1962)
77. J. A. Blauer, M. G. McMath and F. C. Jaye, *J. Phys. Chem.*, 73, 2683 (1966)

78. P. Frisch and H. J. Schumacher, *Z. Physik. Chem.*, **B37**, 1 (1937)
    M. C. Lin and S. H. Bauer, *J. Am. Chem. Soc.*, **91**, 7737 (1969)
79. M. J. Heras, P. J. Aymonino and H. J. Schumacher, *Z. Physik. Chem.*, **NF22**, 161 (1959)
80. (a) E. A. Schuck, E. R. Stephens and R. R. Schrock, *J. Air Pollution Control Assoc.*, **16**, 695 (1966)
    (a) H. W. Ford and S. Jaffe, *J. Chem. Phys.*, **38**, 2935 (1963)
    (a) F. E. Blacet, T. C. Hall and P. A. Leighton, *J. Am. Chem. Soc.*, **84**, 4011 (1962)
    (b) J. Troe, *Ber. Bunsenges. Physik. Chem.*, **73**, 906 (1969)
    (c) F. S. Klein and J. T. Herron, *J. Chem. Phys.*, **44**, 3645 (1966)
81. S. Jaffe and F. S. Klein, *Trans. Faraday Soc.*, **62**, 2150 (1966)
    M. F. R. Mulcahy, J. R. Steven and J. C. Ward, *J. Phys. Chem.*, **71**, 2124 (1967)
    C. J. Halstead and B. A. Thrush, *Proc. Roy. Soc*, **A295**, 363 (1966)
    M. F. Mulcahy, J. R. Steven, J. C. Ward and D. J. Williams, 12th Int. Symp. on Combustion (Combustion Institute, Pittsburgh, 1969), p. 323
82. C. P. Fenimore and G. W. Jones, *J. Phys. Chem.*, **69**, 3593 (1965)
    A. S. Kallend, *J. Phys. Chem.*, **70**, 2055 (1966)
    R. W. Fair and B. A. Thrush, *Trans. Faraday Soc.*, **65**, 1550 (1969)
83. N. Basco and R. G. W. Norrish, *Proc. Roy. Soc.*, **A283**, 291 (1965)
84. E. W. Montroll and K. E. Shuler, *Advan. Chem. Phys.*, **1**, 361 (1958)
    K. E. Shuler and G. H. Weiss, *J. Chem. Phys.*, **38**, 505 (1963)
85. B. Widom, *Advan. Chem. Phys.*, **5**, 353 (1963)
86. H. S. Johnston, Gas Phase Reaction Rate Theory (Ronald Press, New York, 1969
87. J. C. Light, J. Ross and K. E. Shuler, in Kinetic Processes in Gases and Plasmas (A. R. Hochstun, Editor, Academic Press, New York and London, 1969)
88. K. F. Herzfeld and T. A. Litovitz, Absorption and Dispersion of Ultrasonic Waves (Academic Press, New York, 1959)
    B. Stevens, Collisional Activation in Gases (Pergamon Press, Oxford, 1967)
    J. L. Stretton in Transfer and Storage of Energy by Molecules, **2**, Vibrational Energy (G. M. Burnett and A. M. North, Editors, Wiley-Interscience, London, 1969)
89. J. I. Steinfeld and W. Klemperer, *J. Chem. Phys.*, **42**, 3475 (1965)
    J. I. Steinfeld, *J. Chim. Phys.*, **64**, 17 (1967)
90. E. E. Nikitin and G. H. Kohlmaier, *Ber. Bunsenges. Physik. Chem.*, **72**, 1021 (1968)
91. P. G. Bergmann and J. L. Lebowitz, *Phys. Rev.*, **99**, 578 (1955); *Ann. Phys.*, **1**, 1 (1957)
92. T. A. Bak and J. L. Lebowitz, *Phys. Rev.*, **131**, 1138 (1963)
    T. A. Bak and J. L. Lebowitz, *Phys. Rev.*, **131**, 1138 (1963); *Discussions Faraday Soc.*, **33**, 189 (1962)
93. D. G. Rush and H. O. Pritchard, 11th Int. Symp. on Combustion (Combustion Institute, Pittsburgh, 1967) p. 13
    D. L. S. McElwain and H. O. Pritchard, *J. Am. Chem. Soc.*, **91**, 7693 (1969)
94. E. E. Nikitin, Theory of Thermally Induced Gas Phase Reactions (Indiana University Press, Bloomington and London, 1966)
95. B. S. Rabinovitch and D. C. Tardy, *J. Chem. Phys.*, **45**, 3720 (1966)
96. J. Troe and H. Gg. Wagner in Recent Advances in Aerothermochemistry, **1**, (AGARD, Paris, 1967) p. 21
97. J. C. Keck and A. Kalelkar, *J. Chem. Phys.*, **49**, 3211 (1968)
98. J. C. Keck and G. F. Carrier, *J. Chem. Phys.*, **43**, 2284 (1965)
99. C. A. Brau, J. C. Keck and G. F. Carrier, *Phys. Fluids*, **9**, 1885 (1966)
    J. C. Keck, *J. Chem. Phys.*, **46**, 4211 (1967)
    C. A. Brau, *J. Chem. Phys.*, **47**, 1153, 3076 (1967)
100. T. A. Bak and S. E. Nielsen, *J. Chem. Phys.*, **41**, 665 (1964)

101. E. E. Nikitin, *Dokl. Akad. Nauk SSSR*, **116**, 584 (1957); **119**, 526 (1958); **121**, 991 (1957)
102. E. V. Stupochenko and A. I. Osipov, *Zh. Fiz. Khim.*, **33**, 1526 (1959)
103. R. C. Tolman, Statistical Mechanics (Chemical Catalog Co., New York, 1927)
    C. N. Hinshelwood, The Kinetics of Chemical Change in Gaseous Systems (Clarendon Press, Oxford, 3. Ed. 1933)
104. L. S. Kassel, Kinetics of Homogeneous Reactions (Chemical Catalog Co., New York, 1932)
    L. S. Kassel, *J. Phys. Chem.*, **32**, 1065 (1928)
104a. E. V. Waage and B. S. Rabinovitch, *Chem. Rev.*, **70**, 377 (1970)
105. O. K. Rice and R. A. Marcus, *J. Phys. Chem.*, **55**, 894 (1951)
    R. A. Marcus, *J. Chem. Phys.*, **20**, 359 (1952)
106. G. Z. Whitten and B. S. Rabinovitch, *J. Chem. Phys.*, **38**, 2466 (1963)
    B. S. Rabinovitch and R. W. Diesen, *J. Chem. Phys.*, **30**, 735 (1959)
    E. W. Schlag and R. A. Sandsmark, *J. Chem. Phys.*, **37**, 168 (1962)
    E. Thiele, *J. Chem. Phys.*, **39**, 3258 (1963)
    E. W. Schlag, R. A. Sandsmark and W. G. Valance, *J. Chem. Phys.*, **40**, 1461 (1964)
    P. C. Haarhoff, *Mol. Phys.*, **6**, 337 (1963); **7**, 101 (1963)
    S. H. Lin and H. Eyring, *J. Chem. Phys.*, **43**, 2153 (1965)
    W. Forst, Z. Prášil and P. St. Laurent, *J. Chem. Phys.*, **46**, 3736 (1967)
    J. C. Tou and S. H. Lin, *J. Chem. Phys.*, **49**, 4187 (1968)
    J. C. Tou and A. L. Wahrhaftig, *J. Phys. Chem.*, **72**, 3034 (1968)
    M. Vestal, A. L. Wahrhaftig and W. H. Johnston, *J. Chem. Phys.*, **37**, 1276 (1962)
107. B. S. Rabinovitch and D. W. Setser, *Advan. Photochem.*, **3**, 1 (1964)
    M. Wolfsberg, *J. Chem. Phys.*, **36**, 1072 (1962)
108. E. Thiele, *J. Chem. Phys.*, **38**, 1959 (1963)
109. H. M. Rosenstock and M. Krauss, *Advan. Mass Spectry*, **2**, 251 (1963)
110. J. C. Keck, *Discussions Faraday Soc.*, **33**, 173 (1962)
    B. Woznick, *J. Chem. Phys.*, **42**, 1151 (1965)
111. G. M. Wieder and R. A. Marcus, *J. Chem. Phys.*, **37**, 1835 (1962)
    R. A. Marcus, *J. Chem. Phys.*, **43**, 2658 (1965)
112. O. K. Rice and R. C. Ramsperger, *J. Am. Chem. Soc.*, **49**, 1617 (1927)
113. W. Forst, *J. Chem. Phys.*, **48**, 3665 (1968)
    E. Tschuikow-Roux and R. Paul, *J. Phys. Chem.*, **72**, 375 (1968)
    E. Tschuikow-Roux and R. Paul, *J. Phys. Chem.*, **72**, 1009 (1968)
    J. C. Walton, *J. Phys. Chem.*, **71**, 2763 (1967); **72**, 375 (1968)
    B. H. Mahan, *J. Chem. Phys.*, **32**, 362 (1960)
114. E. W. Schlag and H. von Weyssenhoff and M. E. Starzak, *J. Chem. Phys.*, **47**, 1860 (1967)
115. H. A. Kramers, *Physica*, **7**, 284 (1940)
    S. Chandrasekhar, *Rev. Mod. Phys.*, **15**, 1 (1943)
116. S. W. Benson and T. Fueno, *J. Chem. Phys.*, **36**, 1597 (1962)
117. D. L. Bunker and M. Pattengill, *J. Chem. Phys.*, **48**, 772 (1968)
118. A. M. North, The Collision Theory of Chemical Reactions in Liquids (Methuen, London, 1964)
    K. J. Laidler, Reaction Kinetics (Pergamon Press, London, 1963)
    W. J. Le Noble in *Progr. Phys. Org. Chem.*, **5**, 207 (1967)
119. E. Thiele, *J. Chem. Phys.*, **36**, 1466 (1962)
120. N. B. Slater, Theory of Unimolecular Reactions (Cornell University Press, Ithaca, 1959)
121. E. Thiele, *J. Chem. Phys.*, **43**, 2154 (1965); **45**, 491 (1966)
    R. D. Levine, *J. Chem. Phys.*, **44**, 2029, 2035, 2046, 3597 (1966); **48**, 4556 (1968)

80   J. TROE AND H. GG. WAGNER

F. H. Mies and M. Krauss, *J. Chem. Phys.*, **45**, 4455 (1966)
F. H. Mies, *J. Chem. Phys.*, **51**, 787, 798 (1969)
E. E. Nikitin, *Kin. i Kat.*, **6**, 17 (1965)
F. P. Buff and D. J. Wilson, *J. Chem. Phys.*, **45**, 1444 (1966)
D. J. Wilson and E. Thiele, *J. Chem. Phys.*, **40**, 3425 (1964)
W. E. Smyser and D. J. Wilson, *J. Chem. Phys.*, **50**, 182 (1969)
122. M. Polanyi and E. Wigner, *Z. Physik. Chem.*, **A139**, 439 (1928)
123. D. L. Bunker, Theory of Elementary Gas Reaction Rates (Pergamon Press, Oxford, 1966)
D. L. Bunker, *J. Chem. Phys.*, **37**, 393 (1962); **40**, 1946 (1964)
R. C. Baetzold and D. J. Wilson, *J. Chem. Phys.*, **43**, 4299 (1965); *J. Phys. Chem.*, **65**, 3141 (1964)
F. P. Buff and D. J. Wilson, *J. Am. Chem. Soc.*, **84**, 4063 (1962)
E. Thiele, *J. Chem. Phys.*, **39**, 3258 (1963)
E. Thiele and D. J. Wilson, *J. Chem. Phys.*, **35**, 1256 (1961)
N. C. Hung and D. J. Wilson, *J. Chem. Phys.*, **38**, 828 (1963)
124. S. W. Benson and G. C. Berend, *J. Chem. Phys.*, **38**, 25 (1963); **40**, 1289 (1964)
S. W. Benson, G. C. Berend and J. C. Wu, *J. Chem. Phys.*, **37**, 1386 (1962)
R. Dubrow and D. J. Wilson, *J. Chem. Phys.*, **50**, 1553, 1627 (1969)
E. L. Breig, *J. Chem. Phys.*, **51**, 4539 (1969)
125. F. P. Buff and D. J. Wilson, *J. Chem. Phys.*, **45**, 1444 (1966)
126. S. Glasstone, K. J. Laidler and H. Eyring, The Theory of Rate Processes (McGraw-Hill, New York, 1941)
127. M. Jungen and J. Troe, *Ber. Bunsenges. Physik. Chem.*, **74**, 276 (1970), H. Gaedtke and J. Troe, *Ber. Bunsenges. Phys. Chem.*, **76**, (1972) (in press)
128. L. D. Landau and E. M. Lifshitz, Quantum Mechanics (Pergamon Press, London, 1958)
129. V. K. Bykhovskij, E. E. Nikitin and M. Ya. Ovchinnikova, *J.E.T.P.*, **20**, 500 (1961)
E. E. Nikitin, *Optics and Spectry*, **11**, 246 (1961); *Mol. Phys.*, **7**, 389 (1969)
130. J. C. Giddings and H. Eyring, *J. Chem. Phys.*, **22**, 538 (1954)
131. F. P. Buff and D. J. Wilson, *J. Am. Chem. Soc.*, **84**, 4063 (1962)
132. D. J. Wilson, *J. Phys. Chem.*, **64**, 323 (1960)
M. Solc, *Mol. Phys.*, **11**, 579 (1966)
M. R. Hoare and E. Thiele, *Discussions Faraday Soc.*, **44**, 30 (1967)

Chapter 2

Shock Tube Studies of the Hydrogen—Oxygen Reaction System[†]

G. L. Schott and R. W. Getzinger

University of California
Los Alamos Scientific Laboratory
Los Alamos, New Mexico 87544

† Work was performed under the auspices of the United States Atomic Energy Commission

NOMENCLATURE

A	Absorbance
A	Constant factor in expression $k_j^{M_i} = AT^{-m}$
A, B, A', B'	Constants describing induction period lengths
A_j, B_j, C_j	General notation for reactive species
$0, a, b, c, d, e, f, g, h, l, m, n$	Index notation for elementary reactions
C	General chain centre concentration
C_i	Chain centre concentration at end of ignition delay
C	Number of independent chemical elements
E_{act}	Activation energy
F_R, F_L	Functions describing species concentrations in rich, lean mixtures
f	Specific rate of chain branching
$\Delta G°$	Standard free energy change
g	Specific termination rate
$\Delta H°$	Standard enthalpy change
$\Delta H_0°$	Standard enthalpy change at 0 K
I	Chemical symbol for generic primary intermediates OH, H and O (not iodine)
$I_v°$	Incident spectral intensity at v
K_a, K_b, K_c, K_d	Equilibrium constants

K_{II}, K_{III}, K_{IV}	Equilibrium constants
K_a, K_b, K_c, K_f	Product of specific rate coefficient and one or more concentrations
$k, k_{app}, k_j, k_j^{M_i}, k_p, k_s, k_{coll}, k_{prop}$	Specific rate coefficients
L	Path length
M	Chemical symbol for generic collision partner
M_0	Initial mean molecular weight
m	Exponent describing temperature dependence of rate coefficient $k_j^{M_i}$
N	Mole number
N_i	Partial mole number of species i
N_i^0	Initial partial mole number of species i
N_l	Population of absorber in a specific internal energy state
N^{eq}	Value of N at full equilibrium
N_0	Total population of absorber
P_ν	Total absorption coefficient at ν
$P_\nu^{(l)}$	Contribution to P_ν from single line
p	Pressure
p_0	Unshocked gas pressure
R	Chemical symbol for generic reactive species formed in reactions (0) and (h)
R	Gas constant
R_{assoc}	Net rate of recombination
R	Number of independent reaction steps
$R^{(j)}$	Rate of elementary reaction denoted by index j
ΔS°	Standard entropy change
S	Total number of species
T	Temperature
T_0	Unshocked gas temperature
T_l	Explosion limit temperature
T_R	Fractional transmission
t	Time
t_i	Induction time
x, y	General exponents
α	Ratio [OH]/[H]
ε_{eff}	Extinction coefficient
η	Initial hydrogen:oxygen ratio
θ	Initiation rate
ν	Spectral frequency; Dimensionless recombination progress variable
ρ	Gas density
ρ_0	Initial gas density
τ	Laboratory time
ϕ	Exponential growth coefficient
I, II, III, IV	Stoichiometric equations for chemical reaction

2.1. INTRODUCTION AND BACKGROUND

The shock tube is one of the most useful experimental tools currently available to the chemical kineticist for the study of fast reactions in gases, particularly at temperatures above those achievable statically in laboratory apparatus. When coupled with one or more of a great variety of appropriate diagnostic methods, it is well-suited for the study of chemical processes taking place over the approximate range from 10^{-6} to 10^{-3} seconds. Shock wave methods to study chemical reactions and related time-dependent phenomena have been described in the literature throughout the 1950's and 1960's, and the techniques and many of the important experimental results are described in a number of excellent books[1] and recent review articles.[2-4] Notwithstanding the concern for aerodynamic nonidealities in shock tube flows which is increasing conspicuously at the time of this writing, and for the quantitative corrections which these effects make necessary, shock tubes are now a well-established means for obtaining information on chemical reaction rates in thermally perturbed homogeneous systems.

Basically, the shock wave triggers chemical reaction at an elevated temperature and with a precise time origin through an abrupt mechanical compression which raises the enthalpy of the fluid in a controlled way. Chemical experiments are often carried out in the presence of a relatively large concentration of an inert diluent in order to minimize the changes in experimental conditions resulting from the thermochemical effect of the reaction in the adiabatic, compressible flow situation. The earliest, and still by far the majority of fruitful shock tube investigations of fast reactions have been carried out in relatively simple reaction systems which frequently involve only one reactant and even a single chemical element. The objective in the design of such experiments is that only a limited number of elementary reaction steps and chemical compounds play a role in the overall reaction process. The time dependence of any experimental quantity is then more simply related to the rates of a minimum number of elementary reaction steps, and direct reduction of data obtained with only one diagnostic method is rendered possible. The dissociation of a homonuclear diatomic molecule is the simplest example of this type of reaction, and its rate is measurable by such methods as the disappearance of an electronic absorption spectrum of the reactant molecule, the appearance of an emission spectrum from radiative association of the product atoms, or through the direct influence of the endothermic dissociation process upon the density or temperature of the gas. Indeed, the measurement of the previously unobservable rates of thermal dissociation of the diatomic elements was a major part of the developing use of shock waves for chemical studies during ca. 1953–1963. It would appear that part of the incentive for the pursuit of fast reaction studies is the access which they afford to reactions of such formal molecular simplicity.

The more commonly encountered chemical reaction situation, when two or more chemical elements or two different reactant molecules are involved, is generally more complex. The description of the reaction mechanism is in terms of intermediate products, and consists of many elementary steps involving a number of different chemical species. Several steps may make important contributions to the overall rate, and some of the species involved are understood to be present only in exceedingly small concentrations. In such cases the interpretation of data obtained from any single diagnostic method is frequently only possible when one already has a fairly thorough understanding of the reaction system and makes a judicious selection of experimental conditions to isolate a particular process in a rate-determining role. Even then, it may be necessary to forgo conventional considerations of reaction order and graphical or other deductive analyses based on explicit integrated or differential rate expressions. Instead, one employs a trial-and-error series of numerical integrations of the appropriate set of simultaneous reaction rates using adjustable rate coefficient parameters to achieve agreement with the experimental observations and reach a satisfactory understanding of the total reaction kinetics. In fact shock-wave and other reactive flow experiments have been an important developing ground of this computer-assisted approach to complex reaction kinetics since the late 1950's.

Chain reactions belong categorically to the class of complex reactions, in the sense that cyclically regenerated intermediate chemical species are central to the reaction mechanism. When such reactions are characterized by low overall rates and intermediate species concentrations which remain small throughout, the analysis of kinetics is greatly facilitated by the applicability of the quasi-steady state principle. As we shall see, it is characteristic of the fast chain reactions studied in the shock tube that conventional quasi-steady state conditions do not prevail, and that macroscopic chain centre concentrations develop.

The reaction between hydrogen and oxygen is the acknowledged prototype of a branched chain reaction mechanism. No fewer than three elementary reaction steps and three chain carrier species participate, and the reaction is capable of sustained self-acceleration even in the absence of any increase in rate coefficients from thermal feedback. Unravelling the kinetics of a fast reaction of this complexity would seem to pose a formidable problem to the experimental kineticist. However, by providing the capability of time-resolving changes in a variety of system properties over a wide range of mixture proportions, densities and temperatures, shock tube methodology has been successfully applied to the hydrogen–oxygen system and has led to unique contributions to the understanding of this chain reaction occurring under nonsteady-state conditions.

It is the central theme of this chapter to consolidate what we judge to be the more enduring of these contributions, which add to the already substantial body of knowledge previously available from the very extensive work on the hydrogen–oxygen reaction system in lower-temperature, slower reaction regimes. The elucidation of the mechanistic details of the rapid hydrogen–oxygen reaction and the quantitative determination of rate coefficient parameters for many of the key elementary steps stands as one of the major areas of success of the shock tube technique in the fast reaction field. We begin our discussion by considering some general features of chain processes taking place at high speeds and high temperatures and by comparing the shock tube technique with the other experimental techniques which have provided important information about the hydrogen–oxygen reaction. The shock tube diagnostic methods used for the hydrogen–oxygen system and the phenomenological information which they yield are presented next. A major element of the chapter presents the reaction kinetics in terms of the elementary steps of the mechanism, the values of their rate coefficients, and the systematic consequences of the interactions among the elementary steps.

2.1.1. Chain Reactions as Fast Reactions

Chain Reactions

In a chain reaction, consumption of reactants takes place mainly in a cyclic series of elementary steps involving highly reactive intermediate species, or chain centres. In each step, an intermediate is consumed and another is regenerated, thus continuing the propagation of the chain. A simple example of such a process is the following two-step sequence which is operative under certain conditions in the isotopic exchange of elemental hydrogen:

$$H + D_2 \rightleftharpoons HD + D$$
$$D + H_2 \rightleftharpoons HD + H$$

Here the isotopically homogeneous molecules H_2 and D_2 may be the initial reactants, with HD the product, or vice versa, and H and D atoms are chain centre intermediates. The outstanding kinetic feature of the chain mechanism is that once a small amount of either H or D is present in the system, the chain reaction carries the stoichiometric process $H_2 + D_2 = 2HD$ toward equilibrium at a rate which is primarily governed by the H and D atom populations.

Our concern, then, centres upon the kinetics of evolution of the chain centre population throughout the main reaction. In the example introduced above, the hydrogen isotopic exchange occurring slowly under thermal conditions, atoms arise spontaneously from dissociation of the diatomic reactants, and they are removed by the reverse process, mutual association, often appropriately termed recombination.

$$H_2 + M \xrightleftharpoons[\text{reverse}]{\text{forward}} H + H + M$$

In the forward sense this reaction exemplifies a logically necessary aspect of chain mechanisms, namely independent initiation. The reverse step, which is second order in chain centre concentration, represents one of the important subgroups of chain terminating reactions. Their occurrence, as that of chain propagation, necessarily depends upon the presence of the chain centres, and is usually manifested in the kinetics by at least a first order factor. In combination, these particular initiation and termination steps lead to a chain centre concentration which approaches an equilibrium level dependent only upon the reaction conditions and thermodynamics.

Similarly stable levels of chain centre concentration may also arise from the kinetic balance between the rates of thermodynamically independent chain terminating and chain originating processes. A stable chain centre concentration above the equilibrium level can be achieved by externally regulated injection of chain centres, as in the sensitized photochemical stimulation of the above exchange sequence through

$$H_2 \xrightarrow{\text{Hg,h}\nu} H + H$$

when a balance is achieved with the spontaneous three-body recombination step. Alternatively either sort of production step may be counteracted more effectively and the chain centre population held down by an irreversible reaction of H atoms, say with some oxidizing surface, which neither produces another chain centre nor simply catalyzes the $H_2 \rightleftharpoons 2H$ equilibrium. To be sure, isotopically mixed dissociation and recombination make a minor contribution to the exchange reaction.

When the processes which yield such stabilized chain centre concentrations are fast enough to do so prior to any significant consumption of the reactants, either in the chain reaction or simply in the stoichiometric generation of the chain centres, the accelerative phase of the reaction is inconsequential and mathematical description of the rate of the main reaction is greatly simplified. The very powerful quasi-steady state approximation in the rate equations is valid from the outset, and even the stoichiometric analysis is simplified because significant amounts of chain centre intermediates do not accumulate. The rate of the termination process, which appears as a negative term directly involving the chain centre population in the rate equation for the evolution of this population, governs the time scale for achievement of a quasi-steady state. The characteristic lifetime of the chains at any instant is the chain centre concentration divided by the termination rate. When this lifetime is comparatively short, stabilization of the main reaction rate is effective. If this same lifetime is long enough that each chain centre has a high probability of engaging in the chain sequence, the reaction has a high efficiency as measured by the chain length, which is conceptually the ratio of the propagation rate to the initiation rate. Exothermic reaction systems which achieve a prompt quasi-steady

4

state, a high chain length, and a high yield of stable products with a minimum of side reactions are the hallmark of classical chain reaction chemistry.[5,6]

Fast Chain Reactions

We now inquire into how chain reactions may occur on the short time scales of shock tube experimentation, between milliseconds and microseconds. The speed of chain propagation steps is inherently limited by the rate of bimolecular collisions, whose frequency is typically proportional to a rate coefficient, k_{coll}, of the order of 10^{15} cm^3 mole^{-1} sec^{-1}. Chain propagation rate coefficients, k_{prop}, may approach this magnitude, and may typically achieve values near 10^{13} cm^3 mole^{-1} sec^{-1}, provided the Arrhenius activation energy does not exceed $RT \ln (10)$ and the preexponential factor is no smaller than $0 \cdot 1 k_{coll}$. The characteristic rate of the consumption of reactants by the chain process is the product of k_{prop} times the chain centre concentration, and if this rate is to reach 10^4 sec^{-1}, the chain centre concentration must be no smaller than 10^{-9} (to perhaps 10^{-11}) mole cm^{-3}. Now this magnitude is not utterly negligible with respect to the reactant concentrations near 10^{-6} mole cm^{-3} which are typical of laboratory experiments (partial pressures of 20 torr at room temperature).

Achievement of such a minimum chain centre population requires that the chain origination process have a large enough equilibrium constant and a rate which is large enough that irreversible termination does not prevent accumulation of the chain centres. Because chain initiation reactions are most often endothermic, elevated temperatures favor the chain centre concentrations needed for fast chain reaction. Likewise, even if we assume chain centre concentrations as large as 10^{-7} mole cm^{-3}, our attention is still restricted to propagation steps with activation energies no greater than 3 to 4 times $RT \ln (10)$. Thus exothermic or thermoneutral reaction systems with low activation energies in each step are inherently favoured to exhibit high rates by chain mechanisms, just as by direct ones, and elevated temperatures extend the range of positive activation energies that may be encountered.

Chain Termination

The uncommonly large chain centre concentrations necessary for the fast consumption of reactants provide a telling basis for assessing the kinetics of a fast chain reaction. The simple stoichiometric drain on the reactants in providing the chain centres, which may be amplified by the chain reaction during the period of their accumulation if the chain length is large, is enough to suggest that nonsteady kinetics may readily be expected in the macroscopic reaction. Even if the chain centre population during the fast reaction should remain stoichiometrically insignificant, we need to consider the time scale for stabilization of this population. If this is longer than about 10^{-5} sec, the validity of the steady state approximation in a reaction of 10^{-4} to 10^{-3} seconds duration will be substantially compromised. And indeed with an appreciably longer chain lifetime than this, the influence of termination during the bulk of the chain reaction becomes almost negligible.

Neither of the two types of chain termination reactions mentioned above in the isotopic hydrogen example is a fast reaction by these standards. Heterogeneous removal of chain centres requires as a minimum their delivery to a phase boundary, and at typical molecular speeds near 10^5 cm sec^{-1} quite small apparatus dimensions would be necessary to yield a 10^5 sec^{-1} reaction rate even under high vacuum conditions. This argument can be extended to the general observation that heterogeneous steps play no direct part in the interior of the short-lived experiments in shock tubes.

Chain termination by termolecular association is typically described by a rate coefficient which is near 10^{16} cm^6 $mole^{-2}$ sec^{-1} at room temperature, and decreases mildly as the temperature is raised. At ordinary third body concentrations near 10^{-5} mole cm^{-3} (0·25 atm at room temperature) this yields an effective second order rate coefficient of 10^{11} cm^3 $mole^{-1}$ sec^{-1}. This is 10^{-4} times the second order k_{coll} discussed above, and even at a working pressure ten or a hundred times higher, it becomes only comparable to the k_{prop} values we discussed. Thus the rate of mutual association of chain centres does not become large enough to reduce their lifetime to the 10^{-5} second range until their concentration reaches the 10^{-6} mole cm^{-3} range, which is the same as our presumed concentrations of the main reactants. From this we conclude that a fast chain reaction terminated only by steps which are second order in the chain centres surely proceeds under nonsteady conditions and accelerates continuously until the reactants become depleted. Given such a fast chain reaction, though, the termolecular termination steps may be influential in the ultimate removal of chain centres on a time scale comparable to that of equilibration of the system through other mutual reactions between pairs of chain centres or between chain centres and products. These arguments lead to the generalization that in the fast reaction domain it is of greatest importance to consider together with initiation reactions those processes which are first order in the chain centres, on the grounds that it is these reactions which govern the early evolution of the chain centre population. For by the time second order reactions between chain centres can become competitive, the consumption of reactants will already be set in its course.

Termination reactions which are first order in chain centres thus emerge as the favoured candidate to be influential in shaping the kinetics of fast chain reactions. These may occur by association of the centres with a major component of the reacting system in a bimolecular or termolecular step, or by some other bimolecular or even unimolecular step which consumes a chain centre without regenerating another. The termolecular association case is exemplified by the step

$$H + O_2 + M \longrightarrow HO_2 + M \quad \text{reaction } (f)$$

which we shall see is central to the hydrogen–oxygen chain mechanism. The same magnitude of rate coefficients discussed above, 10^{16} cm^6 mole^{-2} sec^{-1}, is appropriate, and with [O$_2$] near 10^{-6} mole cm^{-3} and [M] at least 10^{-5} mole cm^{-3}, a mean lifetime of H atoms of 10^{-5} seconds or shorter is attained.

Exothermic bimolecular reactions may also remove chain centres. As an example we may cite

$$H + I_2 \longrightarrow HI + I$$

which has a rate coefficient equal to $0 \cdot 4k_{coll}$ over a substantial range of temperatures near 700 K.[7] Addition of quite minor amounts of iodine to hydrogen–oxygen systems results in marked inhibition[8] of otherwise explosive reaction, even when H atoms are already limited to a lifetime near 10^{-5} sec by the rate of HO$_2$ formation. The effects of known chain reaction inhibitors have been explored in flames, but they have not received appreciable attention in quantitative shock tube work.

Chain Branching

Having thus introduced the class of first order chain terminating reactions which are most effective in limiting the lifetimes of chain centres early in a fast chain reaction, we now turn to the matter of generation of large chain centre populations.[6,9–11] To achieve even 10^{-9} mole cm^{-3} of chain centres having a lifetime of 10^{-5} seconds, a generation rate of 10^{-4} mole cm^{-3} sec^{-1} is needed, and such a rate would consume our 10^{-6} mole cm^{-3} of reactants in 10^{-2} seconds. Initiation reactions such as the H$_2$ dissociation process mentioned earlier do achieve such characteristic rates under ordinary density conditions, but only at high enough temperatures that they reach a large degree of dissociation at equilibrium.

Besides independent initiation, we have to consider another important source of chain centres, namely processes involving the chain centers themselves in a first order fashion. The one-for-one regeneration of chain centres which occurs so naturally and commonly in abstraction steps involving univalent atoms and radicals, and which is the necessary minimum regeneration for chain continuation, is by no means the maximum possible. A chain reaction which regenerates, say, two chain centres per cycle yields a net generation rate which is equal to the propagation rate. If the latter is in the speed range we are considering, its consequences may readily compete with or even greatly overshadow those of rapid homogeneous chain termination and lead to rapid chain reaction.

The competition between first order chain termination and branching is revealed clearly in the formal rate equation for the population of chain centres;

$$\frac{dC}{dt} = \theta + fC - gC$$

$$= \theta + \phi C. \tag{2.1}$$

C represents the chain centre concentration, θ the positive initiation rate, f the specific rate of chain branching and g the specific termination rate. ϕ, the algebraic sum of f and $-g$, emerges as the characteristic frequency governing the evolution of C, and its sign is of paramount importance. To be sure, ϕ, f and g are functions of the macroscopic condition of the system, and are not constants throughout the entire reaction with whose rate we are concerned. However, these quantities have particular values at the outset, when we suppose C to be negligibly small, and these determine the early evolution of C before second order terms can become important or other changes occur. The behaviour of C within this domain is given directly as

$$C = \theta/\phi(e^{\phi t} - 1) \tag{2.2}$$

If ϕ is negative, termination dominates and C tends toward the stable level $-\theta/\phi$; if ϕ is positive, branching dominates and C grows geometrically with time.

The contrast between these two cases is striking. In the former, the gross chain reaction proceeds at a regulated rate at which its underlying initiation rate is amplified in quasi-steady balance with termination. In the latter, the chain reaction accelerates continuously until the reactants become consumed. Provided that the branching is not a completely minor component of the chain mechanism, the chain centre concentrations rise to the 10^{-9} to 10^{-7} mole cm^{-3} levels at which a fast reaction may occur. Such quantities of chain centres might be quite thermodynamically unattainable independently from the reactants, but the coupled formation of more stable products by the branching chain mechanism temporarily shunts such a constraint. Moreover, when f and g are each of the order of 10^5 sec^{-1}, only those conditions which are balanced to within $\sim 1\%$ have resultant time constants smaller than 10^3 sec^{-1}, and the competition is resolved very promptly. The positive ϕ systems react explosively, and those with negative ϕ react very much more slowly even though the changes in conditions which tip the balance may be quite small.

In the hydrogen–oxygen system the primary mechanism of fast chain branching is firmly understood to be the coupled set of bimolecular steps:

$$H + O_2 \longrightarrow OH + O \qquad \text{reaction } (a)$$
$$O + H_2 \longrightarrow OH + H \qquad \text{reaction } (b)$$
$$OH + H_2 \longrightarrow H_2O + H \qquad \text{reaction } (c)$$

H atoms, O atoms and OH radicals are the chain centres, and the first two of these reactions each generate two of these centres in return for one. The termolecular association of H with O_2 discussed above, forming the comparatively inert HO_2 radical, is tantamount to termination.

High Temperature Reactions

The experimental variable which most influences chain reaction speed is the temperature at the outset. The shock tube studies of the hydrogen–oxygen reaction have involved high temperatures, between about 1000 and 3000 K, and there are several characteristics of this reaction, and of other chain reactions at high temperatures, that stem from this fact.

The most direct effect of elevated temperature on the course of reaction arises thermodynamically through the influence of temperature on the equilibrium relationships among reactants, intermediates and products. Dissociative reactions involving small molecules, that is reactions which form a larger number of molecules than they consume, are generally endothermic but have a positive entropy change, of the order of 30 cal mole^{-1} degree^{-1} per additional molecule. Thus the standard free energy change, $\Delta G° = \Delta H° - T \Delta S°$, passes from significantly positive to significantly negative values as T is raised. In the hydrogen–oxygen system, H_2O_2 is stable with respect to H_2 and O_2 at partial pressures of one atmosphere only up to about 1300 K. HO_2, with a 47 kcal/mole bond, loses its stability with respect to H and O_2 above about 2000 K, and reversibility of the primary chain-terminating reaction becomes a factor. At about 2500–3000 K, the stability of H_2O with respect to the diatomic elements and OH, and of these with respect to the atoms, begins to disappear. At these high temperatures the reaction exothermicity diminishes, and the limitation to the temperature attainable through adiabatic chemical combustion comes to the fore.

Just as the equilibrium constants of dissociative reactions tend toward large numbers of atmospheres as temperature is increased, those of equimolecular reactions, like the mutually bimolecular atom transfer reactions which occur in the H_2–O_2 chain mechanism, tend toward values near unity. The effects of energy differences are diminished, and the residual entropy differences are only a few cal mole^{-1} degree^{-1}. When combined with the substantial chain centre concentrations which accumulate by the nonsteady branching, this effect makes the equilibrium position of the chain propagation steps a much more significant aspect of the chain reaction course than under lower-temperature or quasi-steady reaction conditions.

There are additional kinetic effects in the high-temperature chain reaction regime, besides the much higher and nonsteady gross reaction rate, which arise from the Arrhenius temperature dependence of the rate coefficients of most of the initiation and propagation (including branching) steps. As all the E_{act}/RT factors become smaller, highly endothermic initiation steps such as dissociation of H_2 tend to become more important in relation to less endothermic steps which may have rather smaller preexponential factors or a disadvantageous concentration product in a diluted mixture. H_2 dissociation is an independently observable reaction in shock waves at temperatures down to about 2300 K.[12] However, it is the branching propagation, rather than initiation, which primarily governs the unsteady macroscopic H_2-O_2 kinetics between 1000 and 3000 K. Here, there is the possibility of different propagation steps, with somewhat higher activation energies, becoming prominent at elevated temperatures and making profound changes in the chain reaction mechanism. Such effects might be characterized as more disruptive routes which lead more directly to the stable products. They are not so conspicuous in the hydrogen–oxygen system, where the number of possible molecular species is limited, as they are in hydrocarbon oxidation chain reactions where many partially oxidized products are possible. We may observe, though, that chain continuation by HO_2, as through the step

$$HO_2 + H_2 \longrightarrow H_2O_2 + H$$

with an activation energy significantly above those of the direct branching sequence, becomes comparatively favourable at higher temperature.

Within the direct chain branching mechanism presented above, the ratios of the rate coefficients become significantly closer to unity as temperature is raised appreciably above 1000 K. Below this temperature, the higher activation energy of reaction (a) endows it with a distinct rate-limiting role, even in a modest excess of O_2, and H is far the most abundant of the three chain centres. In fact, even in the explosively accelerating regime at such a relatively low temperature, the turnover of the O and OH chain centre populations is satisfactorily describable by the quasi-steady state approximation, within the outer governance of the nonsteady growth of the H atom branching chain! This particularly simple state of affairs tends to vanish under the same increase in temperature which brings the gross reaction within the optimum range of shock tube study.

Branched Chain Stoichiometry

Besides the gross speed and kinetic order of the propagation and termination steps in explosively accelerated chain reactions such as that between hydrogen and oxygen, and the elevated temperatures at which such reactions are observed, their most prominent phenomenological feature is their stoichiometric course. We have seen how the formation of nontrivial chain centre populations is necessary for the ultimate rapidity of chain propagation. Examination of the three chain reaction steps asserted above as constituting the branching mechanism of the hydrogen–oxygen reaction shows that propagation of this mechanism, comparatively unrestrained in speed and correspondingly unmodified in gross stoichiometry by accompanying termination reactions, has profoundly simple stoichiometric consequences. The three-to-two reduction in number of molecules which is needed to produce the net reaction

$$2H_2 + O_2 = 2H_2O \qquad \qquad (I)$$

cannot occur by these bimolecular chain steps alone. That is, chain termination in the form of termolecular recombination is a necessary component of the complete reaction between hydrogen and oxygen below about 3000 K. Nevertheless the chain ignition occurs; the departure of the chain reaction from quasi-steady behaviour decouples the consumption of reactants from the corresponding formation of terminal products, and the total reaction assumes a particular sequential character. The chain reaction runs its course, substantially consuming the reactants and forming an incomplete measure of stable product, but leaving in its wake a major departure from ultimate thermodynamic equilibrium. Insofar as the macroscopic amounts of atoms and OH radicals which are formed are really intermediates and not long-term metastable products, they are subsequently removed in a second stage of the overall reaction.

Such a stagewise course of reaction, in one degree or another, is a pronounced characteristic not only of the hydrogen–oxygen reaction, but of other fast chain reactions within the scope of shock wave studies. The experimental characteristic by which this phenomenon is recognized is a macroscopically measurable maximum, or kinetic overshoot, in the concentration of one or another chain centre or other intermediate product.

In more general perspective, the hydrogen–oxygen case is peculiar in that the excess of chain centres is produced during the ignition stage without a macroscopic increase in the number of molecules. Rather more common, and correspondingly more complex, are high-temperature non-steady chain reactions in which dissociation steps occur in the mechanism of sustained chain-centre formation. Branching results when the dissociation of one or more unstable intermediates occurs in the chain sequence and contributes to the multiplication of chain centres. Many chain reactions, particularly the oxidation of hydrocarbons by oxygen, involve a net increase in the number of molecules just to achieve the ultimate stoichiometry; for example

$$2C_2H_6 + 7O_2 \longrightarrow 4CO_2 + 6H_2O$$

Formation of CO and of the chain centres OH, H and O demands an even greater contribution from dissociation. The overshoot in number of particles which occurs in branched chain ignition in these systems is then determined by the particular intermediates which are formed, and is not stoichiometrically inherent in the original reactants, as it is in the hydrogen–oxygen system.

Résumé

In the foregoing we have introduced some of the underlying principles of chain reaction kinetics in physical chemistry, and indicated the nature of their extension to describe rapid reactions at elevated temperatures. Clearly, these extensions could have been made entirely theoretically and the description of the processes worked out but for some quantitative parameters. However the actual development has occurred largely in a surge of activity that began in the middle 1950's, in conjunction with fast reaction experiments which reveal the phenomenology and provide access to the quantitative parameters. In place of the conventional quasi-steady state principle, other comparably useful and mathematically approximate aids to the evaluation and comprehension of data have arisen in conjunction with the nonsteady ignition behaviour and the subsequent removal of the residual super-equilibrium population of chain centres. The detailed consideration of these experiments and their interpretation form sections 2.2 and 2.3 of this chapter.

2.1.2. Environments of Fast Exothermic Reactions

The study of fast reaction kinetics entails adequate speed and sensitivity in determining changes in chemical composition and adequate knowledge of the conditions under which these changes occur. Foremost among these conditions for homogeneous reactions are the thermal state of the medium and the basis for reckoning time. In a macroscopic, initially irreversible reaction, the experimental objective is a sequence of finite samples having a common initial condition and subsequent history after a time origin which, for each sample, is well-defined in relation to the ensuing reaction speed. In addition, the reaction conditions should be capable of controlled variation over considerable ranges. In this section we consider and contrast three experimental situations whose applications to the hydrogen–oxygen reaction system have contributed most to the present structure of our understanding of this reaction and similar ones.

Shock Waves in Tubes

The gas shock waves generated in bursting-diaphragm shock tubes are advantageous in meeting the above objectives in the following fashion and to the following degree. Shock waves compress the gaseous medium irreversibly, raising the temperature, specific enthalpy, pressure and density significantly. The equation of state and the conservation relationships for energy, momentum and mass interrelate these changes in state variables and the phase and mass velocities of the shock wave. In the so-called incident shock wave the medium is accelerated along the tube, propelled by the expanding driver gas from the high pressure reservoir, which acts as a piston. When the compression wave reaches the tube end it reflects, and the end wall becomes a piston. The original acceleration is abruptly reversed and the doubly compressed gas again becomes stationary.

Shock wave fronts are microscopically thin, effecting their changes between thermal equilibra in molecular translation and rotation over the space of a few mean free paths. Incident shock fronts more than a few tube diameters ahead of the driving gas are also flat across the tube section, typically to within a few tenths of a millimetre over a 10 cm tube width. Reflected shock waves generated from a flat wall are thus also initially plane.

Shock waves propagate supersonically, which means that in familiar laboratory gases their speeds range from several tenths to a few times 1 mm μsec^{-1}. Moreover, the propagation of incident waves is steady as evidenced by a substantially constant velocity along the test portion of the tube. Thus successive layers of initially homogeneous gas have substantially the same history subsequent to the shock wave, and there is an axial gradient in the time origin which is simply related to the volumetric compression ratio and the shock front speed.

In a planar cross-sectional slice of the tube volume subjected to examination for chemical change, as by *in situ* instrumental observation, one thus has continuously presented differential samples of the recently perturbed gas at successive distances behind the shock wave front. The thickness of the slice and the time inhomogeneity of the sample are mutually proportional, and these can be made typically 1 mm and a few μsec, respectively. In the shock wave reflected at the tube end, the same gas sample may be viewed by stationary apparatus throughout an experiment; in the incident wave flow, gas with a substantially common history flows through the observation region.

This situation is clearly somewhat different from a completely steady flowing fluid experiment, although it is similar in the way time resolution is related to space resolution. The difference, though, is that the shock wave is transient, and the experimenter cannot scan the spatial profile on a time scale of his own choosing, but must accept and adapt to that of the progressive wave phenomenon. Thus the resolution of space and time is directly involved with considerations of signal to noise optimization.

Physical Effects in Shock Tubes

Shock tube studies of fast reactions are subject to several commonly recognized physical effects, some advantageous and others not. The spatial gradient in the time origin of a chemical reaction in the post-shock volume makes for a reaction zone profile with accountable axial gradients in molecular concentrations, temperature, and flow speed. Fortunately, however, the transport processes of diffusion, thermal conduction, and viscous dissipation are so slow in comparison with the near-sonic velocities of the flow that their effects on the reaction zone are quite negligible. Even the very steep gradients at the shock front do not broaden it macroscopically. Radiative transfer of energy to the colder surroundings is also of negligible importance in shocked gases at the temperatures of chemical interest.

The complete accommodation of the thermal state of the sample gas to the shock wave perturbation may be less instantaneous than its translational and rotational components, in the same way that chemical changes are slow enough to be resolved in the postshock profile. The vibrational relaxation of molecular gases is a part of this accommodation whose rate may be measured very well in shock waves, but whose slowness can interfere with the thermal conditions governing simultaneous chemical changes. For example, nitrogen exhibits the largest known vibrational relaxation time, and even when it is a chemically inert diluent its thermal condition influences the temperature of the entire medium. Electronic adjustment is a less conspicuous problem; nearly degenerate components of ground electronic states equilibrate promptly, and the density of highly excited states in most simple molecules and atoms is so small that these states are thermally negligible except at temperatures where ionization is also significant.

Fast reactions, certainly those occurring in the times of shock wave flows, are not subject to much mechanical intervention. Beyond the evident difficulty of mixing reactants at the outset or withdrawing discrete samples during the reaction for an external chemical analysis, there are internal complications when the reaction to be observed has thermochemical consequences. Although the steady shock wave equations do permit proper account to be taken of the cooling or heating effect, the effect itself may be awkwardly large and thermostatting by conduction is impossible. Thermal ballast in the form of an excess of inert gas, usually monatomic, provides as effective a measure as is available, and is a valuable aid to fast reaction studies using shock tubes.

Exothermic reactions generally, and the hydrogen–oxygen reaction specifically, give rise to another more or less conspicuous complication in shock waves. This is instability of the laminar, one dimensional flow which is basic to the controlled environment sought in the experiment. Instead of merging entirely with the axial flow field, some of the energy released in the reaction spontaneously excites uncontrolled transverse mechanical disturbances whose intensity and spatial extent may be ruinous. These effects may distort the shock front macroscopically, as they evidently do in propagating gaseous detonations. In less promptly exothermic systems, however, they may disturb only the heart of the chemical reaction zone, in which case the instability may be quite insidious. Endothermic reactions, by comparison, simply derive their energy uniformly from the laminar flow field of the smooth shock wave. Dilution of exothermically reactive systems with inert gas reduces the tendency for overt instability, besides lowering attendant temperature change. Lower densities of reactants, with the same proportions of diluent, also reduce the tendency for instability, evidently by making the reaction slower and delocalizing the deposition of energy. The result is that many chemical rate measurements of exothermic reactions in plane, steady shock waves are feasible with practical working pressures and extents of dilution. There also exist techniques to generate transient laminar shock wave flows in more severely exothermic mixtures.[13]

The interior flow in shock tubes is also subject to finite influence from the frictional forces at the boundary between the moving gas column and the stationary tube wall. A boundary layer of gas directly influenced by the wall through viscosity and thermal conduction grows progressively behind the shock wave front, and the resulting displacement of material at the edge of the flow field influences the conditions governing the progress of chemical reaction throughout the tube cross section. Gradual attenuation of the shock wave strength with increased distance of travel is an indirect consequence of the growing length over which this effect is operative. In general, the departure of conditions in the shock wave profile from their steady, one dimensional relationships is greater at larger distances from the wave front, and at a given distance, the departure is more severe at lower working pressure. This subject has recently been reviewed from the standpoint of the chemical kineticist, and criteria for evaluating the magnitude of such effects on experimental data have been suggested.[4] Suffice it here to say that in most of the early chemical studies, including those on the hydrogen–oxygen reaction, corrections for this nonideality have not been made; the corrections needed may be substantial in some cases and only a few percent in others. Elevated working densities arising from dilution of the reactants in inert, monatomic gas are beneficial in insulating the shock wave conditions from wall effects, besides having the other benefits discussed above.

On the whole, the conditions attainable in shock tubes for the controlled measurement of fast reaction rates have a number of advantages. The proportions of different reactants and of an inert diluent can be varied over wide ranges, in order to enhance the importance of particular elementary reactions for detailed examination. The temperature range can be selected almost at will, at least in diluted mixtures, although exothermic reactions do give rise to a certain minimum (Chapman-Jouguet) wave speed which makes for a finite gap between the preshock temperature and the postshock temperature prior to the exothermic reaction. In combination with the usable time span, the accessible ranges of temperature and initial conditions facilitate the isolation of several different elementary reactions of a complex mechanism in rate-determining roles under different conditions, and thereby the testing and refinement of a postulated set of elementary steps and their rate coefficients. This is the approach whose pursuit has led to a number of fruitful investigations of the complex hydrogen–oxygen reaction system.

Flames

Of the other experimental methods which have been employed to investigate fast chain reactions at high temperatures, combustion studies in premixed gas flames have been the most generally useful. As with shock waves, profiles within the reaction zone of these flames are strongly dependent on the chemical processes governing the rate of energy release. As has been frequently demonstrated in the past, detailed examination of flame structure can yield significant amounts of kinetic information.

In comparison to the reactive flows generated in shock tubes, flames stabilized on an appropriate burner offer the major advantage of being time invariant. The laminar, essentially one-dimensional combustion flows produced by the flat flame burner have proven to be the most useful for kinetic purposes, as they are readily amenable to diagnostic traversals along the direction of flow. Profiles obtained in this manner indicate that the usual structure of these flames is made up of at least three separate zones: a reactionless region dominated by transport properties, a primary reaction zone, and a secondary, or postflame, reaction zone. The last two are of course of principal interest to the kineticist. The one-dimensional character of the flow for a few cm away from the burner, and the resulting temperature and time profiles, are more readily subject to direct verification than are the corresponding properties of flow behind transient shock waves in tubes.

Because of the relatively low burning velocities of flames, many important structural features are compressed within the region immediately above the burner. Even in the hot gases downstream of the primary reaction zone, flow speeds seldom exceed a few metres sec^{-1}. Spatial resolution is a limiting factor in many kinetic flame investigations of fast reactions, and the range of assessible reaction times lies between about 10^{-4} and 10^{-1} second, some 10^2 times slower than the optimum range for shock waves. Consequently, the earliest successful studies of flame structure were focused upon the postflame region where the reaction velocity is considerably lower than in the primary reaction zone, and the axial extension is correspondingly greater. Analysis of data obtained in the secondary reaction zone is simplified by the fact that the temperature is essentially constant within this region and the influence of transport properties is minimal. It has long been known that the pressure is essentially constant across the entire reaction zone of such flames.

More recently reliable kinetic data have been obtained within the primary reaction zone by studying low pressure flames. The thickness of this zone is roughly proportional to the inverse of the pressure. Even so, interpretation of data from this zone is greatly complicated by the presence of steep concentration and temperature gradients and the influence of heat transfer and diffusion on the measured profiles. However, these experiments have also yielded meaningful mechanistic and quantitative rate coefficient information.

An important additional restriction on flame structure studies of kinetics is provided by the practical problem of maintaining one dimensional flame stability over a wide range of compositions, which limits the ranges of fuel/oxidizer ratios and percentage dilution by inert additives accessible to experimental study. The advantages to the kineticist of freedom in the selection of these quantities has been stressed above. The choice of experimental conditions in flames is also limited by the direct relationship between flame temperature and initial composition. In the shock tube where heating is provided by an outside source, i.e., the shock wave, these two variables can be selected independently of one another. The instability problem with exothermic reactions in shock waves is of a different character from that in flames.

Fast reactions in flames are dealt with in another chapter of this volume, and there are several previous books describing the detailed study of flame structure,[14-16] so that the important contributions of these investigations to the present understanding of selected chain reactions are now available. Later in this chapter (section 2.3.4.) attention will be given some of the more significant kinetic results from H_2-O_2 (—diluent) flames in order to show how they supplement and support kinetic information derived in shock tube studies.

The Explosion Limit Domain

Observable thermal reaction between hydrogen and oxygen in the gas phase occurs only at temperatures of several hundred degrees Kelvin, and above about 1000 K, the reaction is invariably in the fast reaction domain. Between about 700 and 900 K, however, there are circumstances under which vessels in furnaces can be filled uniformly with mixtures of H_2 and O_2 at controlled temperature and pressure, and the course of reaction determined. In this regime of temperature, at each pressure from below 1 torr to above 1 atmosphere, there can be observed an abrupt temperature boundary, $T_i(p)$, between regimes of slow or negligible reaction below T_i and explosive ignition above it. This explosion limit is amenable to quite controlled determination over a very wide range of mixture compositions and other pertinent conditions. The simplest of diagnostic methods often suffice. The extensive body of knowledge on the explosion limit phenomena, dating from the late 1920's, is treated authoritatively elsewhere.[10,11,17,18]

Briefly, in the so-called lower or first limit regime, T_i drops from high temperatures at very low pressure toward a minimum near 700 K at several torr, and then rises with increasing pressure in the upper or second limit regime. The explosive domain between these limiting pressures is termed the explosion peninsula. A maximum in T_i near 900 K is reached at a few hundred torr, from which T_i again tends toward lower temperatures near 1 atmosphere and above, in the third limit regime. The first and second limit data in particular have been interpreted quite successfully on the basis of isothermal branched chain kinetics in which the frequencies of chain centre formation by branching reactions and destruction by termination reactions are just balanced, in the manner indicated in section 2.1.1.

In addition to the limit parameters, the same sorts of experiments also afford measurements of the kinetics of the reaction in the non-explosive regime, particularly adjacent to the second and third limits. The early course of isothermally accelerating reaction in the explosive regime adjacent to the first limit is similarly resolvable. These latter measurements span several hundredths to several tenths of a second, and they confirm quantitatively the exponential growth of the reaction and the finite induction period during which the exponential growth amplifies the reaction rate from an inobservably small initial level.[10,19,20] The reaction mechanism by which we have come to understand the above phenomena includes a number of heterogeneous processes and second order interactions between chain centres and products, but its central core is the set of reactions (a), (b), (c) and (f) introduced in section 2.1.1. These reactions, together with the information which continues to accumulate on their rate coefficients as functions of temperature and the theory of the kinetics to which they lead, form the basis from which our understanding of the fast hydrogen–oxygen reaction in shock waves and flames has evolved.

2.2. Investigation of Reacting H_2-O_2 Systems

Over the broad range of temperatures where the rate of reaction between H_2 and O_2 has been subjected to examination in shock tubes, H_2O is the principal equilibrium product, according to the stoichiometric relationship:

$$2H_2 + O_2 = 2H_2O \qquad \Delta H_0^\circ = -114\cdot2\ \text{kcal/mole} \qquad (I)$$

H atoms, O atoms and OH radicals also appear in the equilibrium state, along with residual O_2 and/or H_2, in amounts which are quite negligible near 1000 K and are progressively larger as temperature is raised, but remain minor in comparison to H_2O up to 2500 K or higher, depending on the density. Any H_2 or O_2 originally in excess of the two-to-one stoichiometric proportions remains essentially unreacted, and exerts a determining influence upon the relative proportions of the minor products.

Formation of these four products, H_2O, H, O and OH from H_2 and O_2 can be represented by reaction (I) and three independent processes such as:

$$3H_2 + O_2 = 2H_2O + 2H \qquad \Delta H_0^\circ = -11\cdot0\,\text{kcal/mole} \qquad \text{(II)}$$

$$H_2 + O_2 = H_2O + O \qquad \Delta H_0^\circ = +1\cdot9\,\text{kcal/mole} \qquad \text{(III)}$$

$$H_2 + O_2 = 2OH \qquad \Delta H_0^\circ = +18\cdot5\,\text{kcal/mole} \qquad \text{(IV)}$$

which may be regarded as side reactions of the H_2 and O_2. The equilibrium concentrations of the six reactive species are determined by the initial H_2 and O_2 fractions and the final temperature and pressure, through the four equilibrium constant relationships representing the reactions (I) through (IV).

This formulation of the composite reaction stoichiometry emphasizes several facts relevant to the measurement of reaction progress. First, reduction in the total number of moles and most of the reaction exothermicity are both associated directly with reaction (I). Second, consumption of H_2 and of O_2, and formation of H_2O, are not singularly tied to reaction (I); and third, the formation of the intermediates, when this can be measured, offers independent information about the course of reaction which complements the determination of progress of reaction (I).

As intermediates in the overall reaction, H, O and OH, along with HO_2 under many conditions of interest, are understood to play a central kinetic role in effecting the conversion (I). Moreover, the concentrations of these intermediates during the reaction are not bounded between zero and their ultimate equilibrium values described above. Instead, as a manifestation of the nonsteady chain ignition behaviour discussed in section 2.1.1, they pass through maximum values which, between 1000 K and 2000 K, are orders of magnitude greater than at equilibrium. With such a large dynamic range, these intermediate species concentrations are sensitive functions of the course of the total reaction leading to reaction (I).

Many investigations of reacting H_2–O_2 systems are approached from the standpoint of combustion phenomena. Accordingly hydrogen is regarded as the *fuel*, and oxygen as the *oxidizer*, and the initial composition is thought of as stoichiometric when the initial $H_2:O_2$ molar ratio is equal to 2, in correspondence with reaction (I). The equivalence ratio is thus one-half the molar $H_2:O_2$ ratio. A *fuel-rich* (often simply termed a *rich*) composition is one with initial $H_2:O_2$ ratio greater than 2, and a *fuel-lean* mixture has this ratio less than 2. A *near-stoichiometric* mixture, for our purposes, is one with $1 \leqslant H_2:O_2 \leqslant 3$, such that it is neither so *rich* as to correspond with reaction (II) nor so lean as to correspond with reactions (III) and (IV).

2.2.1. Electronic Absorption Spectrometry

The determination of the stoichiometric and kinetic role of the free radical intermediates by quantitative spectroscopic means has been central to the early and continuing use of shock tubes to study the high-temperature hydrogen–oxygen reaction. This is especially true of OH, which is found macroscopically in reacting H_2–O_2 systems over a wide range of temperatures, pressures, and particularly, initial $H_2:O_2$ ratios. Moreover, OH is unique among the six species H_2, O_2, H_2O, OH, H and O in exhibiting electronic absorption from its ground state which is suitably intense and sufficiently isolated from interfering spectra for analytical use and which occurs in the readily accessible visible or near ultraviolet spectral region where air is transparent and ordinary photomultiplier tubes and photographic emulsions operate favourably.

The quantitative detection of OH populations by ultraviolet absorption spectrometry was already rather developed for use with the hot gases in flames before it was adapted for shock tube work. The spectroscopic transition used is the $A\,^2\Sigma^+ \to X\,^2\Pi_i$ electronic system, whose strongest vibrational component, (0, 0), has its conspicuous band head at $\lambda = 306\cdot4$ nm (in air). Under ordinary gaseous combustion conditions, this band consists of distinct lines which are a few tenths of a cm^{-1} in combined Doppler and Lorentz width and are separated by as much as ten cm^{-1}, owing to the small moment of inertia of the molecule. The absorption coefficient is appreciable only within these lines, and constant only over a small fraction of their width. Instrumental isolation of a line centre, particularly with adequate luminous intensity for the time scale of shock tube experiments, is a formidable problem, and photographic kinetic spectroscopy on shock tube systems is both uncertain and tedious. Such essential operational difficulties have been overcome by using a narrow line source coincident with the OH absorption spectrum coupled with continuous photoelectric recording. This provides much greater analytical sensitivity than measurement of net absorption from a continuum source with coarse instrumental resolution, and the technique is useful even though quantitative interpretation requires that both the source and absorption line shapes be known independently.[21]

OH *Line Sources*

Electrically excited emission from OH itself provides an appropriate source, and individual lines isolated by a low-resolution monochromator from a low-intensity electric discharge through H_2O vapour are quite suitable for photoelectric work in flames. Kaskan has exploited this method with a d.c. discharge source to detect [OH] overshoots in both fuel-rich[22] and fuel-lean[23] H_2–air flat flames. Absolute concentrations of OH were derived from independently determined spectroscopic parameters for the transition and the approximation that only the line centres were involved. The rate of decay of [OH] was measured in the postflame gases very near equilibrium, and treated in terms of the order of disappearance with respect to [OH].

The much greater transmitted intensity required for a suitable signal-to-noise ratio in shock tubes necessitated some compromises. First, a stronger discharge having broader as well as more intense emission lines was used; and second, a group of lines was subtended by the monochromator slit function. Rather recently, Houghton and Jachimowski[24] have reported the use of a group of narrow lines from a cooled, flowing-gas lamp excited by a 28 MHz discharge. The experimental assembly used with minor variations for a series of shock tube studies of OH formation and disappearance in hydrogen–oxygen systems[25–32] beginning with rate and equilibrium measurements on the thermal decomposition of H_2O[25], is diagrammed in Fig. 2.1. The lamp emitting the OH line spectrum is a smoothly pulsed discharge of several amperes lasting about 5 msec, through a capillary containing H_2O vapour at a pressure (0·9 torr) regulated by a thermostatted salt-hydrate couple. A beam of radiation from the anode region is defined by adjustable slits and passes diametrically through the 10·2 cm i.d. shock tube to the entrance slit of a thermally stabilized monochromator. Collimating lenses and plane windows in the shock tube are of fused

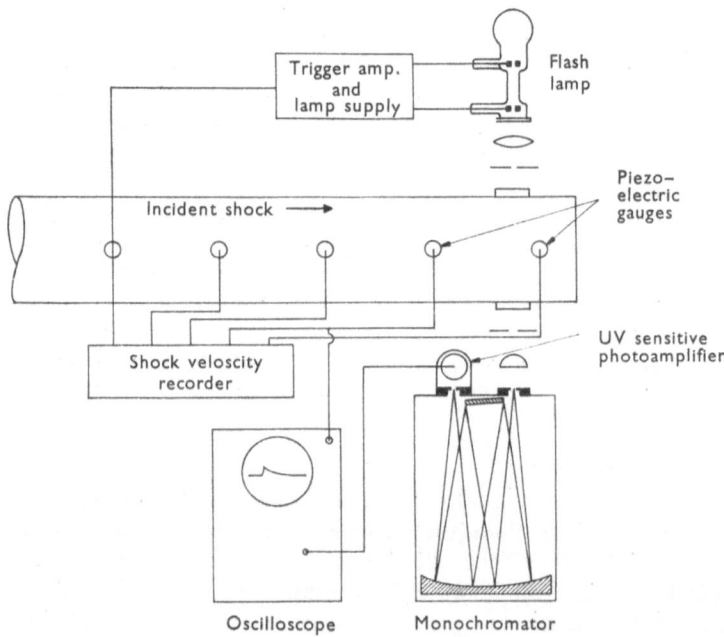

FIG. 2.1. Schematic of the shock tube assembly employed in References 25–32 to monitor [OH] during chemical transformations in hydrogen–oxygen systems.

silica, and radiation selected by the monochromator is received by a photomultiplier tube penetrable by ultraviolet light.

A series of accurately located and precisely responding thermal or piezoelectric sensors monitor the shock front motion along the tube past the optical measurement station. These serve to determine the shock wave speed which is needed to relate the preshock and postshock conditions, and to synchronize the flash lamp pulse with the wave and with the oscilloscope trace recording the time-dependent photo-multiplier output signal over the shock-wave reaction zone. The relationship between the OH radical concentration in the tube and the fractional transmission of the source radiation is monotonic, although it is quantitatively complicated and subject to some uncertainty, as is described below. Empirically, interfering ultraviolet emission from the hot, reacting or equilibrium shocked gases is negligible in relation to the source intensity for all conditions experienced below about 3200 K.

[OH] *Profile in High Temperature* H_2–O_2 *Reaction*

An OH line absorption oscillogram from a representative hydrogen–oxygen reaction experiment containing an excess of argon diluent is shown in Fig. 2.2. The qualitative course of [OH] during the fastest part of the reaction is revealed in this absorption versus time record. Following the shock front, there is a distinct delay, about fifteen μsec in this case, preceding an abrupt rise in [OH], and then a maximum and a rather slower decay which is incomplete at the end of the experiment.

The induction period delay at the beginning of the H_2–O_2 reaction was first measured and studied systematically in the high-temperature, submillisecond regime by this method.[26] Its existence had been antici-pated from the explosion limit phenomena and branched chain reaction theory, and experiments in detonating H_2–O_2 mixtures had sought unsuccessfully to identify it.[33-35] In flames, of course, the induction period has no distinct beginning, nor is the prereaction zone isothermal. As the study of H_2–O_2 ignition in shock waves has been expanded to embrace several techniques and a wide range of reaction conditions, the induction time has become the most commonly measured feature of the process. However, the unambiguous extraction of detailed kinetic information from this sort of measurement has met with limitations.

The nature of the kinetics of the slow disappearance of the radical intermediates in the H_2–O_2 reaction became recognized in flames in the late 1950's, and since 1964 a large amount of the quantitative kinetic information derived from absorption spectrometric work on OH in shock waves has come from analysis of the rate of OH disappearance following its maximum concentration.[28-32] These kinetics measurements are necessarily based on quantitative [OH] information, and their completion was contingent upon reliable calibration of the OH line absorption method. Meanwhile, however, semiquantitative interpreta-tion of the overshoot behaviour which produces the maximum in [OH] led to unification of the whole fast reaction mechanism at high temperatures and established a framework in which shock tube observations by other diagnostic techniques and under more diverse conditions could be related.[27]

OH *Absorption Calibration*

The problem of quantitatively relating [OH] to the transmission of a particular incident radiation spectrum over an extended range of operating conditions is not easily solved with great certainty. The photoelectrically measurable fractional transmission, T_R, is given in terms of the incident spectral intensity, I_ν°, and the variable absorption coefficient, P_ν, by:

$$T_R = \frac{\int I_\nu^\circ \exp(-P_\nu)\, d\nu}{\int I_\nu^\circ\, d\nu} \tag{2.3}$$

The contribution to P_ν from each absorption line, which we may denote $P_\nu^{(l)}$, is related to a particular fraction N_l/N_0, of the total [OH] through an integral proportionality,

$$\int P_\nu^{(l)}\, d\nu \propto N_l \tag{2.4}$$

and the shape of each line is affected by a combination of thermal Doppler broadening and collisional Lorentz broadening. The latter depends upon gas composition and pressure in addition to temperature, and even for a monochromatic source, the relevant value of P_ν depends upon these environmental conditions. Still, a unique relationship between T_R and [OH] exists for each particular environment when N_l/N_0 has the thermal equilibrium distribution. No serious departure from this distribution among the angular momentum states in the ground vibrational and electronic configuration of OH is experienced on the time scale and under the thermally achieved conditions of the hydrogen–oxygen systems which have been studied in shock waves.

The groups of lines selected for multiple-line quantitative absorption spectrometry on OH in shock waves have lower state rotational quantum numbers in the range 1–10, so that over the 1000–3000 K range, some of the N_l/N_0 values involved increase and others decrease with changing temperature. There is thus only a minor effect of temperature upon the effective extinction coefficient,

$$\varepsilon_{eff} = -([OH] \times L)^{-1} \log_{10}(T_R) \tag{2.5}$$

relating particular values of [OH], T_R, and the sample path length, L. The most serious variation of ε_{eff} is found to be with [OH] itself, at

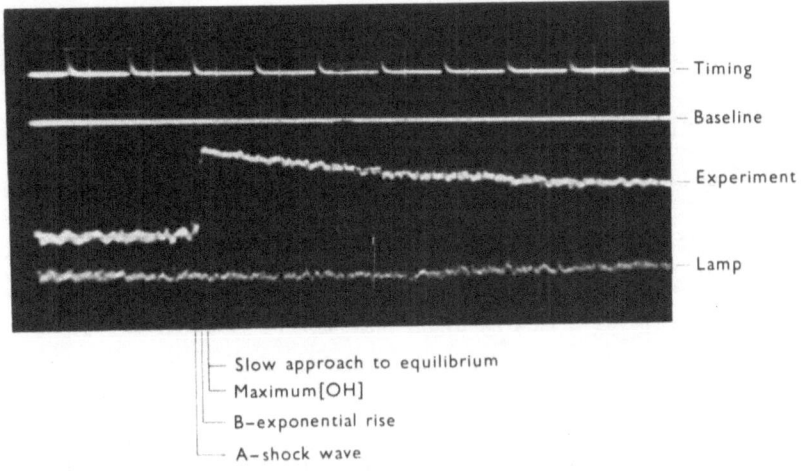

Timing

Baseline

Experiment

Lamp

Slow approach to equilibrium
Maximum[OH]
B–exponential rise
A–shock wave

A–B, induction period

FIG. 2.2. Experimental oscillogram showing absorption by OH in 306·4 nm band during shock-initiated reaction in a 1·0% H_2–2·0% O_2–97·0% Ar mixture. Transmission increases downward from baseline, and time increases from left to right. Time marks: 100 μsec. Initial pressure: 10·0 cm Hg. Shock velocity: 1·215 km sec⁻¹. Mean reaction zone temperature: 1601 K.

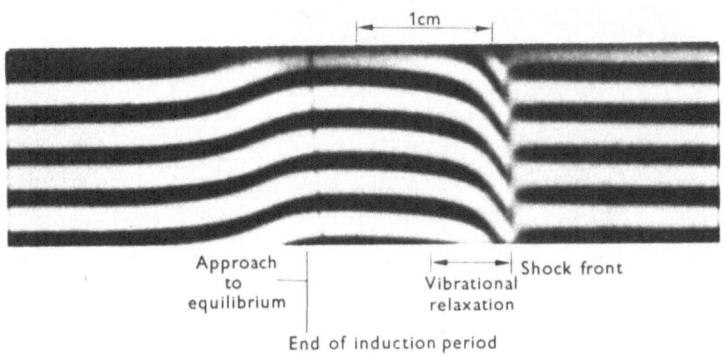

1cm

Approach
to
equilibrium

Vibrational
relaxation

Shock front

End of induction period

FIG. 2.4. Interferogram showing instantaneous gas density profile (Reference 54) near the shock front in a 1·5% H_2–98·5% O_2 mixture. Shock wave propagating from left to right with velocity of 1·412 km sec⁻¹. Initial pressure: 6·0 cm Hg. Increasing density displaces fringes upward. Mean reaction zone temperature: 1266 K.

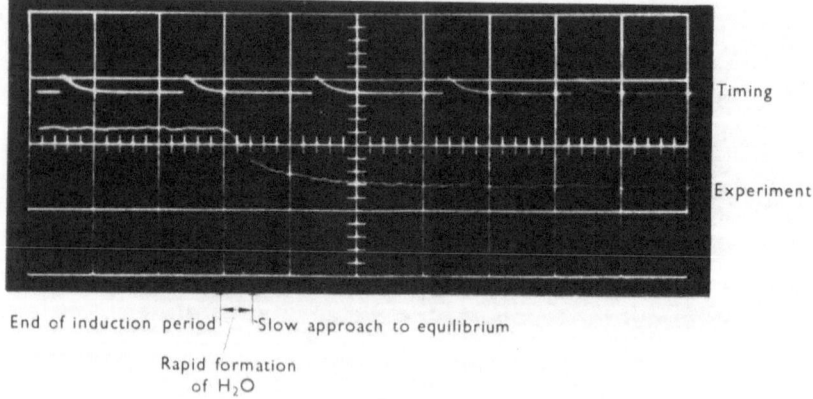

End of induction period ┤├ Slow approach to equilibrium

Rapid formation
of H_2O

FIG. 2.5. Profile of infrared emission at 2·7 μm during shock-initiated reaction in 2·0% H_2–2·0% O_2–96·0% Ar mixture. Emission is downward. Time increases from left to right. Time marks: 100 μsec. Initial pressure: 10·0 cm Hg. Shock velocity: 1·121 km sec^{-1}. Mean reaction zone temperature: 1435 K. Field of view defined by slits opened to 5 mm (after Blair and Getzinger[60]).

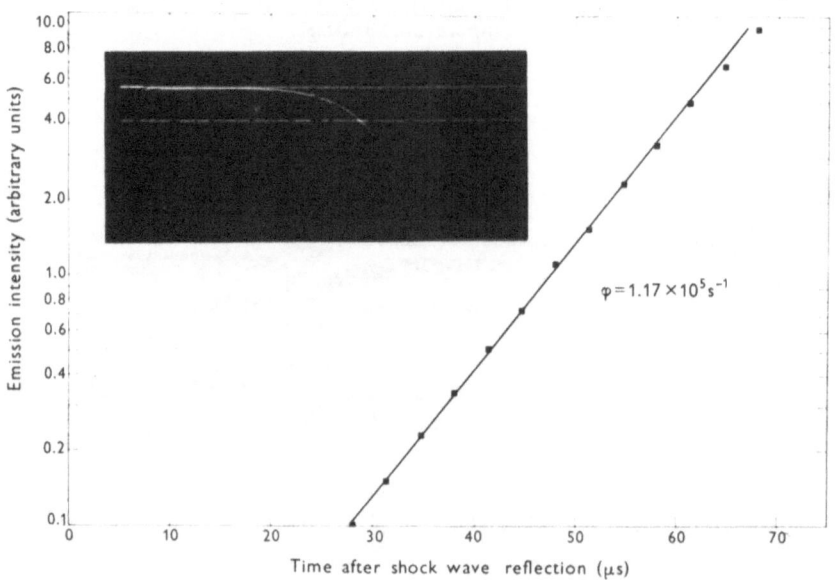

FIG. 2.6. Representative oscillogram and exponential growth plot for intensity of O + CO chemiluminescent emission during induction period in end-on reflected shock experiment. 1·5% H_2–0·5% O_2–3·3% CO–94·7% Ar. Initial pressure: 12·2 cm Hg. Reflected shock temperature: 1412 K. 10 μsec timing marks on oscillogram (after Gutman and Schott[62]).

fixed L, owing to the nonuniform value of P_v over the incident spectral interval. The customary exponential absorption relationship with constant ε, the Beer–Lambert law, is not followed at lower values of T_R. Rather, curves of absorbance, $A = -\log_{10}(T_R)$, versus ([OH] × L) are steepest at the origin and become less sloped as the more strongly absorbed spectral components are selectively removed. The precise trajectories of these curves depend upon temperature, pressure and composition of the system, as well as upon the particular incident spectrum. This behaviour is shown graphically in Fig. 2.3.

The earliest determination of such a curve for OH in shocked gases

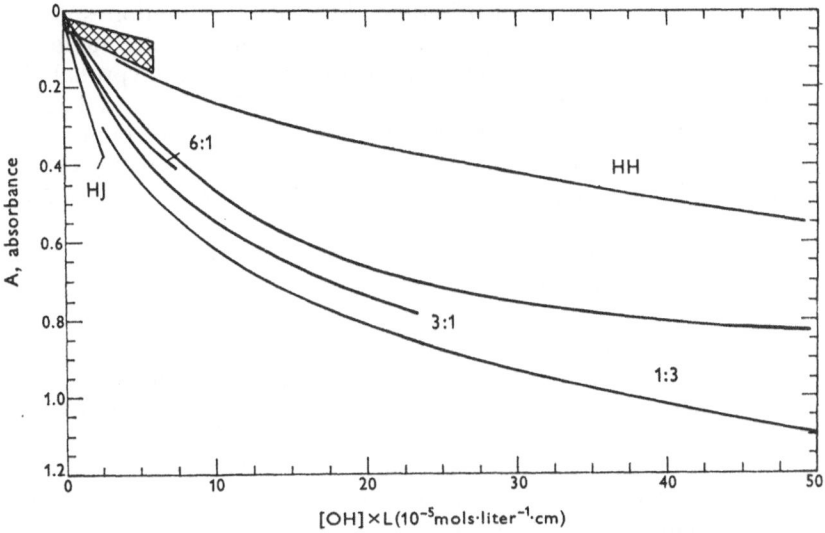

Fig. 2.3. Absorbance as a function of optical density for selected shock tube investigations employing OH electronic absorption spectrometry. The unmarked curve represents the semi-empirical relationship derived in Reference 37, evaluated at a pressure (5·1 atm) and temperature (1520 K) typical of recombination experiments in an argon diluent. The curves labelled 6:1, 3:1 and 1:3 were empirically determined over a selected range of recombination pressures and temperatures for mixtures dilute in argon with those particular initial H_2/O_2 ratios (Reference 32). The curve identified by HJ (Reference 24) was empirically determined in a 1% H_2–1% O_2–98% Ar mixture at 1300 K for a selected range of pressures. The cross-hatched area represents the approximate range of absorbances and optical densities observed with an atomic bismuth line source (Reference 41). Also shown are the line HH derived from photographic spectroscopy using instrumental definition of absorption line centres on a continuum (Reference 48), and a solid circle (beyond the range of the abscissa) denoting the photoelectric absorbance reported in Reference 47 for a continuum source at an optical density of 750 × 10^{-5} moles liter^{-1} cm.

was based upon computed equilibrium populations of OH between 10^{-8} and 10^{-7} mole cm^{-3} attained both from H_2O–Ar and $(2H_2 + O_2)$–Ar mixtures observed in reflected shock waves near 2830 K.[25] A series of determinations have subsequently been made for different source lamps and with particular attention to the 1400–2000 K temperature range and to pressure and composition effects in the absorbing gas.[36,37] All have been based upon shock tube measurements under conditions where a steady or nearly steady [OH] level was ascertained for equilibrium or other confidently calculable conditions.

Relation of such empirical calibration to quantitative spectroscopic theory was pursued with two of the different source lamps by determining their spectral distributions from high resolution spectrographic plates made by repeated flashes, combined with numerical evaluation of T_R via equation (2.3) using the band transition probability factor or f-number, and the pressure broadening factor, as well as the absorber temperature, as selectable parameters.[36] Uncertainty concerning the presence of continuum radiation between the OH lines in the source spectrum ultimately limited the definiteness of this calibration procedure.

With a more recent source lamp, employed after deterioration of the ones whose spectral output had been thoroughly examined, somewhat different calibration curves were experienced for OH formed in reacting H_2–O_2–Ar mixtures in the 1300–1900 K range, and in particular there was a noticeable effect of the initial $H_2 : O_2$ ratio[32] which had not arisen in the earlier work. This is illustrated in Fig. 2.3. Shown for comparison is the previously experienced calibration relationship for comparable temperature and total pressure conditions,[37] which had not been perceptibly dependent upon particular constituents other than the major diluent, argon. Also included, for comparison of gross analytical sensitivities, are representations of particular experience reported for other source spectra and monitoring procedures, as discussed below.

Bismuth Line Source

Ultraviolet absorption by OH has been used in other ways for quantitative shock tube investigations of the high temperature hydrogen–oxygen kinetics. In an important series of papers and unpublished dissertations,[38–45] Gardiner and co-workers used a source lamp emitting the 306·77 nm resonance line of atomic bismuth in low pressure, incident shock wave experiments on dilute H_2–O_2–Ar mixtures as well as on reactions of O_2 with C_2H_2 and with mixtures of H_2 and CO or CO_2. This line is coincident with the $R_2 10$ and $R_2 9$ lines of the $A\,^2\Sigma^+ \rightarrow X\,^2\Pi(0, 0)$ band, and although ε_{eff} is smaller and less reproducible than with the pulsed OH spectrum source (see Fig. 2.3), much higher signal-to-noise ratios and higher overall sensitivity were achieved.

Typical oscillograms obtained in H_2–O_2 mixtures by this method resemble Fig. 2.2, but differ primarily because of the influence of the lower total gas pressures on the reaction kinetics. The decay of [OH] following its maximum is too slow to be evident, and after its initial rapid rise, the absorption remains essentially constant for the duration of the experiment. In stoichiometrically balanced mixtures, this constant level is the maximum; under some fuel-lean conditions, a short-lived maximum precedes the essentially flat level.[42] Gardiner and co-workers have used these low-pressure experiments to study in detail both the induction period kinetics over a large range of initial $H_2:O_2$ ratios[39–41,43,44] and the [OH] maxima.[42] Where calibration data were necessary, experimental absorbances were converted to concentration ratios by assuming the validity of Beer's Law, or into estimates of absolute concentration with an empirically derived and somewhat erratic calibration which indicated variations in the Bi emission spectral distribution. Significant results from these investigations will be discussed further in section 2.3.

Continuous Sources

Absorption by OH radicals from a continuum ultraviolet source has also been used as a quantitative diagnostic to study the hydrogen–oxygen reaction under conditions producing high enough OH concentrations. In one investigation employing a xenon discharge lamp and monochromator, induction times to rapidly rising OH absorption were measured in H_2–O_2–Ar mixtures containing between 10% and 30% reactants following the passage of a reflected shock front through the source beam, and experiments were carried out at temperatures extending below those of earlier shock tube studies and approaching the explosion limit regime.[46]

In another study, OH absorption in the 310 nm region from a continuum source was measured in the reaction zone of a low-pressure, undiluted H_2–O_2 detonation.[35,47] It was possible to detect OH overshoots preceding an assumed Chapman–Jouguet (equilibrium) state, but not to resolve an induction period for the formation of OH in the early stages of the reaction.

More recently, sensitivity quite comparable to that afforded by the line source method has been obtained by timed flash photographic spectroscopy, by which the rate of appearance of OH in the shock-wave decomposition of H_2O vapour was reinvestigated.[48] Values of A over the instrumental slit function, averaged for several lines, are included in Fig. 2.3.

Absorption Spectrometry of Other Intermediates

Line-source absorption spectrometric methods quite analogous in their essentials to the method described above for OH have also been developed for H and O atoms.[12,49] Highly allowed atomic resonance transitions in the vacuum ultraviolet are employed, and in the case of

H atoms absorbing the Lyman-α line at 121·5 nm, sensitivity to [H] values in the 10^{-12} mole cm^{-3} range is achieved with a shock tube setup. These methods have not been applied to near-stoichiometric H_2–O_2 systems, but the rate of formation of H atoms from the H_2–O_2 chain reaction in H_2–Ar mixtures containing traces (0·1 % and 0·01 %) of O_2 has been measured and interpreted successfully.[12]

Absorption spectrometry of decomposing H_2O_2 in shock waves has been performed with a continuum source and monochromator, operating at 230 nm and 290 nm.[50] Transient absorption at 230 nm was identified with HO_2 radicals formed during the reaction in amounts estimated as about 10^{-8} mole cm^{-3}.[50,51] Although HO_2 is believed to be important kinetically in rapidly reacting H_2–O_2 mixtures studied in shock tubes, its concentration remains miniscule in the high temperature and low density systems that have received the most attention, and the presence of this intermediate, or of H_2O_2, has not been demonstrated by optical spectrometric methods.

2.2.2. Shock Wave Densitometry

One of the most versatile classes of diagnostic methods for reactive shock wave profiles is based on the thermal effect of the postshock reaction. In the case of an exothermic reaction there is both temperature rise and volumetric expansion, and the expansion of transparent gas is particularly amenable to instantaneous, noninterfering optical measurement. In a particular gas sample, especially one containing an inert diluent, the departure of the refractive index from unity is substantially proportional to the density, although changes in chemical composition do exhibit a determinable effect upon the proportionality coefficient which calls for independent calibration.[52,53] Because this diagnostic method depends upon a bulk change in the postshock fluid, high dilution with inert gas diminishes its sensitivity to chemical phenomena.

Interferometry

The most usual setup to determine postshock density profiles by their refractivity uses a Mach–Zehnder interferometer with a short-duration spark light source to map the instantaneous density distribution adjacent to the shock front by means of the pattern of parallel fringes in the photointerferogram made through a pair of optically plane windows in opposite sidewalls of a rectangular shock tube. These interferograms span several centimetres both along the shock tube and across it, and in addition to the profile along the shock wave path, they also afford direct evidence of the presence or absence of transverse disturbances in the flow field that is not provided by the measurements through small apertures or slits common in other diagnostic methods.

Shock wave interferograms have been obtained in a very wide range of fuel–oxygen–diluent mixtures by White and co-workers.[13,54–56] In comparatively dilute mixtures these interferometric experiments were successfully carried out in a conventional, constant-area shock tube. However, since one of these authors' objectives was the structure of the reaction zone of a detonation wave, many of the experiments were done in mixtures with little or no dilution. In such mixtures laminar shock waves of the appropriate strength for kinetics measurements were unstable. To overcome this difficulty the detonation wave front was first strongly accelerated and then strongly decelerated by passage through a converging-diverging rectangular nozzle formed by a wedge in the shock tube.[13,54] Interferograms of the cylindrically expanding shock wave during the exit phase showed laminar reaction zone structure and yielded kinetics data for relatively undiluted mixtures.

A representative constant-area shock wave interferogram obtained in a mixture of $1\cdot5\%$ hydrogen in oxygen is shown in Fig. 2.4.[54] The shock front, readily identifiable as the abrupt discontinuity in the fringes, is followed immediately by an increase in the density to a maximum value. Still farther behind the shock the density plateau is terminated by the beginning of a region in which the density is slowly decreasing toward a limiting equilibrium value. This behaviour can be compared qualitatively with that displayed by OH absorption in a different mixture in Fig. 2.2. The gradual compression immediately following the shock is a result of vibrational relaxation of the reactants, principally O_2 in this case. No consequence of such a process is seen in Fig. 2.2, as OH absorption is still negligible at this point. Quantitative interpretation shows that termination of the density plateau corresponds approximately to the end of the induction period for OH formation, and that the subsequent region of most rapid density decrease occurs near the peak in OH absorption. The final, slower density decrease can be identified with the slow decay of [OH] toward equilibrium.

Besides distinguishing between vibrational relaxation and macroscopic chemical reaction zones, these interferometric experiments have been most useful in systematic determination of the induction period length, t_i, in mixtures of extreme compositions, and in exploring directly the maximum rates of chemical energy release under selected circumstances. When applied to shock waves in hydrogen–oxygen mixtures dilute enough and at low enough densities to yield undisturbed, laminar reaction zones, the interferometric method has been a less sensitive indicator of the occurrence of chemical reaction than the OH absorption spectrometric method. Thus the latter diagnostic detects reaction rather earlier in the induction zone and more importantly, follows the approach to equilibrium much farther. A combination of factors influences the relative sensitivities of these methods, but a major determinant is the fact that the reference level in the spectrometric measurement (zero absorption) is experimentally established prior to occurrence of the shock wave and is stable, whereas the induction period plateau and terminal equilibrium density levels are both established by the shock wave itself and are subject to more uncertainty.

Schlieren and Sweeping-Image Photography

Another form of optical refractometric measurement of the exo-thermic zone in shock-wave initiated hydrogen–oxygen combustion has also evolved from the study of the detonation behaviour of concentrated mixtures. Reflected shock waves at the closed end of a constant-area tube are employed to initiate explosive reaction behind an initially planar shock front, and Toepler schlieren and interferometric optics, slit-shaped windows extending along the tube, and a sweeping image camera provide a time-resolved record of (i) the position of the original shock wave before and after reflection, (ii) the densities of particular zones, and (iii) the development of compression waves in the confined exothermic reaction volume adjacent to the end wall. This variant of densitometric technique has contributed to early observations of the induction time, particularly in the exploration of very long ignition delays and nonlaminar phenomena at comparatively low temperatures and high gas densities.[57–59]

2.2.3. Emission Spectrometry

The time or position of occurrence of reaction, and relative or absolute quantitative indications of its extent, can often be examined very simply by means of emission spectra which are characteristic of the reaction system. At sufficiently high temperatures and for suitable values of concentrations, transition probabilities, and optical paths, thermally governed vibrational and even electronic emission spectra can be strong enough for time-resolved determination of either temperature or the chemical concentration of the emitting molecule or atom. In the hydrogen–oxygen reaction system below 3000 K, the infrared spectrum of H_2O is the only thermally excited emission which has been developed as a direct diagnostic of the reaction progress in shock tube experiments. Possibilities exist for use of the 306·4 nm electronic band of OH above about 2500 K.[43]

Another source of quantitatively useful emission spectra is electronic chemiluminescence, which is commonly found in atomic chain reactions generally and is particularly prevalent in combustion reactions, owing to their exothermicity. The emitter of such radiation is recognized qualitatively as one of the chemical constituents, but the upper state population, and thus the radiation intensity which represents it, is very much greater than that which corresponds to thermal equilibrium with the chemical concentration of the emitter, and generally is not proportional to it. Instead, the emission is governed by the populations of those other reactive components which are producing the excited state dynamically, in competition with quenching, during the reaction. Hence the luminosity is generally useful as an indicator of the chemical concentrations of species other than the emitter itself.

Some of the best understood and quantitatively most useful emission mechanisms of this type occur by radiative association of atoms and molecular fragments, in which the excitation energy arises by the incipient formation of a new, strong chemical bond between previously dissociated partners interacting initially through the potential energy field of an excited electronic state. Examples whose uses in studies of the hydrogen–oxygen reaction are discussed below are

$$H + O \longrightarrow OH\ (A^2\Sigma^+) \longrightarrow OH\ (X^2\Pi) + h\nu$$

which gives rise to emission of the characteristic ultraviolet band spectrum of OH, and

$$CO + O \longrightarrow CO_2^* \longrightarrow CO_2 + h\nu$$

which under typical shock tube conditions with H_2–O_2–CO mixtures displays a broad, essentially continuous spectrum over the blue and near ultraviolet regions.

H_2O Infrared Emission

The collisional vibrational relaxation time of H_2O under typical shock tube conditions is very short compared to the vibrational radiative lifetime, and the emission spectrum of this molecule during the H_2–O_2 reaction is quite thoroughly thermal. Figure 2.5 is a representative oscillogram of a profile of emission near $2 \cdot 7\ \mu m$, including contributions from the ν_1, ν_3 and $2\nu_2$ transitions of H_2O, from the incident-shock initiated reaction between hydrogen and oxygen dilute in argon. This record was obtained in a recent investigation of the kinetics of the slow decay of the system to equilibrium following the initial production of large concentrations of intermediates.[60] That the emission intensity at any point following the initial displacement from the baseline is directly proportional to $[H_2O]$ was confirmed in auxiliary experiments which demonstrated (i) optical thinness, and (ii) the absence of significant interfering radiation from species (including OH) other than H_2O.

The passage of the shock front across the slit defining the field of view is not displayed in Fig. 2.5. The initial appearance of a significant concentration of water vapour, which occurs at some time after shock passage, corresponds approximately to the end of the induction period. At a somewhat later time, when the intensity has reached about half of its ultimate level in this experiment, there is a fairly sharp discontinuity in the slope of the emission profile as the system passes from the region in which intermediate concentrations are increasing into the final, slow approach to equilibrium. It was established in independent experiments that the constant deflection at very long reaction times is attributable to the equilibrium H_2O concentration. The fraction of this level which had been reached at the transition from fast to slower H_2O formation was also studied quantitatively in the rate study,[60] and in an independent investigation employing the $4 \cdot 7$ to $10\ \mu m$ (ν_2 and pure rotational transitions of H_2O) spectral band.[61]

The definition of space, and hence of time in shock wave profiles by means of slits is generally more troublesome in emitted light experiments than it is in absorption spectrometry, owing to the adverse effects of reflections and other forms of scattering, combined with the ever-present need for a large enough signal to be satisfactorily free of noise. The resolution achieved in the work which produced Fig. 2.5 was insufficient for the initial, rapid rise in [H_2O], but was quite adequate for the subsequent slow phase of the reaction. Because the fractional change in signal during this part of the reaction is not large, and because the reference levels are only determined within the postshock profile itself, this method compares with the interferometric situation, and is not as sensitive as is OH absorption spectrometry. Nor is it useful for the decelerating phase of the reaction over such a large range of initial $H_2 : O_2$ ratios and densities, owing to the more complete formation of H_2O in the prompt reaction which occurs in mixtures far from the 2:1 stoichiometric proportions, and to the need to reach the equilibrium level with certainty within the time available for measurement.

O–CO Chemiluminescence

It was discovered in flame experiments,[23] and adapted advantageously to shock tube studies of the hydrogen–oxygen reaction, that inclusion of carbon monoxide as an indicator in this reaction system provides a means of detecting oxygen atoms quantitatively through the intensity of chemiluminescent emission which is proportional to the product [CO][O], while introducing minimal and readily accountable perturbation of the H_2–O_2 kinetics. The involvement of CO in the reaction mechanism is particularly inconsequential during the accelerating phase, especially when H_2 is in plentiful supply. Thus the growth of the CO flame spectrum intensity, monitored with a photomultiplier with suitable spectral and spatial definition, has provided a sensitive means of observing the relative oxygen atom population and systematically studying the kinetics of exponential chain branching over a wide range of mixture composition, density and temperature.[62–67] Later in the reaction, this same diagnostic reveals the slower oxidation of CO, the reduction of added CO_2, or effects of overall radical removal through association.[45]

OH Luminescence

In the absence of CO or other additives, the most conspicuous electronic emission from the H_2–O_2 reaction is the 306·4 nm and related $A^2\Sigma^+ \rightarrow X^2\Pi$ bands of OH. At temperatures above about 2000–2500 K the intensity of this emission behaves thermally, but below 2000 K it is abnormally strong, less steeply dependent upon temperature, and subject to variability due to collisional quenching, and it thus represents chemiluminescence.[43] The chemiluminescent intensity of the OH bands is awkwardly weak for time-resolved measurement in many low-density shock wave experiments. Also, a weak continuous emission spectrum accompanies the pure H_2–O_2 reaction and covers the whole 220–600 nm region, with maximum intensity near 450 nm.[68] This is associated with the presence of OH, although the most satisfactory accounting of the dependence of its intensity upon composition[69] indicates the chemiluminescent mechanism:

$$H + OH \longrightarrow H_2O^* \longrightarrow H_2O + h\nu$$

Nevertheless several investigations of H_2–O_2—diluent mixtures reacting in incident or reflected shock waves have employed ultraviolet emission to observe the length of the induction period,[70–72] the time constant of the exponential growth of the emission intensity with fixed quenching and detection conditions,[43,64,73] and the postignition zone under very low density conditions.[43] Many of the results are in accord with proportionality of the OH chemiluminescence intensity to the product [O][H].

End-On Viewing of Reflected Shock Waves

The time constant for exponential acceleration of the H_2–O_2 (and the H_2–CO–O_2) reaction toward the end of the induction period but prior to macroscopic reaction has been examined directly by quantitative emission spectrometry of radiation proportional to the exponentially growing concentration of one or more reaction products.[62–67,74] As we shall see in section 2.3.2, this time constant offers the most direct information yet obtainable about the elementary reaction rate coefficients which govern the branching chain kinetics at high-temperatures.

In order to enhance the sensitivity and time resolution of this approach, the following adaptation of shock tube technique has been developed.[62] The reaction is studied in reflected shock waves, and the luminosity is observed throughout the growing isothermal reaction volume between the reflected shock front and the shock tube end plate, through a large window in the latter. During the measurement, the exponential growth of luminosity is common to each of the successively shocked layers, save for a transient zone of very low luminosity adjacent to the shock front, and so the spatially integrated intensity also exhibits the same exponential growth in time. The problem of excluding reflected or scattered radiation from strongly luminous gas which is already reacted beyond the exponential phase is avoided, and large aperture optics and detectors are usable without degrading the definition of time. Thermochemical effects of the reaction which ultimately accelerate the reflected shock front and may also stimulate transverse compression waves are likewise avoided until after the measurement is finished.

This technique was introduced in the study of exponential growth of O–CO radiation from H_2–O_2 mixtures with CO added.[62] A representative oscillogram from one of these experiments and the resulting semilogarithmic graph of photoelectric intensity against time are shown in Fig. 2.6. The exponential growth constant, ϕ, is interpreted as that of the growing oxygen atom concentration. The kinetics results from these experiments have subsequently been verified and extended by the end-on technique in the absence of CO using the H_2O vibrational emission in the infrared[74] and a combination of the OH band and H_2O continuum emissions in the ultraviolet.[64] Under common conditions of temperature and reactant concentrations, this ultraviolet emission is found to grow at twice the exponential rate of either the O–CO or the infrared spectrum, and the inference is that it is proportional to the product of two exponentially growing constituents.

2.3. INTERPRETATION OF H_2–O_2 KINETICS

2.3.1. Mechanism of the Reaction

The kinetic interpretation of the rapid burning of hydrogen and oxygen in various proportions at elevated temperatures is in terms of the following most important steps:

$$H_2 + O_2 \longrightarrow 2I \text{ or } I + R \qquad (0)$$

$$H + O_2 \longrightarrow OH + O \qquad (a)$$

$$O + H_2 \longrightarrow OH + H \qquad (b)$$

$$OH + H_2 \longrightarrow H_2O + H \qquad (c)$$

$$OH + OH \longrightarrow H_2O + O \qquad (d)$$

$$H + H + M \longrightarrow H_2 + M \qquad (e)$$

$$H + O_2 + M \longrightarrow HO_2 + M \qquad (f)$$

$$H + OH + M \longrightarrow H_2O + M \qquad (g)$$

$$HO_2 + I \longrightarrow 2R \qquad (h)$$

The rate coefficients of these elementary reactions will be denoted as k with the identifying letter as subscript. The reactions are written here in the direction of their most conspicuous contribution to the mechanism; their occurrence in the reverse sense, to be denoted by the same identifying letters with minus signs, is understood to be included as departures from equilibrium may dictate. Reactions (a) through (d) are fully specified. Reactions (0) and (h) are partially specified reactions involving the reactants named, the generic primary intermediates OH, H and O, which are collectively designated as I, and the generic reactive species, comprising H_2, O_2, H_2O, OH, H, O and HO_2, collectively designated as R. In the termolecular reactions (e)–(g), M designates the generic collision partner whose concentration multiplies the third order rate coefficient applicable at total pressures up to at least ten atmospheres for these diatomic and triatomic association reactions. For these reactions the product $k_j[M]$ is to be understood as an abbreviation representing the sum $\sum_i k_j^{M_i}[M_i]$ over all species as collision partners, and [M], used alone will represent the total gas concentration.

The Induction Period

This reaction mechanism (0)–(h) operates in the $k_a > k_f[M]$ regime in the following way. Reaction (0), whose products are often assumed to be two OH radicals, is an initial source of chain centre intermediates which is appreciably faster than the dissociation of H_2 (reaction (−e)) or O_2 in the 1000–2500 K range by virtue of a much lower activation energy. These initial intermediates accumulate and stimulate reactions (a)–(c) in a first order branching chain sequence which converts H_2 and O_2 to H_2O and the intermediates at an accelerating rate which quickly dwarfs the rate of reaction (0) and stabilizes as an exponential increase with a particular time constant, ϕ. During the exponential growth phase, the relative proportions of all the product species hold fixed values determined by the initial composition and thermodynamic state and the several rate coefficients. Competition between reactions (a) and (f), which is enhanced at higher densities and lower temperatures, may partially inhibit chain branching and influence the rate of exponential acceleration without altering its essential nature.

The initiation transient and the period of exponential growth following it make up the induction period, which continues until macroscopic departures from the initial conditions disrupt the exponential growth and alter the course of reaction. This alteration takes place by (i) depletion of the original reactants, leading ultimately to deceleration, (ii) changes in the temperature and density of the gas by the thermochemical effect of reaction, and (iii) the increasing influence of the balance of the reactions in the mechanism, which are second order in the chain reaction products. The speeds of reactions (a)–(c) at high temperatures are great enough, in relation to the bimolecular collision rate, that the second order reactions do not emerge before depletion becomes serious.

As we saw in section 2.2, suitably sensitive experiments penetrate the induction period and reveal the time constant ϕ. In other experiments the time interval between shock wave heating and some detected manifestation of the accelerating reaction measures an induction time, t_i. To a good approximation, t_i is inversely proportional to ϕ. However, the detailed interpretation of this induction time depends upon the initiation rate and also upon the sensitivity of the experiment and whether or not the strict exponential growth has continued to the time t_i.

The Main Reaction

Once the branching chain reaction has become established early in the induction period, the populations of the reactive intermediates dominate the kinetics for the remainder of the reaction. Following the induction period, the first order chain steps proceed to consume a substantial fraction of the original H_2 and O_2, and concurrently the concentrations of the intermediates and the rates of their reactions reach maxima and later subside. Reactions which are of higher than first order in the chain centres and products become as influential as those involving the original reactants and only a single reactive intermediate. As this happens, the reaction evolves from an accelerating phase, which we may term ignition, to a decelerating, completion phase. The roles of the individual steps as initiating, propagating, or terminating components in a chain mechanism become inconspicuous and are supplanted by the roles which these same steps play in the kinetics of attainment of chemical equilibrium.

Where H_2O is the dominant equilibrium product according to

$$2H_2 + O_2 = 2H_2O \qquad\qquad (I)$$

completion of the reaction entails a net decrease in the number of molecules. In the gas phase, the steps which yield this decrease are three-body reactions in which previously separated fragments are united to form a stabilized product molecule in the influence of the collision partner. By common usage the association reactions, whose reverse sense is readily identified as dissociation, have become known as recombination steps, without regard for the origin of the combinable fragments. Reactions (e), (f) and (g) are the main steps which accomplish this essential function in the high-temperature hydrogen–oxygen mechanism. The relationship between these association reactions and others which we might consider as minor contributors to the decrease in number of molecules is shown schematically in Fig. 2.7. The evidence in support of the importance of reactions (e), (f) and (g) is considered later (section 2.3.3), in relation to experimental conditions and the data on their rate coefficients with specific species as M.

Operating simultaneously with these association reactions are the mutually bimolecular reactions among the constituents of the reacting mixture. These include the original first-order chain branching steps, (a), (b) and (c), their reverses, and the reversible reaction (d). Also in this group, in principle, are reactions (0) and (h). After the induction period reaction (h) rapidly removes the unstable HO_2 radicals formed in reaction (f) and limits their concentration to a steady state level which we suppose is stoichiometrically negligible under most high-temperature conditions.

These reactions clearly do not contribute directly to the essential function of reactions (e), (f) and (g), which is creation of additional chemical bonds and release of their energy of formation. From the standpoint of contributing to progress of reaction (I), the function of the bimolecular reactions is the provision of an intermediate reservoir of H atoms and other combinable fragments from which reactions (e), (f)

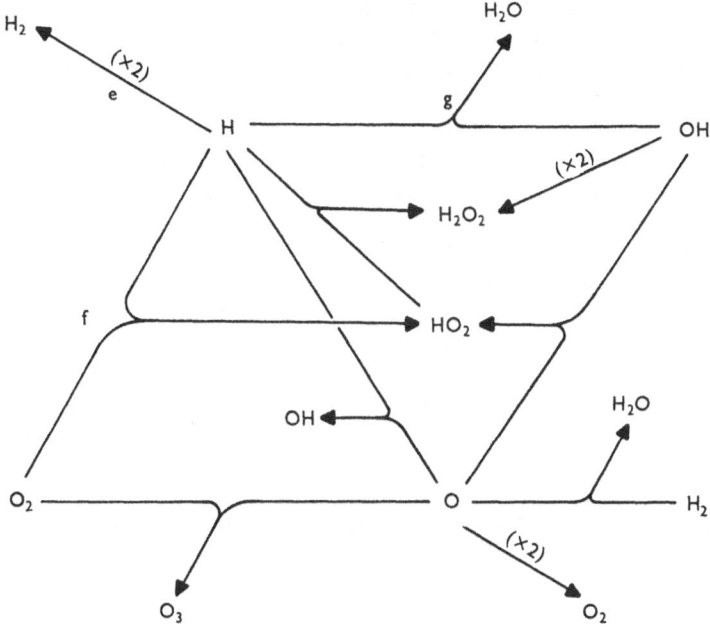

FIG. 2.7. Diagrammatic representation of possible paths for termolecular recombination in reacting hydrogen–oxygen mixtures.

and (g) may draw. Initially this reservoir is fed irreversibly from the
original reactants by the branching chain mechanism, but when
ignition is accomplished the reactants are merged into the reservoir of
combinable intermediates and the role of the bimolecular reactions is to
alter the final course of the ignition process and ultimately to regulate
the kinetics of the association reactions by continuously readjusting the
proportions of all the reactive species. In this readjusting, the equimolecular reactions operate according to their own equilibrium
tendencies, which are independent of the association–dissociation
component of the total reaction.

If there were no rate-determining step in all of this, the fast reaction
following the induction period would be quite complex. The concentrations of as many as four of the six species H_2, O_2, H_2O, OH, H and
O may vary independently and influence the kinetics, subject funda-
mentally only to the independent conservation of each chemical
element.

Termolecular and Bimolecular Rates

The phenomenon of kinetics of small gas molecules by which the
high-temperature hydrogen–oxygen reaction yields its most important
simplification is the third-order kinetics of association reactions at
familiar gas pressures. The proportionality of the rates of reactions

5

(e)–(g) to the concentration of collision partners, combined with the magnitude and temperature dependence of the rate coefficients of these reactions and of reactions (a)–(d), results in the association steps being inherently very much slower than the bimolecular steps in the mechanism, provided only that the density is low enough.

This rather complete separation of the rates of bimolecular reactions from termolecular association reactions, with the latter much slower, is achieved at total gas pressures up to about 0·5 atm near 1000 K and up to 5 to 10 atm near 2000 K. This, we may observe, is the extreme interior of the regime whose extension to lower temperatures manifests itself as the low-pressure explosion peninsula (see section 2.1.2). It is in this regime that the most extensive and satisfactorily interpreted shock tube studies on the H_2–O_2 reaction kinetics have been performed. In fact it is only in this regime that serious attention has been given to the kinetics of the reaction beyond the induction period. We proceed here to enumerate and discuss the several simplifying concepts for treating this separation of rates which have either arisen in shock tube studies or have been generalized from their prior application in the burned gas region of premixed flames.[75]

The Mole Number

When we consider the progress of reaction (I) under high-temperature, low-pressure conditions, reactions (e)–(g) collectively constitute a rate-determining step, even though this step is not isolated in such a way that its progress is stoichiometrically reflected in the concentration of any particular reactant or product. The total exothermic effect of the reaction is closely proportional to the progress of the association reactions, but the analysis of kinetics has led to an exact specification of the net effect of association steps in terms of the total number of moles per unit mass of reacting mixture, which is conceptually the reciprocal of the mean molecular weight. We designate this mole number as N and define it in reduced units as the number of moles per mole of original reactants. Thus it is the sum of the partial mole numbers, N_i, of all species in the reacting mixture. N is unaffected by reactions which produce as many molecules as they consume, and its rate of change thus gives the net rate of dissociation–association reactions, according to the volumetric rate equation:

$$-\frac{\rho}{M_0}\frac{dN}{dt} = -\frac{\rho}{M_0}\frac{d\sum_i N_i}{dt}$$

$$= R_{assoc} = \sum_j \{k_j[M][A_j][B_j] - k_{-j}[M][C_j]\} \quad (2.6)$$

in which ρ is the density, M_0 is the mean molecular weight of the original reactants, and the index j ranges over all association reaction steps which unite species A_j and B_j to form C_j.

The value of N is an integral measure of the net progress of association and dissociation reactions within the overall reaction. Initially, and throughout the induction period when all reaction is macroscopically negligible, N is unity. Subsequently, in the hydrogen–oxygen system below about 3000 K, N is driven to decrease monotonically by the spontaneity of the forward reactions (e)–(g) as ignition progresses. The utility of N as one element in the description of reaction progress is common to all systems in which a change in N occurs, whether by chain reaction or otherwise, whether with net dissociation or association, and whether N changes monotonically or through an extremum.

Partial Equilibrium States

The value of the mole number, N, is only one ingredient in the relationship between reaction progress and the composition of the reacting mixture. The other important matter to consider in the regime where association is rate-limiting is the consequences of the rapid portion of the mechanism. In fact, equation (2.6) is quite independent of the presence or absence of equimolecular reactions in the mechanism, much less of the establishment of equilibrium relationships in such reactions when they are present. But insofar as the association rate is all but negligible, the fast bimolecular reactions in the H_2–O_2 reaction mechanism do tend to reach and maintain a close approximation of their own equilibrium positions This leads to the following inter-relationships between N and the complete copomsition of the system.

The number of independent equilibrium relationships among the partial mole numbers, N_i, is equal to the number of independent steps in the fast equimolecular mechanism,[76] which we call R. Where changes in N are possible, R may not exceed $(S - C - 1)$, where S is the total number of species being considered (the range of the index i), and C is the number of stoichiometric components of these species (the independent chemical elements). When $R = (S - C - 1)$, the composition to which the system tends through the fast reactions alone is a unique function of N, determined by the temperature and the initial composition. This function is conspicuously independent of density.

This state of affairs obtains in the hydrogen–oxygen mechanism we are considering, reactions (0) through (h). Two of reactions (b)–(d) are independent, and these together with reaction (a) give three independent equilibrium relationships, which are a complete set among the six species H_2, O_2, H_2O, OH, H and O at fixed N and initial composition $N_{H_2}^0$, $N_{O_2}^0$ and $N_{H_2O}^0$. Inclusion of HO_2 and the independent equilibrium condition of reaction (h) preserves the completeness of the equilibrium relationship between N and composition, and influences the larger N_i imperceptibly. These independent equilibrium relationships coexisting with a macroscopic departure from equilibrium in some other chemical process are called *partial equilibrium* relationships. The composition which is determined by them in conjunction with a supplementary specification of N (or, in general, of the conditions of any other non-interacting degrees of freedom) is then a *partial equilibrium* composition.

Association and Partial Equilibrium

The sequence of definite *partial equilibrium* compositions which would describe a hydrogen–oxygen reaction mixture at successive values of N in the limiting circumstance of infinitesimally slow decrease in N provide us a very useful approximate description of the trajectory of the composition under the finite but rate-limiting influence of the actual association reactions at low to moderate reaction system pressures, once the ignition transient has abated and the several *partial equilibrium* relationships have become approached.

These relationships provide the means of relating experimental measurements on the reacting system to the quantities in the association rate equation, (2.6). Under *partial equilibrium*, the association reaction progress reckoned by N is a definite though indirect function of any observed species concentration or other system property of adequate sensitivity, such as shocked gas density.

To analyze the rate terms on the right-hand side of equation (2.6), the use of the *partial equilibrium* approximation is extended to permit evaluation of other unmeasured species concentrations which may enter these terms. This use of the measured main reaction progress to evaluate the concentration of a kinetically significant intermediate is closely analogous to the conventional quasi-steady state approximation in kinetics, but free of the usual restriction on its accuracy or utility that the concentration so evaluated be stoichiometrically minor.

Upon the establishment and maintenance of the full set of equimolecular *partial equilibrium* relationships during the completion phase of the hydrogen–oxygen reaction, only one degree of freedom remains in the ongoing reaction. All the parallel association reactions are rendered equivalent in their net effect on the composition, and the reaction trajectory is such that for each association reaction which yields one molecule fewer than it consumes, the ratio of the equilibrium expression to its corresponding equilibrium constant is the same (multiple association reactions obviously involve integral powers of this ratio).[75] Because the equimolecular equilibria are independent of pressure, so is the reaction trajectory, and within the high-temperature, low-pressure regime where the applicability of this approximate description is preserved, the time scale of the association phase of the reaction simply scales with the inverse second power of the gas density.

The occurrence of equimolecular *partial equilibrium* is of course only an approximation, and the equilibria are subject to sustained perturbation by the finite rate of association. The main influence in favour of the equilibria is the relative slowness of termolecular association at low total density. The second general influence is elevated temperature, since all of the bimolecular rate coefficients increase with temperature, whereas the association rate coefficients decrease. Dilution of the reactants in an inert gas is not directly a factor, at a given total pressure, except secondarily in that monatomic gases are noticeably less efficient as collision partners for association than are more complex molecules, notably H_2O.

Within these general influences, the separate components of the bimolecular mechanism exhibit differences in the sizes of their forward and reverse rates at or near *partial equilibrium* which depend upon the reaction steps themselves and the total composition of the system. It is these differences that determine which equilibrium relationships are most surely applicable. The larger the forward and reverse rates of each step are, the more durable is its equilibrium position. Thus the equilibria in reactions (c) and (d) are promoted by the accumulation of H_2O as the reaction proceeds, and reactions (b) and (c) are likewise influenced by an excess of H_2, as is reaction (a) in an excess of O_2. Correspondingly, very small species concentrations on either or both sides of a reversible step work against the preservation of its equilibrium. As reaction (a) involves no major species in a fuel-rich situation, $[O_2]$ may indeed be perturbed rather far from its *partial equilibrium* trajectory. However, unless there is a measurement involving O_2 specifically, this is inconsequential, as reactions (e) and (g) dominate equation (2.6) and N_{O_2} is negligible in the total mole number. Likewise, neglect of HO_2 in the use of the *partial equilibrium* approximation is permissible so long as N_{HO_2} is negligible and no reaction of HO_2 is rate-determining. In an excess of O_2, the participation of the minor intermediate H in the inherently rapid association step (f) poses the most conspicuous limitation on the range of applicability of the *partial equilibrium* approximation.

As association nears completion, all the intermediate species concentrations diminish, but so does the rate of association, which behaves rather like a second order reaction in the excess intermediates. In a fuel-rich system reactions (c) and (−c) remain first order in excess intermediates, and thus this equilibrium condition becomes progressively better established, while reactions (a), (b) and (d) retain the same tendency to be perturbed by reactions (e) and (g). In fuel-lean mixtures, only reaction (d) improves its equilibration capability, as the competitive relationship between the primary association path, reaction (f), and reaction (a) is influenced only by [M].

Ignition without Association—Overshoots

The separation of bimolecular and termolecular rates in the hydrogen–oxygen reaction at high temperatures and modest pressures also gives rise to significant simplification of the kinetics of the ignition phase of the reaction. First, in this regime, interruption of the branching chain by HO_2 formation is insignificant. Because the whole ignition process results from bimolecular reactions, its course for each composition and temperature is the same, with the time scale simply proportional to the inverse first power of the density.

The course of the macroscopic branched chain reaction between the induction period and the approach to partial equilibrium still has three degrees of freedom, even when changes in N and reactions involving HO_2 are negligible. The composition of the system can be partially described, however, by combining the accounting of N with the conservation of the chemical elements.[77] Thus we have:

$$N_{H_2O}^0 + N_{H_2}^0 + N_{O_2}^0$$

$$= (N_{H_2} + N_{O_2} + N_{H_2O} + N_{OH} + N_H + N_O)_{N=1} \quad (2.7)$$

$$2N_{H_2O}^0 + 2N_{H_2}^0 = 2N_{H_2} + 2N_{H_2O} + N_{OH} + N_H \quad (2.8)$$

$$N_{H_2O}^0 + 2N_{O_2}^0 = 2N_{O_2} + N_{H_2O} + N_{OH} + N_O \quad (2.9)$$

Algebraic elimination of N_{H_2} and N_{O_2} from these equations also eliminates N_{OH} and the diatomic N^0's and leaves

$$N_{H_2O}^0 = (N_{H_2O} - N_H - N_O)_{N=1} \quad (2.10)$$

which represents the fact that until the net progress of association becomes appreciable, the purely bimolecular branching chain reaction produces the same amount of atoms, H and O, as triatomic products, H_2O, from its diatomic reactants. The formation of the other intermediate, OH, is independent of this constraint.

Equation (2.10), together with the *partial equilibrium* relationships, determines a *fully ignited* composition which approximately describes the junction between the ignition phase of the reaction and the purely decelerating association phase which follows it. The sizable concentrations of the intermediates OH, H and O in this composition account for the magnitudes of the essential overshoots of these species which arise early in the main reaction simply because N must decrease in order for the reaction to be completed. To be sure, individual excursions of the concentrations of these intermediates above these *fully ignited partial equilibrium* values are possible before the individual bimolecular equilibria are approached, but such excursions are short-lived in comparison to the overshoots that depend upon association for their removal. Under many circumstances, particularly in near-stoichiometric mixtures, no such excursions occur and the *fully ignited* composition represents upper bounds to the observed concentrations of the intermediates.

Résumé

The shock tube investigations of the hydrogen–oxygen reaction described in section 2.2 have revealed the significant qualitative features of the high-temperature mechanism considered above, viz. exponential growth of intermediate and product concentrations, the induction period, overshoots in intermediate concentrations, and the final, slow approach to full equilibrium. These studies have also contributed quantitative information about the relative importance of reactions (0)–(h) under a wide range of conditions, and values of many of the important rate coefficients have been determined. In the remainder of section 2.3 we discuss the significant quantitative results from these experiments. The order to be followed is that determined by the temporal evolution of the reaction.

2.3.2. Kinetics of Shock-Wave Ignition

Exponential Acceleration Rates

The exponential growth coefficients, ϕ, discussed in section 2.2 describe the concentrations of intermediate and product species during the induction period of the explosively rapid reaction between hydrogen and oxygen. Values of ϕ have been measured in shock tubes by several experimental methods. From the standpoint of the quantitative kinetic information which they afford, the most comprehensive and useful of such measurements have been those employing emission spectrometry to determine relative populations of a single product of the branching chain reactions. Emissions proportional to either [O] or [H_2O] have been monitored in such experiments. We now consider how experimental ϕ's determined from graphs like Fig. 2.6 are related to the rate coefficients $k_a - k_c$ and k_f.

As demonstrated by Kondratiev[18] and others,[39,78-79] for the initiation and first order branching mechanism comprising reactions (0), (a), (b), (c), (f) and (−f), with all second order processes negligible, the system of rate equations for the intermediates is linear and until such time as sensible changes in the initial conditions occur, the coefficients are constants. Following the initiation transient, the chain reaction intensity grows exponentially with the time constant ϕ_p, the single positive root of the derived secular equation. With the exclusion of reaction (−f) and of other first order chain-continuing reactions of HO_2 which we shall mention later, this equation is the cubic:

$$\phi^3 + \{K_a + K_b + K_c + K_f\}\phi^2$$
$$+ \{K_bK_c + K_f(K_b + K_c)\}\phi - K_bK_c(2K_a - K_f) = 0 \quad (2.11)$$

where

$$K_a = k_a[O_2] \qquad K_b = k_b[H_2]$$

$$K_c = k_c[H_2] \quad \text{and} \quad K_f = k_f[M][O_2]$$

The constants K_j appearing in equation (2.11) correspond to the thermal conditions achieved between the shock wave front and the macroscopically reacted gas.

To reduce data obtained over a range of gas densities and compositions, equation (2.11) is rearranged to the form

$$\left\{\frac{\phi}{2[O_2]}\right\}^3 + \tfrac{1}{2}\{k_a + (k_b + k_c)\eta + k_f[M]\}\left\{\frac{\phi}{2[O_2]}\right\}^2$$

$$+ \tfrac{1}{4}\{k_b k_c \eta^2 + k_f[M](k_b + k_c)\eta\}\left\{\frac{\phi}{2[O_2]}\right\}$$

$$- \tfrac{1}{8}k_b k_c(2k_a - k_f[M])\eta^2 = 0 \quad (2.12)$$

where $\eta = [H_2]/[O_2]$.

Under the low-density and/or high temperature conditions of many
of the experiments, where the influence of reaction (f) is negligible,
equation (2.12) reduces to

$$\left\{\frac{\phi}{2[O_2]}\right\}^3 + \tfrac{1}{2}\{k_a + (k_b + k_c)\eta\}\left\{\frac{\phi}{2[O_2]}\right\}^2$$

$$+ \tfrac{1}{4}k_b k_c \eta^2\left\{\frac{\phi}{2[O_2]}\right\} - \tfrac{1}{4}k_a k_b k_c \eta^2 = 0 \quad (2.13)$$

The following observations[65] concerning equations (2.11)–(2.13) are
pertinent at this point. First, in equation (2.13) the quantity $\phi/2[O_2]$
is a function only of temperature and η, and not of the total concentration [M], as it is in the case of equation (2.12). A second important
observation is that the coefficients k_b and k_c only enter equations
(2.11)–(2.13) as their sum $k_s = (k_b + k_c)$ and product $k_p = k_b k_c$. This
fact limits the extraction of individual values of k_b and k_c from measured
ϕ's, so that some independent information is needed to interpret
exponential growth rates in terms of k_b and k_c independently. Thirdly,
consideration of the limiting values of $\phi/2[O_2]$ at extreme values of η
indicates the sensitivity of the measurements to the different rate
coefficients. For $\eta \gg 1$, equation (2.13) yields $\phi/2[O_2] \approx k_a$, and for
$\eta \ll 1$, $\phi/2[O_2] \approx \eta\,(k_p/2)^{1/2}$. Sensitivity to k_a and to k_p are thus
attained by measuring ϕ in very hydrogen-rich or hydrogen-lean
mixtures, respectively. Moreover, the rationale of representing measured
$\phi/2[O_2]$ data in Arrhenius form is self-evident.

Experimental ϕ's measured in reflected shock waves with the spatially-
integrated CO–O flame spectrum emission method over a range of
temperatures and for different values of η, and normalized by $2[O_2]$,
are displayed in Fig. 2.8. In accordance with equation (2.13), the data
separate only with η at higher temperatures. Variation of $\phi/2[O_2]$ with
total gas concentration, predicted by equations (2.11) and (2.12)
is seen in $\eta = 0\cdot33$ mixture data at the lowest temperatures. Also shown
are recent infrared measurements[74] which have extended the previously
limited $\eta = 10$ results up to nearly 2200 K, while yielding excellent
agreement with the CO–O data.

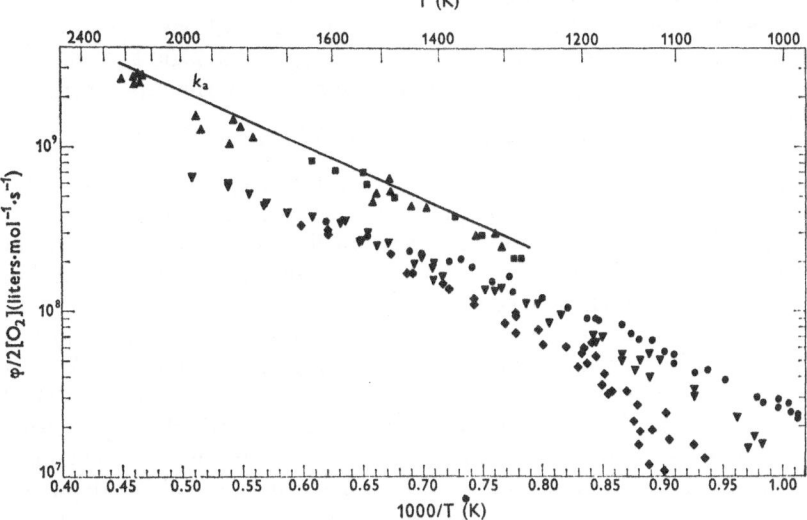

FIG. 2.8. Experimental values of the logarithm of $\phi/2[O_2]$ against reciprocal reflected shock wave temperature for exponential growth data from various sources. $\eta = 10$: ■; CO–O flame spectrum emission measurements from Reference 62, ▲; infrared emission measurements from Reference 74. $\eta = 0.33$: CO–O emission measurements from Reference 63 for various ranges of total reflected shock gas concentration and identical mole fractions of H_2 and O_2. ●; $0.15 - 1.4 \times 10^{-2}$ moles liter^{-1}. ▼; $1.6 - 3.7 \times 10^{-2}$ moles liter^{-1}. ◆; $5.5 - 9.2 \times 10^{-2}$ moles liter^{-1}. Solid line: $k_a = 9.5 \times 10^{13} \exp(-15\,000/RT)$ cm^3 mole^{-1} sec^{-1} (Reference 74).

Various courses have been followed in applying equations (2.11)–(2.13) to the extraction of rate coefficients and their Arrhenius parameters from data of the type shown in Fig. 2.8. These include such procedures as (i) assuming that $k_f[M]$ is small enough that equation (2.13) can be used,[62,63] (ii) assuming that k_f and/or k_c are well enough known independently and that ϕ is sufficiently insensitive to them that their values can be inserted into equation (2.12) and the data for two or more values of η then used to determine k_a and k_b,[62,63,80] (iii) determination of k_a, $k_b k_c$ and $(k_b + k_c)$ from data at three or more values of η,[65,74] and (iv) massive variation of the CO mole fraction and inclusion in the analysis of the effect of the reaction

$$OH + CO \xrightarrow{k_l} CO_2 + H \qquad (l)$$

in parallel with reaction (c), thereby permitting simultaneous extraction of k_a, k_b, k_c and k_l from data at four compositions.[66]

The rate coefficient most conclusively determined by any of these procedures is k_a, and closely concordant values have been obtained in overlapping temperature ranges in several investigations of hydrogen-rich mixtures. One of these[74] covered the range 1300–2200 K at $\eta = 10$ and lower values of η and yielded the empirical Arrhenius expression

$$k_a = 9 \cdot 5 \times 10^{13} \exp\left(-15{,}000 \, cal/RT\right) cm^3 \, mole^{-1} \, sec^{-1}$$

whose relation to the data is displayed in Fig. 2.8. We take this result as the most satisfactory determination of the value of k_a over its temperature range, and observe that earlier results based on lesser ranges of data differ little from it in magnitude and provide rather less significant Arrhenius coefficients.

There has been greater difficulty in obtaining concordant data on ϕ in the $\eta \leqslant 0 \cdot 3$ regime, where k_b and k_c are most influential. Higher values of ϕ, and thereby of k_b and k_c, are supported by end-on CO–O emission[65] and OH absorption measurements,[43,80] which suggests that the less sensitive infrared[74] and side-viewed CO–O[66] measurements may be affected by declining chain branching rates caused by depletion of reactants. There are also serious unresolved questions about the influence of boundary layer growth behind incident shock waves upon the time scale and temperature in the experiments to determine ϕ.[67]

Notwithstanding these and other possible difficulties, the results from the lean mixture regime do suffice to indicate that k_b and k_c have nearly equal values throughout the $T \geqslant 1500$ K regime. At 1600 K, the value of both these coefficients derived from end-on CO–O emission data[65] is $k_b \approx k_c \approx 4 \times 10^{12}$ cm^3 mole^{-1} sec^{-1}. These values can be compared with some of those derived from independent sources; for example the 1968 University of Leeds compilation[81] gives $k_b = 8 \cdot 9 \times 10^{11}$ cm^3 mole^{-1} sec^{-1} at 1600 K, while an Arrhenius extrapolation of particular data covering the range $1000 \geq T \geq 350$ K[82] gives $k_b = 1 \cdot 3 \times 10^{12}$ cm^3 mole^{-1} sec^{-1} at 1600 K. Similarly, k_c at 1600 K is $4 \cdot 3 \times 10^{12}$ cm^3 mole^{-1} sec^{-1} from the Leeds work,[81] and an extrapolation from $500 \geq T \geq 300$ K[83] gives $k_c = 1 \cdot 1 \times 10^{12}$ cm^3 mole^{-1} sec^{-1}.

Finally, the observable dependence of $\phi/2[O_2]$ on total gas concentration at low temperatures in $\eta = 1 \cdot 0$ and $\eta = 0 \cdot 33$ mixtures has been analyzed with equation (2.12) to deduce a value $k_f^{Ar} = 3 \cdot 3 \times 10^{15}$ cm^6 mole^{-2} sec^{-1} in the neighbourhood of 1100 K.[63] The $\eta = 0 \cdot 33$ data upon which this result is based are included in Fig. 2.8; its relationship to other termolecular association results is considered in section 2.3.3.

Induction Times

As was indicated in sections 2.2.1 and 2.3.1, the occurrence of an induction period is a prominent feature of shock-wave experiments sensitive to the macroscopic hydrogen–oxygen reaction. The acceleration at the beginning of the reaction is so marked that in any linear measure of the reaction profile there is a perceptible period of negligible reaction which is followed by an abrupt S-shaped portion. We now consider the quantitative interpretation of this behaviour in relation to the reaction mechanism.

Quantitative description of the induction period by a definite time interval, or *induction time*, t_i, is aided by the precise time origin provided by a shock wave. The specification of the end of the induction period is conceptually more difficult, however, and no universally preferred criterion has been found. One approach, which is appealing when one wants to emphasize the induction time as a determinant of the time scale of the main reaction, is to reckon t_i by the occurrence of an inflection or maximum in such quantities as the rate of exothermic expansion or in the formation of some product, byproduct (such as chemiluminescent photons), or intermediate. This approach suffers conceptually from the fact that such phenomena are generally not quite simultaneous with each other, even though the differences are minor in comparison with the major effects of varying gas density, composition and temperature.

Clearly, though, such induction times include a portion of the macroscopic reaction. Thus another approach has been prevalent, in which one seeks to confine the measurement of t_i to the initial constant-state conditions of exponential acceleration, in order to avoid departures from equation (2.2), which present major complications for most gas compositions. This is achieved by arbitrarily defining t_i by the attainment of some small but reproducible manifestation of the accelerating reaction which is experimentally resolvable. Such t_i values are of course systematically shorter than those based upon an inflection or maximum in an experimental signal, but nevertheless are characteristic of the reaction conditions and lend themselves to systematic, albeit often approximate correlation.

The rationale for such correlation of t_i data with the reaction kinetics stems from equation (2.2), applied to concentrations C_i at time t_i under conditions where $\exp(\phi t_i) \gg 1$, so that one has:

$$\phi t_i = \ln C_i - \ln(\theta/\phi) \approx \text{constant} \qquad (2.14)$$

C_i is taken as fixed by experimental standardization, whereas the effective value of C at time zero, θ/ϕ, is subject to variation with experimental conditions as the reaction kinetics may determine. However, θ/ϕ enters only as the argument of a logarithm, and provided that $\theta/\phi \ll C_i$, ϕt_i is affected little even by quite large changes in θ/ϕ. Thus, for an induction period in which branched chain acceleration dominates, it is the time constant ϕ which mainly governs t_i, and the product ϕt_i may be regarded as invariant, to an acceptable approximation.

This $\phi t_i \approx constant$ principle was first employed extensively in the work of Nalbandyan[10,84] near the explosion limits as the basis for correlating t_i values over ranges of reactant density and temperature. For high-temperature shock wave conditions, consider first the regime where reaction (f), and hence a diluent which only acts as a collision partner, M, has negligible influence upon ϕ. Here ϕ is directly proportional to the density of reactants for fixed composition and temperature, and t_i should thus exhibit the inverse of this proportionality. For the same reasons that were applied in the previous section for correlating the direct data on ϕ, we select $[O_2]$ as the measure of reactant density, and represent the product $[O_2]t_i$ in the form

$$\log [O_2]t_i = A + B/T \qquad (2.15)$$

Of course, the coefficient B involves a combination of the activation energies of the reactions, (a)–(c), which determine ϕ, together with the temperature dependence of $\log (\theta/\phi)$, which brings in reaction (0) as well. Also, it is clear that the neglect of reaction (f) and its higher-order dependence upon gas density makes equation (2.15) totally unsuitable for extension to conditions which approach the second or third explosion limits.

In the high-temperature shock wave investigation by Schott and Kinsey, using ultraviolet absorption by OH as the diagnostic and the fixed $[OH]_i$ cutoff criterion, the utility of equation (2.15) over the ranges $1100 \leqslant T \leqslant 2600$ K, $p \leqslant 2$ atm was demonstrated.[26] A sample of these data for stoichiometric H_2–O_2 mixtures ($\eta = 2 \cdot 0$) diluted with argon, over a range of reactant density, is reproduced in Fig. 2.9. The greatest shortcoming of the Schott and Kinsey study has been that, owing to insufficiency in both experimental precision and range of $H_2:O_2$ ratios ($0 \cdot 5 \leq \eta \leq 5$), the dependence of ϕ upon η in near-stoichiometric compositions was not recognized correctly, and in its place an appealing rationalization in terms of reaction (a) being totally rate-limiting was offered. However, for stoichiometric and near-stoichiometric mixtures, the essential soundness of the t_i data and equation (2.15) has been confirmed by several other diagnostic means.[39,54,57] Moreover, in spite of the crudeness of the kinetics analysis which was used, the data did yield an estimate of k_a near 1650 K which is within 50% of that given by the extensive direct data on ϕ in hydrogen-rich mixtures.

The utility of such an operationally simple experiment as the measurement of t_i to investigate the kinetics of hydrogen–oxygen ignition was exploited considerably in the early 1960's, and more precise high temperature data covering wide composition ranges ($0 \cdot 0075 \leq \eta \leq 100$ in one investigation)[54] were reported. Some representative data are included in Fig. 2.9. In fact it was the clear recognition of the effect of $H_2:O_2$ ratio in these t_i data that led to productive consideration[39,78] of equations (2.11)–(2.13), and not until afterward[62] were direct measurements of ϕ developed. For simple functional representation of induction time data where ranges of temperature, density and composition are involved, the $\phi t_i \approx constant$ rationale was adapted either by considering separate curves of $\log [O_2]t_i$ vs $1/T$ for each $H_2:O_2$ ratio,[39] or,

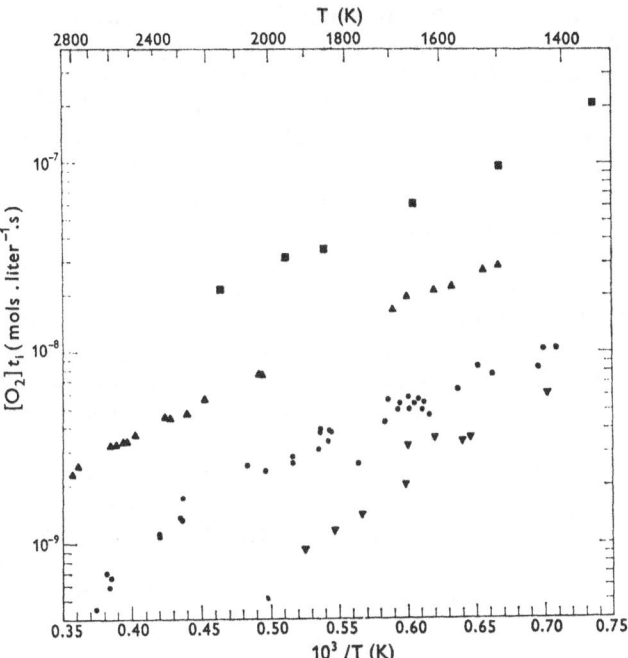

FIG. 2.9. Graph of the logarithm of $[O_2]\, t_i$ against reciprocal post-shock temperature for representative selection of experimental induction time data. ■: Interferometric measurements in $3 \cdot 0\%$ H_2–$97 \cdot 0\%$ O_2 ($\eta = 0 \cdot 031$) from Reference 54; ▲: OH absorption measurements in 1% H_2–10% O_2–89% Ar ($\eta = 0 \cdot 10$) from Reference 41; ●: OH absorption measurements in $1 \cdot 0\%$ H_2–$0 \cdot 5\%$ O_2–$98 \cdot 5\%$ Ar ($\eta = 2 \cdot 0$) from Reference 26; ▼: Interferometric measurements in $65 \cdot 4\%$ H_2–$1 \cdot 3\%$ O_2–$33 \cdot 3\%$ Ar ($\eta = 50$) from Reference 54.

upon observing that such curves are closely parallel to each other over the $1400 \leqslant T \leqslant 2300$ K range, by approximate representation of the density and composition effects by the semi-empirical equation:

$$\log\, [H_2]^x [O_2]^y t_i = A' + B'/T$$

Here x and y are taken as positive rational[54-55] or decimal[39] fractions whose sum is near unity and whose individual values are determined, with tolerances of perhaps $\pm 0 \cdot 1$, to fit particular data or results of numerical simulation of the kinetics.

Further quantitative interpretation of induction times for the extraction of information about individual rate coefficients, pursued particularly by Gardiner and co-workers, has involved explicit attention to the values of C_i, ϕ and θ/ϕ, and has led to such discoveries as:

(i) Reaction (0), whatever its mechanism may be, must be appreciably faster and have a much lower Arrhenius activation energy than either reaction ($-e$) or its oxygen counterpart, in order to account for the magnitude and temperature dependence of t_i in the 1400–2200 K range and to be compatible with independently credible magnitudes of k_a, k_b and k_c;[40]

(ii) although the vibrational relaxation of O_2 is independently measurable early in the induction period under certain circumstances,[85] there is as yet no discernible effect of the vibrational condition of either O_2 or H_2 upon ϕ,[39,86] and the interaction between vibrational relaxation and reaction (0) is very difficult to disentangle from other uncertain effects;[41,44] and

(iii) the accuracy of t_i data needed to extract meaningful values of the several individual rate coefficients k_a, k_b, k_c and k_0 from these data alone is such that the interfering effects of nonideal shock tube flow cannot be ignored and must be controlled and correctly assessed.[87]

In arriving at these deductions, the complete solution of the coupled set of rate equations for all the intermediates and reactions in the model mechanism has been used, either in analytic form for situations in which depletion of reactants was not important, or through numerical integration for conditions where secondary effects entered.

Low Temperature Induction Periods

Shock wave techniques have also been extended to studies of delayed ignition in hydrogen–oxygen mixtures at temperatures in the neighbourhood of 1000 K, where the effects of proximity to the explosion limit phenomena become strongly evident.[46,57,58,59,71,72,88,89] Reflected shock waves have mostly been employed, to avoid influence by the sorts of nonsteady shock wave behaviour discussed in section 2.1.2 and associated with the spontaneous detonation tendencies of even rather dilute mixtures. Studies of undiluted mixtures have also been pursued at these low temperatures, where their ignition delays at manageable total pressures are not too short to be resolved.

The main difference between the kinetics in this regime and in the one considered above is the noticeable competition between reactions (a) and (f), as was illustrated in the case of the $\eta = 0.33$ exponential growth data in Fig. 2.8. By equations (2.11) and (2.12), this results in ϕ being keenly dependent upon the product of $[O_2]$ and the difference $(2k_a - k_f[M])$. As a function of gas pressure at a given temperature, t_i departs radically from inverse proportionality to density and in fact exhibits a minimum, approximately at $k_f[M] = k_a$. Correspondingly the temperature dependence of t_i is different for each density of reactants and diluents, and its correspondence with equation (2.15) and the activation energies of elementary chain reaction steps vanishes. Data demonstrating this behaviour are exhibited in Fig. 2.10, as individual curves of log t_i vs $1/T$ for each postshock pressure.

Induction time data from moderately dilute hydrogen–oxygen and

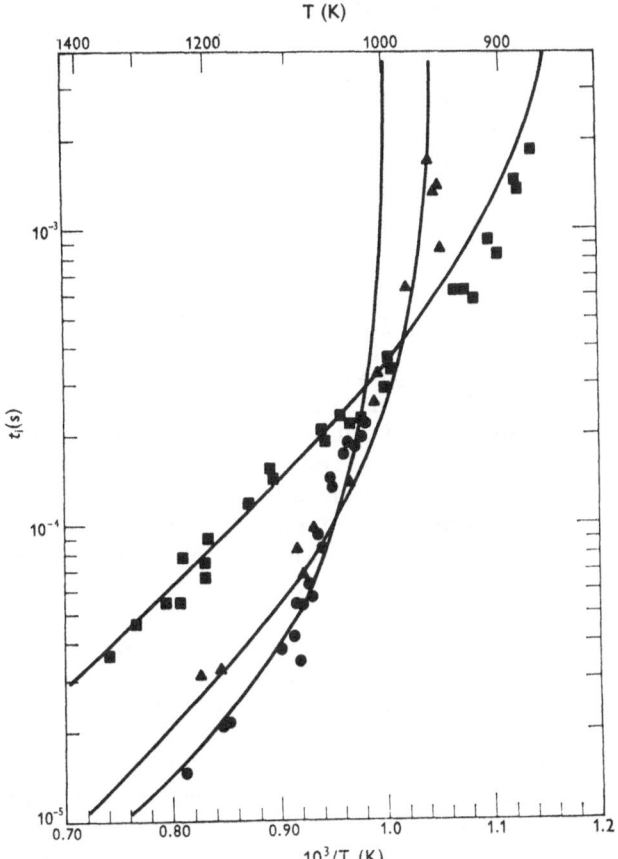

FIG. 2.10. Ignition delay as a function of postshock temperature and pressure for a stoichiometric hydrogen–air mixture (29·6% H_2–14·8% O_2–55·6% N_2) near 1000 K. ■: 0·41–0·49 atm; ▲: 1·4–1·47 atm; ●: 1·95–2·37 atm. The solid curves were calculated[89] with the low temperature mechanism discussed in the text and a selected set of rate coefficients (after Schmalz[89]).

hydrogen–air mixtures within the approximate regime $(2k_a - k_f[M])/k_a \geqslant 0·2$, $t_i \leqslant 0·5 - 1$ msec are more or less satisfactorily accountable by equation (2.11) and the existing body of knowledge concerning reactions (0), (a), (b), (c) and (f). Many experiments have also been done at postshock temperatures below or pressures above the $2k_a = k_f[M]$ condition. At low enough pressures ($p \leqslant 0·5$ atm), this condition is associated with the second explosion limit, which prevails at temperatures up to about 850 K. But at the rather higher pressures of reflected shock wave ignition experiments, the same condition is encountered at temperatures above 1000 K and pressures of several atmospheres. From the mechanism of the third explosion limit, the equality of $2k_a$ and $k_f[M]$ defines an *extended second limit* which is predicted to divide the high-temperature region of rapid ignition, with only minor impedance by HO_2 formation, from another region which extends downward in temperature to the third limit. In this high-pressure, low-temperature region, exponentially accelerating reaction is still expected to occur, with ϕ governed by reactions (*a*) and (*f*), together with a chain-continuing reaction of HO_2, variously proposed as

$$HO_2 + H_2 \longrightarrow H_2O_2 + H \qquad\qquad (m)$$

or

$$HO_2 + H_2 \longrightarrow H_2O + OH \qquad\qquad (n)$$

The rate equations for this regime, in which the counterpart of equation (2.11) is a quartic which again has one and only one positive, real root, have been thoroughly worked out and their solutions examined with numerous sets of hypothetical rate coefficients.[58,59,72,79,88,89] The correspondence with the experimental t_i data, however, has been unsatisfactory, in that the latter are much shorter and less sensitive to temperature than can be reconciled with credible chemical kinetics.

The explanation of the anomalously short induction periods appears now to lie in the domain of gasdynamics in shock tubes.[58,79] One knows independently that the gas behind reflected shock waves experiences additional heating, which is progressively more severe away from the end plate but not negligible adjacent to it.[90,91] This effect is caused primarily by interaction between the reflected shock wave and the consequences of the side-wall boundary layer of the incident shock flow, and it is more severe the lower the specific heat ratio, γ, of the gas. Recognition of the importance of this form of shock-wave nonideality in shortening what otherwise might be unobservably long ignition delays stems most directly from densitometric (schlieren photographic) experiments in undiluted hydrogen–oxygen mixtures, where it was discovered that the earliest ignition did not occur in the gas adjacent to the end plate, where the earliest shock-wave heating had occurred![58,59] The conclusion, for the present at least, is that shock tubes do not provide a valid means of measuring the slow ignition of the hydrogen–oxygen system in the high-pressure, low-temperature regime, owing to a very unfavourable combination of long chemical delays and extreme sensitivity to temperature. Hence no useful information about reactions (*m*) or (*n*) is yet available from shock-wave studies, and even the indications about k_f have been clouded in many experiments in small shock tubes and concentrated H_2–O_2 mixtures, which have failed to show even the effects seen in Fig. 2.10.[46,71]

Maximum Intermediate Concentrations—Overshoots

The portion of the high-temperature hydrogen–oxygen reaction that has received least detailed experimental attention is the period of most rapid reaction following the induction period and preceding the orderly decrease in intermediate species concentrations through termolecular association under the close governance of *partial equilibrium*. The experimental study of this region has been almost exclusively confined to observations of the amplitude and qualitative shape of the maximum in [OH]. The complexity of the kinetics in this postinduction or transition portion of the reaction is considerable, and the understanding of the phenomena governing the maximum in [OH], and in [H] and [O] as well, has been greatly advanced by computational and approximate analytical studies of the reaction mechanism.[42,92]

The bulk of the pertinent experimental data are profiles of the ultraviolet absorption by OH in incident shock waves, made either for deliberate study of $[OH]_{max}$,[24,42] or in conjunction with systematic studies of induction times[27] or of recombination kinetics.[28-32] To date no measurements of $[H]_{max}$ and only qualitative observations of $[O]_{max}$ made by uncalibrated measurements of O–CO chemiluminescence under conditions of high postshock pressure have been reported.[92] Experiments like that in Fig. 2.2, over the ranges $0.33 \leq \eta \leq 13$, $1300 \leq T \leq 2000$ K, $0.8 < p < 4.6$ atm have consistently yielded [OH] maxima that develop rapidly and decay comparatively slowly and lie between about 0.5 and 1.0 times the ideal, $N = 1.0$ *partial equilibrium* values.

These results are quite consistent with the conclusion that prior to the [OH] maximum the progress of termolecular recombination is negligible or accountably small, depending systematically on the total density, and that *partial equilibrium* conditions prevail at the time of the maximum and thereafter. However the absolute accuracy of the [OH] measurements over the necessary range of conditions is not great enough for conclusions about either the net change in N preceding $[OH]_{max}$, or the departure of [OH] or other species concentrations from *partial equilibrium* conditions thereafter, to be reached from the departure of $[OH]_{max}$ from the ideal, $N = 1$ *partial equilibrium* overshoot value in systematic experiments over an extend range of density. Rather, the most convincing method of examining these questions has been by computation, using the mechanism (0)–(h) and the progressively refined knowledge of the rate coefficients. The consistency of the measured [OH] maxima with the expected behaviour attests most directly to the long-term stability and accuracy of the spectrometric measurement of [OH] using the H_2O discharge lamp. In fact, theoretically justified confidence in the attainment of $N = 1$, *partial equilibrium* conditions at low enough gas densities is great enough that the quasi-steady OH absorption level following the ignition transient serves as an internal calibration point in many experiments.[24,32,36,40,42]

Thus the impact of the [OH] overshoot measurements upon our understanding of the H_2-O_2 reaction kinetics in shock waves has been, first, through the stimulus which the first such measurements[27] provided for the recognition of the substantial absence of association during rapid, high-temperature ignition, and second, through the confidence in the *partial equilibrium* description of the course of the subsequent OH disappearance, and in the diagnostic method for following it, that the $[OH]_{max}$ measurements afford.[28-32]

Brief Concentration Maxima at Low Density

Computer studies of the course of macroscopic ignition in shocked H_2-O_2 mixtures containing an excess of either reactant have revealed the occurrence under certain conditions of short-lived excursions of intermediate species concentrations significantly above the quasi-stable $N = 1$ *partial equilibrium* levels.[92] These so-called *spikes* occur between the time of the peak chain reaction rate, which occurs when the lesser reactant is about 50% depleted, and the time of establishment of *partial equilibrium* proportions among the major intermediates. In an excess of H_2, the *partial equilibrium* O-atom concentration is quite

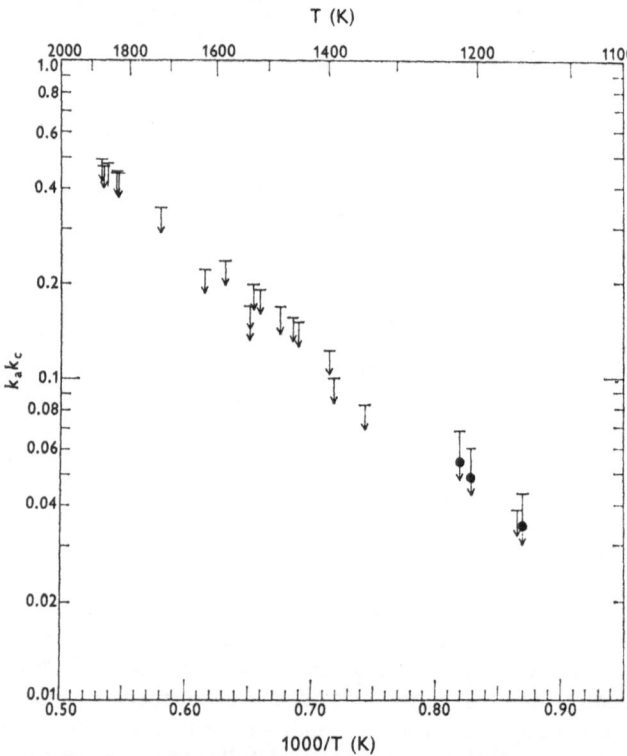

FIG. 2.11. Upper bounds to (k_a/k_c) imposed by failure of [OH] to exhibit spikes in hydrogen-rich mixtures above 1300 K, and indicated values of (k_a/k_c), with upper error bounds, from apparent [OH] maxima in three of four recombination experiments below 1300 K (after Hamilton and Schott[92]).

small, particularly at lower temperatures. It is smaller, in fact, than the transient [O] level which can be reached kinetically before much of the O_2 is consumed, particularly if k_a/k_b is large enough in relation to $(K_bK_c)^{-1}$ and the original $O_2:H_2$ ratio. The same effect is possible in [OH] if k_a/k_c is large enough in relation to $(K_c)^{-1}$, and similar behaviour of the minor radicals [H] and [OH] is calculable for very lean initial compositions.

The experimental requirements for exploring such spikes are clear enough. One needs to work at low total density in order for the super-imposed progress of reactions (e), (f) and (g) to be minor, and to have high enough analytical sensitivity for the resulting small [OH], [H], or [O] value, together with high enough spatial resolution to observe the transient spike preceding *partial equilibrium*.

The existing body of $[OH]_{max}$ data extends to sufficiently hydrogen-rich mixtures to establish the absence of discernible spikes above about 1300 K, and this is interpreted[92] to establish the upper bound to the ratio k_a/k_c shown graphically in Fig. 2.11. This information, combined with the value of k_a displayed in Fig. 2.8, leads to the conclusion that at 1600 K, $k_c \geq 3 \cdot 4 \times 10^{12}$ cm^3 mole^{-1} sec^{-1}, which is more definite than, yet compatible with, the conclusions derived from the exponential branching rate data in lean mixtures discussed previously. Below 1300 K, data from a few high-density experiments originally done for recombination rate measurements have given evidence of small spikes (less than a factor of two above the subsequent quasi-steady level and perhaps better described as bumps) which indicate values of k_a/k_c somewhat below the upper bound.[92]

Distinct spikes in O atom concentration are expected to occur throughout the hydrogen-rich regime, on the basis of k_a/k_b values compatible with the exponential branching kinetics.[92] However no quantitative determination of k_a/k_b by this method has yet been completed. Further use of the [OH] and [O] spikes in shock waves in rich H_2-O_2—diluent mixtures, to determine ratios among k_a, k_b and k_c independently of the exponential rate measurements, is to be expected.

In hydrogen-lean mixtures ($H_2:O_2 \leq 0 \cdot 2$), [OH] excursions up to 50% above the subsequent quasi-steady, *partial equilibrium* level have been reported.[42] The phenomena which give rise to this behaviour are evidently more complex than those in hydrogen-rich mixtures, owing to the involvement of reaction (d) and to the earlier emergence of reverse reactions on account of the smaller equilibrium constants governing the more endothermic formation of the major chain centres, O and OH. Satisfactory reconciliation of the $[OH]_{max}$ data with the low-temperature data on k_d and with the exponential branching data then available was not possible.[42] Resolution of this puzzle will probably have to await fuller study of either the exponential branching rate parameters for lean mixtures or the [O] and [OH] spike behaviour in rich mixtures, or both.

2.3.3. Kinetics of Shock-Wave Recombination

Next to the kinetics of branched chain ignition in hydrogen-rich mixtures at low densities, termolecular association is presently the best understood portion of the high-temperature hydrogen–oxygen reaction mechanism, in spite of the persisting uncertainties in many of the rate coefficients for individual collision partners. Several factors have combined to produce this situation. The utility of the shock tube method for studying this process over a wide range of reaction conditions has provided a means of acquiring extensive data. The general soundness of the *partial equilibrium* approximation throughout the decelerating phase of the reaction has been of considerable assistance in the analysis of these data. Finally, the comparatively small number of termolecular steps which have been found to be important contributors to reducing the mole number to its full-equilibrium value in any given experiment, and the occurrence of these steps as independent, parallel paths and not in the sort of chain sequence which occurs in the ignition phase of the reaction, have greatly simplified the interpretation of the recombination data. As a result, the sufficiency of reactions (*e*), (*f*) and (*g*) to describe the recombination kinetics has been demonstrated, and specific third-order rate coefficients have been obtained for these reactions with each of three or four different collision partners.

Shock tube studies of recombination kinetics in reacting hydrogen–oxygen mixtures have used OH ultraviolet absorption,[28-32] optical interferometry,[54,56] and H_2O infrared emission.[60] Of these, OH absorption has been the most extensively and systematically employed and has yielded the most definitive results, and we concentrate on this method in developing the kinetics results. The other two, less sensitive diagnostics have provided largely confirmatory evidence over narrower ranges of conditions, and we present their results in the form of comparisons with those from OH absorption work. The occurrence of recombination can be seen in the representative records shown in Figs. 2.2, 2.4 and 2.5.

Recombination Rate Treatment

Kinetic interpretation of the recombination portion of the [OH] profile proceeds by means of the *partial equilibrium* approximation, by which the course of change of the entire system composition with recombination is computed, using the measured shock wave speed and initial gas composition and known thermochemical properties of the expected species. To place the progress of recombination in perspective, the change in N from its original value of unity to the value at full equilibrium, N^{eq}, is reckoned in terms of the normalized progress variable defined as

$$\nu = \frac{N - N^{eq}}{1 - N^{eq}},$$ (2.16)

which decreases from unity to zero as recombination proceeds. Also, for the flowing gas behind incident shock waves, the laboratory time differential, $d\tau = \rho_0/\rho \, dt$, is incorporated into the rate equation, (2.6), which then becomes

$$\frac{-p_0}{RT_0}(1 - N^{eq})\frac{dv}{d\tau} = R_{assoc} = \sum_j k_j[M][A_j][B_j] \qquad (2.17)$$

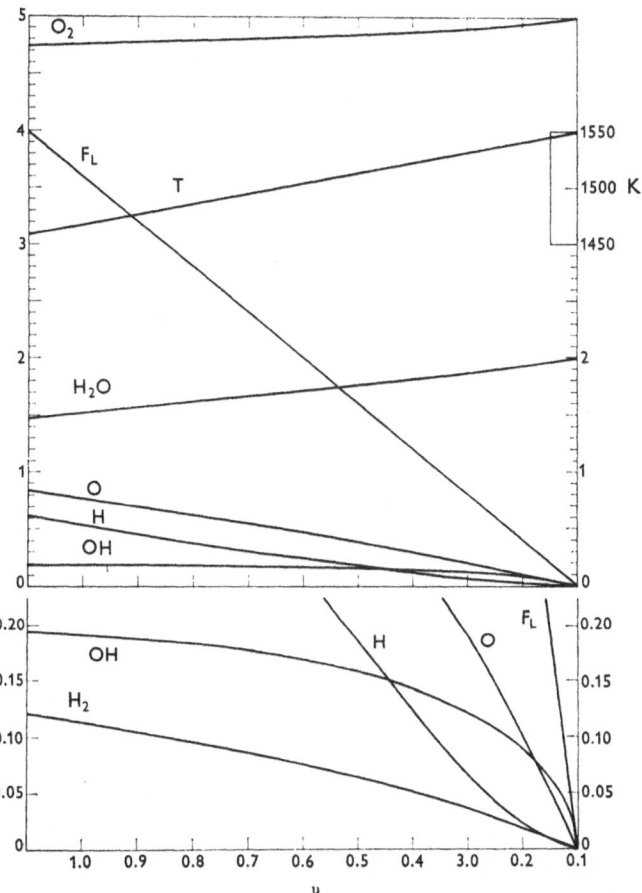

FIG. 2.12. Computed partial equilibrium trajectory as a function of recombination progress, v, for representative experiment in $1\cdot0\%$ H_2–$3\cdot0\%$ O_2–$96\cdot0\%$ Ar ($\eta = 0\cdot33$) mixture. Shock velocity: $1\cdot174$ km sec^{-1}. Plotted with dimensionless scale at left are $N_i/(1 - N^{eq})$ for each of the six stoichiometrically significant reactants, and F_L, a measure of the pool of recombinable species in hydrogen-lean mixtures. $F_L = (N_{OH} + 2N_O + 2N_{H_2} + 3N_H)/(1 - N^{eq})$. Plotted with scale on right is the temperature. An expansion of the lower portion of the ordinate scale is presented at the bottom of the figure.

when the reverse reaction (dissociation) is ignored. ρ_0, p_0 and T_0 are respectively the density, pressure and temperature of the unshocked gas.

The computed mole numbers of the six stoichiometrically significant reactants as a function of ν for recombination under *partial equilibrium* conditions in particular experiments are shown in Figs. 2.12, 2.13 and 2.14. Table 2.1 systematically summarizes the *partial equilibrium*

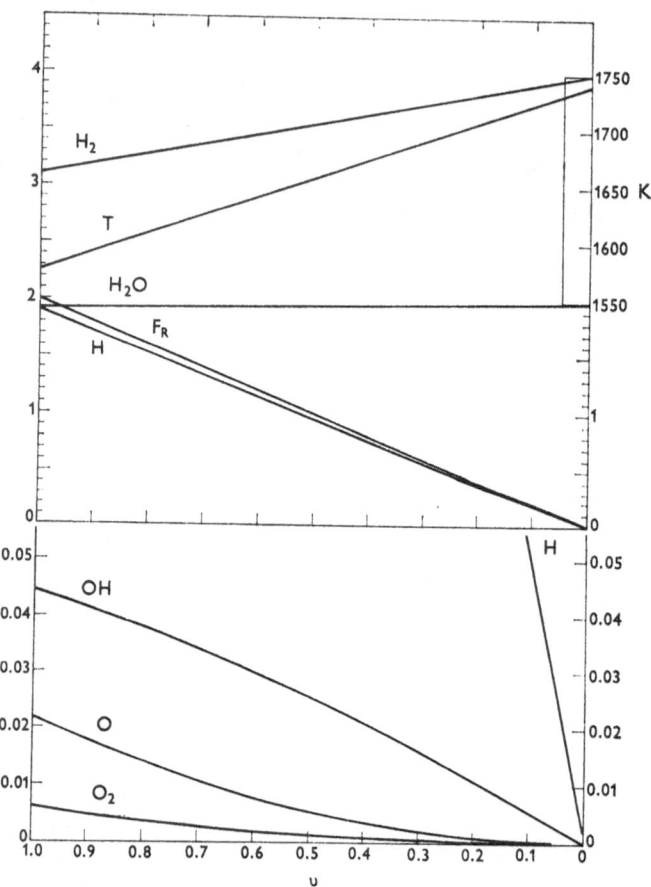

FIG. 2.13. Computed partial equilibrium trajectory as a function of recombination progress, ν, for representative experiment in 6·0% H_2–1·0% O_2–93·0% Ar ($\eta = 6·0$) mixture. Shock velocity: 1·256 km sec^{-1}. Plotted with dimensionless scale at left are $N_i/(1 - N^{eq})$ for each of the six stoichiometrically significant reactants, and F_R, a measure of the pool of recombinable species in hydrogen-rich mixtures. $F_R = (N_H + N_{OH} + 2N_O + 2N_{O_2})/(1 - N^{eq})$. Plotted with scale on right is the temperature. An expansion of the lower portion of the ordinate scale is presented at the bottom of the figure.

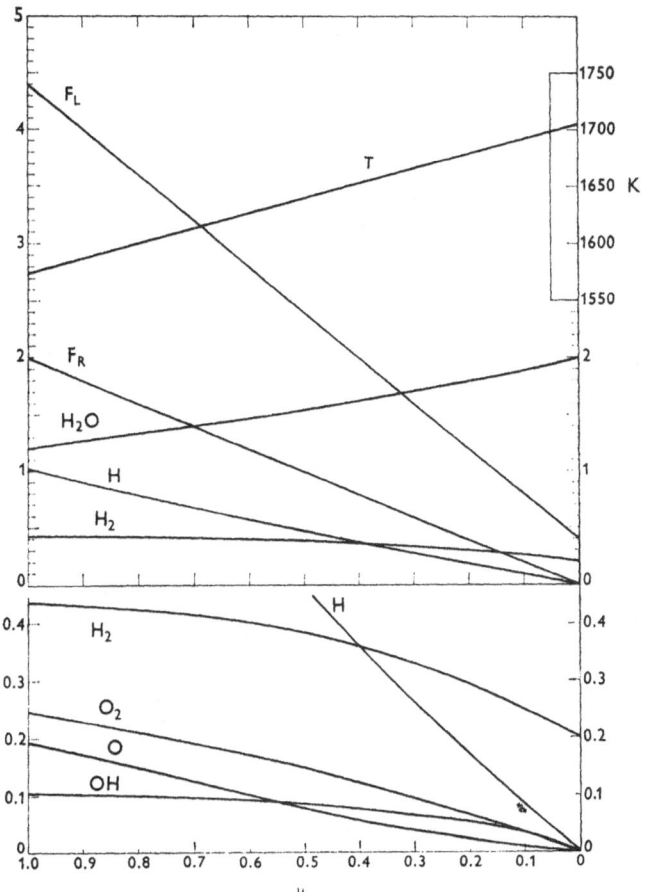

FIG. 2.14. Computed partial equilibrium trajectory as a function of recombination progress, v, for representative experiment in 1·65% H_2–0·75% O_2–97·60% Ar ($\eta = 2·20$) mixture. Shock velocity: 1·222 km sec^{-1}. Plotted with dimensionless scale at left are $N_i/(1 - N^{eq})$ for each of the six stoichiometrically significant reactants, and F_L and F_R, as defined in Figs. 2.12 and 2.13, respectively. Plotted with scale on right is the temperature. An expansion of the lower portion of the ordinate scale is presented at the bottom of the figure.

relationships between the concentrations of pertinent pairs of the species which are ultimately consumed rather completely, in terms of equilibrium constants and the comparatively constant concentrations of H_2O and either H_2 or O_2, whichever may be in excess.

From the primary [OH] vs τ profile and the computed *partial equilibrium* trajectory, one has the wherewithal to apply equation (2.17). Because the functional dependence upon v of the recombinable species concentrations, [H], [OH] and [O$_2$], which enter the right-hand side

TABLE 2.1

Equimolecular equilibrium relationships between pairs of intermediates and consumable reactants

	OH	H	O
H	Excess H₂ $$[H] = \frac{K_c[H_2]}{[H_2O]}[OH]$$ Excess O₂ $$[H] = \frac{K_d}{K_a[H_2O][O_2]}[OH]^3$$		
O	$$[O] = \frac{K_d}{[H_2O]}[OH]^2$$ $$K_d = K_c/K_b$$	Excess H₂ $$[O] = \frac{[H_2O]}{K_bK_c[H_2]^2}[H]^2$$ Excess O₂ $$[O]^3 = \frac{K_a^2 K_d[O_2]^2}{[H_2O]}[H]^2$$	
H₂, O₂	$$[H_2][O_2] = \frac{1}{K_{IV}}[OH]^2$$ $$K_{IV} = K_a K_b$$	$$[H_2]^3[O_2] = \frac{[H_2O]^2}{K_{II}}[H]^2$$ $$K_{II} = K_a K_b K_c^2 = K_a K_b^3 K_d^2$$	$$[H_2][O_2] = \frac{[H_2O]}{K_{III}}[O]$$ $$K_{III} = K_a K_c$$

of this equation is generally not straightforward, and because the [OH] profiles have been well enough resolved to permit it, equation (2.17) has been treated directly in its differential form. Smoothing procedures used in deducing v and $dv/d\tau$ at selected points over the range $0 < v < 1$ are given in detail in Refs. 29–32.

Lean, Rich and Stoichiometric Regimes

The several shock-wave studies of OH radical disappearance[28–32] have established the following general features of recombination in the hydrogen–oxygen reaction. For the range of temperatures, densities and hydrogen:oxygen ratios included in the experiments, essentially all recombination, beginning at $v \approx 1$, is satisfactorily accounted for by termolecular reactions proceeding in a *partial equilibrium* environment. Only three reactions, (e), (f) and (g) contribute significantly to recombination for $8 \geq \eta \geq 0.33$. The relative importance of reactions (e)–(g) is dependent on η. Reaction (f) is the only termolecular reaction of importance over a range of hydrogen-lean mixtures. Reaction (e) is dominant in hydrogen-rich mixtures, although the contribution of reaction (g) cannot be ignored except in an extreme excess of H_2, and over an extended range of hydrogen-rich compositions the two reactions (e) and (g) together govern the recombination rate, to the exclusion of reaction (f) and other possible paths involving O or O_2. The principal reaction in near-stoichiometric mixtures is reaction (g), with both reactions (e) and (f) also playing significant roles.

The lean regime where reaction (f) dominates recombination is approximately specified by $\eta \leq 1$. As Fig. 2.12 illustrates, $[O_2]$ and $[H_2O]$ greatly exceed the other reactive species concentrations, and change comparatively little between $v = 1$ and $v = 0$. In an excess of Ar as diluent, terms representing the effects of the very minor constituents as collision partners are ignored, and the form of equation (2.17) to be compared with experiment then becomes

$$R_{assoc} = R^{(f)} = [H][O_2]\{k_f^{Ar}[Ar] + k_f^{H_2O}[H_2O] + k_f^{O_2}[O_2]\}$$

$$= k_f[M][H][O_2] \quad (2.18)$$

Sufficiency of this expression is illustrated by Fig. 2.15, which shows $\ln(-dv/d\tau)$ versus $\ln[H]$ for three separate experiments carried out under nearly identical conditions in the 96% Ar, 1% H_2, 3% O_2 mixture to which Fig. 2.12 pertains. Each data point corresponds to a value of [OH] deduced from the original oscillogram. The least squares line best describing all of these points has slope 1.01 ± 0.01, in accordance with expectation. More detailed consideration of data covering substantial ranges of T, v and $0.33 < \eta < 1$ has confirmed that termolecular reactions other than reaction (f) have no discernible influence on the deduced $dv/d\tau$ values.

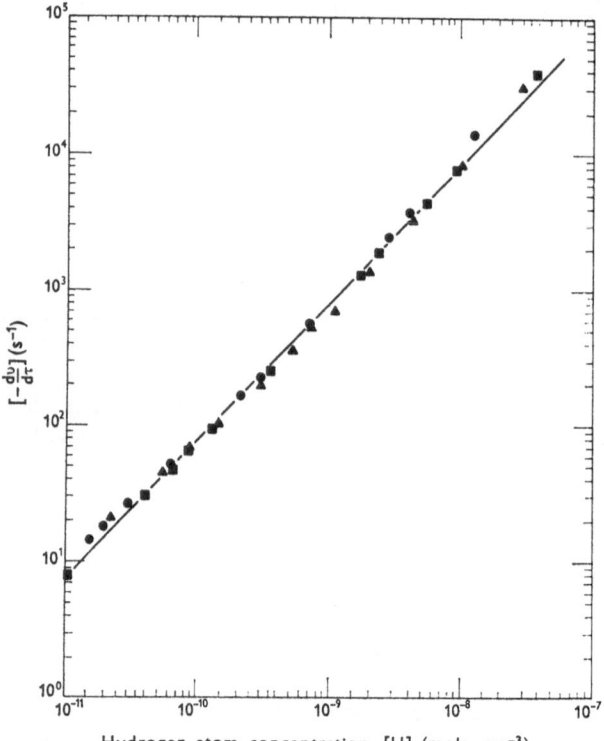

Hydrogen atom concentration, [H] (mols · cm⁻³)

FIG. 2.15. Graph of $\ln(-dv/d\tau)$ versus $\ln$ [H] for recombination rate data of three experiments (■, ▲, ●) carried out under nearly identical conditions in a $1\cdot0\%$ H_2–$3\cdot0\%$ O_2–$96\cdot0\%$ Ar ($\eta = 0\cdot33$) mixture. Initial pressure: 15 cm Hg. Mean reaction-zone temperature: ~1415 K. v values range between $0\cdot55$ and $0\cdot004$, and cover a period, $\Delta\tau$, of about 450 μsec. Slope of line determined by least-squares fit $= 1\cdot01 \pm 0\cdot01$ (after Getzinger and Schott[29]).

Mixtures sufficiently hydrogen-rich ($\eta \gtrsim 3$) that [H] and [OH] substantially exceed $[O_2]$ and [O], especially as $v \to 0$, and that $[H_2]$ and $[H_2O]$ change relatively little with v, lead to the other comparatively simple form of equation (2.17). Reaction (f) is inconsequential, and the contributions of reactions (e) and (g) are in the ratio $R^{(g)}/R^{(e)} = k_g[M][OH]/k_e[M][H]$, which is practically constant between $v = 1$ and $v = 0$, owing to the well-maintained partial equilibrium relationship $\alpha = [OH]/[H] = [H_2O]/K_c[H_2]$. Thus we may write

$$R_{assoc} = R^{(e)} + R^{(g)} = k_{app}[M][H]^2 \qquad (2.19)$$

where $k_{app}[M] = k_e[M] + \alpha k_g[M]$. Again considering an excess of Ar as diluent, and inconsequential changes in $k_{app}[M]$ during a given shock-wave run, equation (2.19) predicts that $(-dv/d\tau)$ is proportional

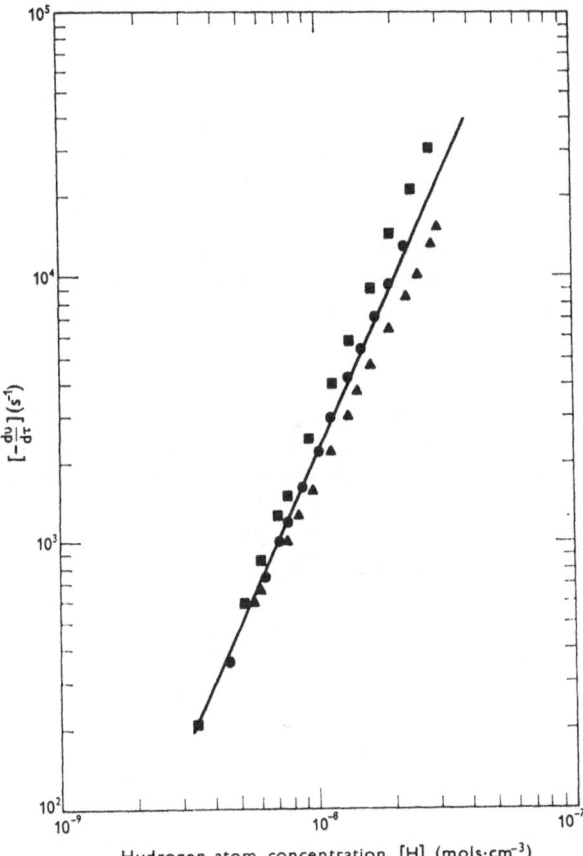

FIG. 2.16. Graph of ln $(-d\nu/d\tau)$ versus ln [H] for recombination rate data of three experiments (■, ▲, ●) carried out under nearly identical conditions in a $6\cdot0\%$ H_2–$1\cdot0\%$ O_2–$93\cdot0\%$ Ar ($\eta = 6\cdot0$) mixture. Initial pressure: $12\cdot1$ cm Hg. Mean reaction-zone temperature: $\sim$1635 K. ν values range between 0·75 and 0·09, and cover a period, $\Delta\tau$, of about 240 μsec. Slope of line determined by least-squares fit $= 2\cdot10 \pm 0\cdot08$.

to $[H]^2$. A logarithmic plot of $(-d\nu/d\tau)$ versus [H] for three similar experiments in a 93% Ar, 6% H_2, 1% O_2 mixture, to which Fig. 2.13 pertains, is presented in Fig. 2.16. As the ν values involved in this plot are generally higher than those in Fig. 2.15, and the sensitivity of [OH] to ν is less, the value of k_{app} is not so precisely determined by these experiments as was k_f. However the mean slope of the three sets of data is $2\cdot10 \pm 0\cdot08$, demonstrating approximately the effective second-order kinetics.

Recombination in near-stoichiometric compositions ($1 \leqslant \eta \leqslant 3$), such as the $\eta = 2\cdot2$ mixture to which Fig. 2.14 applies, has significant contributions from all three of reactions (e), (f) and (g), and there is no instructively simple form of equation (2.17). Experiments done in this regime have been analyzed by subtracting from equation (2.17) the values of $R^{(e)}$ and $R^{(f)}$ measured in the same diluent at richer and leaner conditions and examining the residuum. These experiments afford the greatest obtainable sensitivity to $R^{(g)}$ and, by yielding results which are consistent among themselves over ranges of T and ν and with values of k_g derived from richer mixtures where reaction (f) is insignificant, serve to exclude other imaginable recombination steps (refer to Fig. 2.7) from the necessity of further consideration.

Reaction Orders and Pools of Excess Species

Here we digress to expand upon the approximate but informative simplifications of equation (2.17) for rich and lean compositions and to relate them to the rather more conventional "order-of-reaction" forms which one encounters in the recombination work in postflame gases.[14] In rich mixtures, with which the vast bulk of the flame work has dealt, and as Fig. 2.13 illustrates, [H] and [OH] are not only proportional to each other, but also to ν, once ν becomes small enough that [O] and [O_2] are utterly negligible. Thus $(-d\nu/dt)$ and $(-d[H]/dt)$ or $(-d[OH]/dt)$ are simply related, and by equation (2.19), one sees that the disappearance of either of these species is described by an apparent second order rate equation. This means of data analysis has also found use in the early shock wave work.[28]

Again referring to Fig. 2.13, the precise relationship between ΔN_H, $\Delta\nu$ and the corresponding changes in the other species concentrations is illustrated by the function F_R. This function is the simplest of an infinite family of equivalent linear combinations of the changes in selected species concentrations that can be constructed[77] from the right-hand side of equation (2.7) by elimination of N_{H_2O} and either N_{H_2} or N_{O_2}, using equations (2.8) and (2.9), in order to leave an expression for the numerator of ν in equation (2.16) in which all the remaining N_i^{eq} terms are substantially zero. These functions have been taken to specify a *pool of excess radicals* which disappears in the course of recombination.[14]

The function of corresponding utility for lean compositions, F_L, is represented in Fig. 2.12, where one also sees that the asymptotic behaviour, $\lim\limits_{\nu\to 0} \dfrac{\Delta N_{OH}}{\Delta\nu(1 - N^{eq})} = 4$ is approached closely only at such low values of ν that it has not been useful in the shock wave recombination work.

Finally, both these functions are exhibited in relation to the trajectory of *partial equilibrium* recombination in a near-stoichiometric mixture in Fig. 2.14. Here it is clear that F_R and F_L both provide linear bases for reckoning Δv, irrespective of η, although except in a precisely stoichiometric mixture, only one yields proportionality to v.

Individual Collision Partner Effects

Experiments such as those represented in Figs. 2.15 and 2.16 yield values of the quantities $k_f[M]$ and $k_{app}[M]$ which appear in equations (2.18) and (2.19). Systematic variation of reactant mixture compositions then provides a means of reducing these effective rate coefficients to the substantially temperature-independent termolecular rate coefficients, $k_j^{M_i}$, for the elementary reactions (e)–(g) with specific collision partners. In this manner selective dilution with argon[28-32] and nitrogen[31,32] has permitted extraction of 6 such elementary coefficients, viz. $k_e^{M_i} - k_g^{M_i}$ for M_i = Ar and N_2. These are presented in Table 2.2.

Addition of significant amounts of water vapour[31,32] to reactant mixtures has yielded important information about the remaining coefficients in equations (2.18) and (2.19), including specific values of $k_f^{H_2O}$ and $k_g^{H_2O}$. The accessible ranges of $[H_2]$ and of $[O_2]$ have proven insufficient for determination of the effects of $k_e^{H_2}$ or $k_f^{O_2}$, respectively. Therefore, independent observations of $k_e^{H_2}$ were invoked to subtract the small contribution it makes to k_{app} in the rich mixtures, and an upper bound to $k_f^{O_2}$ was estimated from the lean mixture data. These results are also presented in Table 2.2.

Isolation of $k_e^{H_2O}$ and $k_g^{H_2}$ individually from k_{app} data in rich mixtures is not possible simply by composition variation. This limitation is recognized by writing out the expression for $k_{app}[M]$ from equation (2.19), viz.

$$k_{app}[M] = k_e^{Ar}[Ar] + k_e^{H_2O}[H_2O] + k_e^{H_2}[H_2]$$

$$+ \alpha k_g^{Ar}[Ar] + \alpha k_g^{H_2O}[H_2O] + \alpha k_g^{H_2}[H_2] \quad (2.20)$$

and observing, from $\alpha = [H_2O]/K_c[H_2]$, that the terms in $k_e^{H_2O}$ and $k_g^{H_2}$ exhibit the same dependence upon composition, to wit:

$$k_e^{H_2O}[H_2O] + \alpha k_g^{H_2}[H_2] = (k_e^{H_2O} + K_c^{-1}k_g^{H_2})[H_2O] \quad (2.21)$$

Thus, only the sum $(k_e^{H_2O} + K_c^{-1}k_g^{H_2})$ can be isolated, unless a large enough range of K_c could be covered, but the range of temperature needed to do so is beyond that readily accessible to shock-wave (or flame) recombination experiments.[32] Consequently, only upper limits to these two coefficients have been determined.

The eight termolecular rate coefficients deduced from shock-wave OH absorption measurements, and the upper bounds for three others, which are summarized in Table 2.2 are believed to be valid over the approximate temperature range 1300–1900 K. The value of k_f^{Ar} ($2\cdot2 \times 10^{15}$ cm^6 mole^{-2} sec^{-1}) compares favourably with the previously cited result deduced by Gutman et al. ($3\cdot3 \times 10^{15}$ cm^6 mole^{-2} sec^{-1})[63]

TABLE 2.2

Selected termolecular coefficients determined between 1300 and 1900 K in OH absorption experiments (after Reference 32)†

Reaction ($j =$)	$k_j^{N_2}$	k_j^{Ar}	$k_j^{H_2O}$	$k_j^{H_2}$	$k_j^{O_2}$	$k_j^{H_2O}/k_j^{Ar}$	$k_j^{H_2}/k_j^{Ar}$	$k_j^{O_2}/k_j^{Ar}$
e	6	7·5	≤100	15*	—	≤13	2	—
f	21	22	540	—	≤43	25	—	≤2·0
g	86‡	33	660	≤1600	—	20	≤50	—

† Units: 10^{14} cm^6 mole^{-2} sec^{-1}. All values, except those for $k_j^{N_2}$, were determined with the more recent of two source lamps (see section 2.2.1). (Based on the data in Ar diluent determined previously— References 29–31—with the same equipment and procedures, the nitrogen values were found to be approximately 1·5, 1·5 and 1·6 times as large as the corresponding k_j^{Ar} for reactions (e), (f) and (g), respectively.)

* Determined by extrapolating the expression appearing in Table I of Reference 93 to 1700 K, and used in evaluating $k_e^{H_2O}$, $k_g^{H_2O}$ and $k_g^{H_2}$.

‡ Evaluated by dividing k_g[M] by the total gas concentration, [M], and hence not corrected for the undetermined, but presumably small, contributions of other collision partners.

from exponential growth measurements at approximately 1100 K. Not surprisingly, the relative values indicate the presence of a small negative activation energy for k_f^{Ar}. The coefficients k_e^{Ar} and k_g^{Ar} can be compared with extrapolated values deduced from recent shock tube measurements of the rates of dissociation of H_2[12,93] and H_2O[48,94] in an argon environment. At 1700 K the H_2 dissociation coefficients yield $k_e^{\text{Ar}} = 7 \cdot 4 \times 10^{14}$ [12] and $5 \cdot 9 \times 10^{14}$ [93] cm^6 $mole^{-2}$ sec^{-1}. Similarly, from H_2O dissociation measurements, k_g^{Ar} (1700 K) $= 1 \cdot 9 \times 10^{15}$ [94] and $3 \cdot 2 \times 10^{15}$ [48] cm^6 $mole^{-2}$ sec^{-1}. The ratio $k_g^{\text{H}_2\text{O}}/k_g^{\text{Ar}}$ was found to be approximately 20.[48] The agreement between these values and those in Table 2.2 is quite satisfying. Comparisons with coefficients derived in other, nonshock-wave studies of the high temperature hydrogen–oxygen reaction will be discussed in section 2.3.4.

Densitometric and Infrared Emission Studies of Recombination

The primary emphasis in shock tube interferometric studies of the hydrogen–oxygen reaction has been on induction period phenomena. Recently, however, the entire postshock density profiles of a selection of rich, lean and near stoichiometric H_2–O_2–Ar mixtures have been studied by numerical integration of an assumed reaction mechanism.[56] In this manner it was shown that the characteristic features of the profile prior to the end of the density plateau are essentially independent of the recombination kinetics. Thereafter, however, the shape of the profile is largely accounted for by termolecular reactions (*e*)–(*g*). Systematic variation of the termolecular rate coefficient values in experimental regimes where recombination is most sensitive to reactions (*f*) or (*g*), respectively, has yielded temperature-dependent expressions of the form $k_f^{\text{M}_i} = AT^{-m}$ for k_f^{Ar} and k_g^{Ar} believed valid over the range 1400–3000 K. The expression of Jacobs et al.[93] was found satisfactory for k_e^{Ar}. In all three cases, variation with temperature is small $(1 \cdot 0 \geq m \geq 0 \cdot 5)$. Values at 1700 K, $k_e^{\text{Ar}} = 5 \cdot 9 \times 10^{14}$ (cited above), $k_f^{\text{Ar}} = 1 \cdot 9 \times 10^{15}$, and $k_g^{\text{Ar}} = 3 \cdot 6 \times 10^{15}$ cm^6 $mole^{-2}$ sec^{-1}, are in excellent accord with those listed in Table 2.2.

Experimental limitations restricted the quantitative analysis of H_2O infrared emission profiles to hydrogen-lean ($\eta = 1 \cdot 0$ and $0 \cdot 5$) H_2–O_2–Ar mixtures.[60] Under these conditions an empirical calibration, combined with the *partial equilibrium* approximation and equation (2.18), made possible an evaluation of $k_f[\text{M}]$. Correction for the small contribution of $M_i = H_2O$ using the ratio $k_f^{\text{H}_2\text{O}}/k_f^{\text{Ar}} = 25$ from Table 2.2 yielded a series of values of k_f^{Ar} that average $3 \cdot 0 \times 10^{15}$ cm^6 $mole^{-2}$ sec^{-1}. As in the OH absorption experiments, k_f^{Ar} is essentially independent of ν, indicating that reaction (*f*) is the sole termolecular reaction of significance throughout the recombination regime.

An additional, qualitative comparison of OH and infrared results was reported for other mixtures within the range $2 \cdot 2 \geq \eta \geq 0 \cdot 33$.[60] The coefficients listed in Table 2.2 were incorporated in numerical syntheses of typical infrared recombination profiles. The agreement was believed to be quite satisfactory.

As with the interferometric measurements, the principal significance of the infrared emission experiments lies in the confirmatory evidence they provide regarding the essential correctness of our present understanding of the recombination kinetics in reacting hydrogen–oxygen mixtures. Excellent agreement with the conclusions of the OH absorption experiments has been observed for a significant range of temperatures, densities, initial H_2/O_2 ratios, and ν values. An essential feature of the analysis of $[H_2O]$ and $[OH]$ profiles was the assumed validity of the *partial equilibrium* approximation at all ν. The agreement between the two independent sets of results is convincing experimental evidence for the validity of this assumption for a wide range of reaction conditions under which it had not been previously tested.

2.3.4. Kinetics in Nonshock Environments

Shock tube contributions toward unraveling the kinetics of the high temperature hydrogen–oxygen reaction were predated by, and have been continually supplemented by, results derived with other experimental methods. Specifically considered in the following paragraphs are significant quantitative observations derived from studies of laminar, premixed flames and from determination of first and second explosion limit parameters.

Flame Structure Studies

Numerous studies of the structure of hydrogen–oxygen (—diluent) flames employing a great variety of experimental techniques have been reported in the literature. A detailed discussion is beyond the scope of this chapter, and the interested reader is referred to the texts cited previously.[14–16] What follows is by necessity a very restricted discussion, whose primary emphasis will be on comparing kinetic results derived from such studies with those obtained in the shock tube.

Measurements of the time-invariant temperature and composition profiles by axial traversal through the flame provide the data required for a quantitative investigation of the high temperature hydrogen–oxygen reaction. Typically, thermocouples, mass spectrometers (for stable component concentrations), and various optical and ESR spectroscopic techniques (for intermediate concentrations) have provided the means for determining flame structure in both the primary and secondary reaction zones. Subatmospheric pressures, and dilute mixtures which burn to produce relatively low temperatures, have been employed to slow the overall reaction, and thus provide adequate spatial resolution for experimental measurements in the primary zone.

The chemically interesting portion of a hydrogen–oxygen (—diluent) flame is bounded upstream by a plane marking the point where diffusion of intermediates and/or heat transfer has produced perceptible reaction in the mixture. The primary reaction zone, which follows, contains steep temperature and concentration gradients, evidence of significant reaction progress and release of the bulk of the reaction exothermicity. Further downstream, passage into the secondary reaction zone is not clearly defined, but corresponds approximately to the beginning of a region where the temperature and concentrations of the stable components are substantially constant. Thereafter, reaction progress is most apparent in the falling concentrations of intermediates, marking the approach to complete system equilibrium.

All of the principal kinetic features of the high temperature hydrogen–oxygen reaction described previously, with the exception of initiation reaction (0), supplanted by diffusion of intermediates and conduction of heat into the preignition region, are, in principle, amenable to study by detailed probing of the flame structure. The primary experimental limitations in such studies are difficulties in accurately measuring concentration profiles for the intermediates, uncertainties in the magnitude of corrections for diffusion, and, as in shock-wave experiments, lack of sensitivity to certain important elementary reactions. The recent development[95] and application[95,96] of numerical procedures for computing flame velocities and detailed flame structure for such relatively simple flames provide a means of assessing the effect of these and other limitations on the accuracy of kinetic information derived from experimental profile measurements.

Structure studies of the primary and secondary reaction zones of hydrogen–oxygen (diluent) flames have yielded values for the bimolecular coefficients $k_a - k_c$, and important conclusions about the recombination kinetics, including values for some of the coefficients $k_e^{M_i} - k_g^{M_i}$. Representative values of k_a and k_c, determined in a recent investigation of subatmospheric, low-temperature hydrogen–oxygen flames,[96] can be compared with the shock tube values given previously. At an approximate mean flame temperature of 900 K k_a was found to be $1 \cdot 9 \times 10^{10}$ cm^3 mole^{-1} sec^{-1}. This is in excellent agreement with the value $2 \cdot 2 \times 10^{10}$ cm^3 mole^{-1} sec^{-1} computed at this same temperature from the shock tube expression for k_a given in Fig. 2.8. For comparison with the shock tube value for k_c ($\geqslant 3 \cdot 4 \times 10^{12}$ cm^3 mole^{-1} sec^{-1} at 1600 K) quoted earlier, it is necessary to utilize the Arrhenius expression derived in Reference 96 for the temperature range 500–1500 K. Extrapolation to 1600 K yields $2 \cdot 9 \times 10^{12}$ cm^3 mole^{-1} sec^{-1}. A similar comparison with the shock tube value of k_b ($\sim 4 \times 10^{12}$ cm^3 mole^{-1} sec^{-1} at 1600 K) is possible by extrapolation of the consensus Arrhenius expression given by Fristrom and Westenberg.[15] At 1600 K k_b is computed to be $5 \cdot 6 \times 10^{11}$ cm^3 mole^{-1} sec^{-1}. The differences between flame and shock tube values for k_b and k_c reflect the difficulty of determining these coefficients in either type of experiment, and emphasize the need for more reliable high temperature measurements of both coefficients.

6

The influence of termolecular association reactions (e)–(g) on flame structure will now be considered. These three are the only recombination steps whose contribution to stable flame propagation has been confirmed experimentally.

Structure studies in the primary zone of both hydrogen-rich[95–97] and hydrogen-lean[98] flames have demonstrated that reaction (f) plays a significant role in determining concentration and temperature profiles under conditions prevailing before significant depletion of the reactants. This is a consequence of the relatively low temperatures and large concentrations of molecular oxygen present early in the primary regime. At such temperatures the importance of (f) relative to the chain branching step (a) is considerably enhanced. With increased reaction progress, the influence of (f) relative to other elementary steps is substantially reduced in hydrogen-rich mixtures. However, in the presence of excess oxygen, reaction (f) continues to influence measured profiles, especially that of the temperature, which is particularly sensitive to the rate at which chemical energy is liberated, throughout the primary zone.

The significance of quantitative determinations of k_f in such studies is limited by uncertainties concerning the relative contributions of various collision partners, M_i. No systematic effort to separate these effects has been reported. In a recent study of profiles in one hydrogen-rich flame, the mean value of k_f at 600 K for the particular gas composition was estimated to be 3×10^{16} cm^6 mole^{-2} sec^{-1}.[96] Similarly, concentration and temperature measurements within the primary zone of a hydrogen-lean flame yielded $k_f^{M_i}$ (1000–1350 K) = 1×10^{17} cm^6 mole^{-2} sec^{-1}, with M_i assumed to be H_2O.[98] Both values are in reasonable agreement with those in Table 2.2, if allowance is made for the estimated influence of composition.

Studies of recombination in postflame gases have made particularly significant contributions to the understanding of the high temperature hydrogen–oxygen reaction kinetics.[22,75] This is true in spite of the experimental limitations, particularly with respect to composition variation, mentioned in a previous section. Equilibration of reactions (b) and (c) throughout the postflame region of hydrogen-rich mixtures was established as early as 1956.[75] The rate of decay of H atoms was seen to be third order in total gas concentration and second order in [H],[75] in agreement with the presently accepted mechanism of recombination via reactions (e) and (g) in a *partial equilibrium* environment. Subsequent work in slightly hydrogen-lean mixtures[14,23] confirmed the equilibration of reaction (d) late in the postflame region, and yielded a rate of decay of [OH] compatible with an [OH]^3 dependence, again in agreement with current expectations. Both of these studies predated similar investigations in shock tubes.

Efforts to evaluate the important termolecular rate coefficients from data in postflame gases have been restricted by: (i) the comparative complexity of the recombination environment; (ii) the appearance of flame instabilities for η below approximately 0·5–1·0;[23] and (iii) the paucity of established spectroscopic diagnostics suitable for work in hydrogen-lean mixtures.[99] Item (i) is a consequence of the relatively large percentages of reactants required to support a stable flame. At least three third bodies M are of potential importance in promoting recombination via reactions (e)–(g) in the postflame gases of non-stoichiometric mixtures.

In spite of these difficulties, a considerable amount of information about the coefficients $k_e^{M_i}$ and $k_g^{M_i}$, particularly where $M_i = N_2$, H_2 and H_2O, has been deduced from hydrogen-rich flames. Table 2.3 lists values reported in three detailed studies[100–102] of recombination in such mixtures. As in comparable shock tube experiments, evaluation of $k_e^{H_2O}$ and $k_g^{H_2}$ is further complicated by the presence of the equilibrated reaction (c), and, in general, only upper limits can be determined for these two coefficients.

There is considerable scatter in the coefficients presented in Table 2.3, and, where comparison is possible, between these values and those in Table 2.2. The differences are probably indicative of the corresponding experimental uncertainties. There is no reason to believe that either flames or shock tubes provide a superior method for determining these quantities. In both environments the major complexity, and principal source of uncertainty, in the evaluation of the termolecular coefficients lie in their extraction from experimentally-determined total recombination rates. Evidence that this is the case can be seen in a recent comparison in which bulk recombination rates under conditions typical of postflame gas mixtures were satisfactorily reconstructed with the shock tube coefficients in Table 2.2.[32] The values in Tables 2.2 and 2.3 represent the best that can be achieved from studies of the high temperature combustion of hydrogen–oxygen mixtures, without some significant advancement over existing experimental techniques.[102]

Isothermal Explosion Limit Studies

The explosion limit phenomenology observed in hydrogen–oxygen mixtures has been semi-quantitatively described in section 2.1.2. For an initial pressure below approximately one atmosphere there exists one threshold temperature for explosion, typically between 700 and 900 K, whose magnitude is determined by the appropriate isothermal branched chain kinetics. More specifically, this boundary, separating régimes of slow, quasi-steady state and explosively rapid reaction, is defined by an equality between the rates of chain centre formation by branching steps and destruction by termination steps.

For the case of a single chain centre subjected to competitive, first-order chain branching and termination, equations (2.1) and (2.2), this equality is simply identified with the condition that ϕ be equal to zero. Kondratiev[18] and others have shown that equations (2.1) and (2.2)

TABLE 2.3

Recombination coefficients from postflame region of hydrogen-rich flames

$(10^{14} \, cm^6 \, mole^{-2} \, sec^{-1})$

Reaction $(j =)$	k_j^{Ar}	$k_j^{N_2}$	$k_j^{H_2O}$	$k_j^{H_2}$	$k_j^{N_2}/k_j^{Ar}$	Reference	Reference temperature (K)
e	—	5·4	≤36	86	—	101	1400
e	—	8·5	8·5	55	—	100	1072
e	18	19	≤30	—	1·0	102	1900
g	—	470	1020	≤3630	—	101	1400
g	—	550	550	—	—	100	1072
g	32	32	270	≤430	1·0	102	1900

provide a satisfactory description of the time-evolution of the hydrogen–oxygen reaction in the explosion limit domain, where the quantity C is identified with [H] and the branching frequency f with $2k_a[O_2]$, provided that behaviour of the O and OH chain centre populations is governed by quasi-state steady considerations (cf. section 2.1.1) and termination is controlled either by reaction (f) or by effective removal of H at the vessel walls. There is ample experimental evidence that these restrictions are satisfied under all conditions typically employed in determining subatmospheric explosion limits, except in very hydrogen-lean mixtures.[103] It follows that experimental parameters defining the location of the explosion peninsula also satisfy the equality $\phi = 0$. One then needs only to know the detailed expression for the termination frequency g in equations (2.1) and (2.2) to translate the limit parameters into useful kinetic information.

The first or lower explosion limit is determined by competition for hydrogen atoms between reaction (a) and wall termination. The appropriate expression for g, a function of such factors as vessel geometry and the mechanism (kinetic or diffusion) controlling the rate of wall removal of H,[18] can be evaluated on the basis of independent experimental and theoretical studies, to give values for k_a. As an illustration, a recent paper[104] reports careful measurements of limit pressures and temperatures, and evaluations of the k_a's required to satisfy the equality $\phi = 0$. At 900 K k_a was found to be $2 \cdot 0 \times 10^{10}$ cm^3 mole^{-1} sec^{-1}, a value in excellent agreement with results from shock tube and flame experiments cited above.

At the larger pressures typical of the second or upper explosion limit, termolecular recombination reaction (f) competes effectively for H with reaction (a), and g takes the simple form $k_f[M][O_2]$. Thus if k_a is known at the limit temperature, $k_f[M]$ can be evaluated from the relationship $\phi = 0$. Selective variation of mixture composition, and determination of the corresponding limit conditions, then provides a set of simultaneous algebraic equations which can be solved for the individual $k_f^{M_i}$. For a value of k_a (813 K) $= 6 \times 10^9$ cm^3 mole^{-1} sec^{-1} (compared with $8 \cdot 7 \times 10^9$ cm^3 mole^{-1} sec^{-1} computed from the extrapolated shock-tube Arrhenius expression in Fig. 2.8) Baldwin[105] has obtained $k_f^{Ar} = 1 \cdot 7 \times 10^{15}$ cm^6 mole^{-2} sec^{-1} from second explosion limit measurements. This value of k_f^{Ar} is close to, but below all of the higher temperature measurements. Second explosion limit measurements also yield ratios $k_f^{M_i}/k_f^{Ar}$, where $M_i = N_2$, H_2O, H_2 and O_2, of $2 \cdot 2$, 32, $5 \cdot 0$ and $1 \cdot 8$, respectively.[106] Where a comparison is possible, these ratios are seen to be in approximate agreement with values listed in Table 2.2.

2.3.5.   Status and Prognosis

In closing this section we summarize the present state of knowledge of the high-temperature hydrogen–oxygen reaction kinetics. It can be safely asserted that the principal kinetic features of the reaction are now quite well understood. With some modification the basic mechanism is the same as that operating in the low-temperature region below the explosion limits. The principal differences are in the mode of termination: wall reactions are replaced by homogeneous termolecular recombination in the fast-reaction situation brought about by large, nonsteady concentrations of the atomic and reactive radical intermediates, and peroxide species do not accumulate in such significant amounts as they do at lower temperatures.

The rate coefficients of many of the important elementary steps at high temperatures are now well established, particularly k_a and the functions $k_f[M]$ and $k_{app}[M]$ which describe the recombination kinetics for those gas compositions which have been studied directly. Improved experimental accuracy is apparently needed in the determination of the rate of recombination before more definitive values of the coefficients for individual collision partners, $k_e^{M_i}$, $k_f^{M_i}$ and $k_g^{M_i}$, can be anticipated. Also, more quantitative information is desirable concerning the high-temperature rate coefficients of the other important bimolecular steps, k_b, k_c and k_d. Some of this can be provided, without major advances in experimental technique, by further study of the nonequilibrium excursions of intermediate species concentrations toward the end of the ignition process under selected conditions in nonstoichiometric mixtures, and from further resolution of the exponential branching behaviour of lean mixtures, as discussed in section 2.3.2.

Some mechanististic details of the high-temperature reaction have yet to be elucidated. Most notable among these are determination of the specific rates and identification of the remaining reactants or products in reaction (0), by which the branching chain process is independently initiated, and reaction (h), by which HO_2 formed in reaction (f) is prevented from accumulating after ignition. Experiments to explore these questions effectively will have to meet stringent criteria, because the rate of bulk reaction progress is not very sensitive to the contributions made by these steps. Reaction (0) may be examined indirectly by quite exact measurements of induction times, or more directly by very sensitive experiments which detect the development of reaction before the exponential branching phase. Either approach requires quite controlled temperature and composition conditions, particularly the exclusion of impurities which might contribute to initiation. Thus far, the induction period measurements and their interpretation pursued by Gardiner and his co-workers have shed the most light on this process. Two approaches to reaction (h) can also be envisioned: (i) a very sensitive measurement of the concentration of HO_2 during recombination in lean mixtures, with interpretation by means of the quasi-steady state condition $R^{(h)} = R^{(f)}$; and (ii) a study of lean mixture recombination kinetics above about 2000 K, where one may expect the net rate of recombination to be influenced by competition between reactions (h) and $(-f)$.

The kinetics of HO_2 as a chain centre in the first-order branched chain ignition process at pressures above the extended second explosion limit condition, i.e., where $k_f[M] \geq 2k_a$, also remain to be elucidated at high temperatures. As was discussed in section 2.3.2, shock wave experiments exploring the so-called high-pressure, low-temperature regime near 1000 K have not yielded satisfactory results because the times needed have exceeded those for which well enough controlled temperatures are provided in reflected shock waves. However, it would seem that success might be realized with little departure from existing shock wave induction time or exponential acceleration methods by working with an order of magnitude higher pressures (near 10^2 atmospheres) and high enough temperatures, perhaps 1400 K or above, that rapid ignition is inevitable because reaction $(-f)$ prevents the complete sequestering of H atoms in the form of HO_2. Under such conditions, there may or may not be a large enough contribution from reactions (m) or (n) to enhance the branching rate measurably.

REFERENCES

1. J. N. Bradley, Shock Waves in Chemistry and Physics (John Wiley and Sons, New York, 1962)
   A. G. Gaydon and I. R. Hurle, The Shock Tube in High Temperature Chemical Physics (Reinhold Publishing Corporation, New York, 1963)
   E. F. Greene and J. P. Toennies, Chemical Reactions in Shock Waves (Academic Press, New York, 1964)
2. I. R. Hurle, Reports on Progress in Physics, Vol. 30, Part 1, Institute of Physics and the Physical Society, London, p. 149 (1967)
3. R. A. Strehlow, *Progress in High Temperature Physics and Chemistry*, 3, 1 (1969)
4. R. L. Belford and R. A. Strehlow, *Ann. Rev. Phys. Chem.*, **20**, 247 (1969)
5. F. S. Dainton, Chain Reactions: An Introduction (Methuen and Co. Ltd., London, 1956)
6. S. W. Benson, The Foundations of Chemical Kinetics (McGraw-Hill Book Company, Inc., New York, 1960) Chapters 3, 13, 14
7. J. H. Sullivan, *J. Chem. Phys.*, **30**, 1292 (1959)
8. W. L. Garstang and C. N. Hinshelwood, *Proc. Roy. Soc.*, **A130**, 640 (1931)
9. C. N. Hinshelwood, The Kinetics of Chemical Change (Oxford University Press, Oxford, 1940) Chapter 7
10. N. N. Semenov, Some Problems in Chemical Kinetics and Reactivity (Princeton University Press, Princeton, New Jersey) Volumes 1 (1958) and 2 (1959)
11. V. N. Kondratiev, Comprehensive Chemical Kinetics, Volume 2, The Theory of Kinetics, C. H. Bamford and C. F. H. Tipper (Editors)—(Elsevier Publishing Company, Amsterdam, London and New York, 1969) Chapter 2
12. A. L. Myerson and W. S. Watt, *J. Chem. Phys.*, **49**, 425 (1968)
13. D. R. White and K. H. Cary, *Phys. Fluids*, **6**, 749 (1963)
14. C. P. Fenimore, Chemistry in Premixed Flames (Pergamon Press, New York, 1964)
15. R. M. Fristrom and A. A. Westenberg, Flame Structure (McGraw-Hill Book Company, New York, 1965) Chapter 14
16. J. N. Bradley, Flame and Combustion Phenomena (Methuen and Company, Ltd., London, 1969) Chapters 3, 5–6

158   G. L. SCHOTT AND R. W. GETZINGER

17. B. Lewis and G. von Elbe, Combustion, Flames and Explosions of Gases (Academic Press, New York, 1961) Chapters 1–4
18. V. N. Kondratiev, Chemical Kinetics of Gas Reactions (Pergamon Press, London, 1964) Chapters 9–10
19. A. A. Kovalskii, *Phys. Z. Sow.*, **4**, 723 (1933)
20. L. V. Karmilova, A. B. Nalbandyan, and N. N. Semenov, *Zhur. Fiz. Khim.*, **32**, 1193 (1958)
21. O. Oldenberg and F. F. Rieke, *J. Chem. Phys.*, **6**, 779 (1938)
22. W. E. Kaskan, *Combustion and Flame*, **2**, 229 (1958)
23. W. E. Kaskan, *Combustion and Flame*, **2**, 286 (1958); *ibid.*, **3**, 29, 39 (1959)
24. W. M. Houghton and C. J. Jachimowski, *Appl. Opt.*, **9**, 329 (1970)
25. S. H. Bauer, G. L. Schott, and R. E. Duff, *J. Chem. Phys.*, **28**, 1089 (1958)
26. G. L. Schott and J. L. Kinsey, *J. Chem. Phys.*, **29**, 1177 (1958)
27. G. L. Schott, *J. Chem. Phys.*, **32**, 710 (1960)
28. G. L. Schott and P. F. Bird, *J. Chem. Phys.*, **41**, 2869 (1964)
29. R. W. Getzinger and G. L. Schott, *J. Chem. Phys.*, **43**, 3237 (1965)
30. R. W. Getzinger, Eleventh Symposium (International) on Combustion (The Combustion Institute, Pittsburgh, 1967) p. 117
31. R. W. Getzinger and L. S. Blair, *Phys. Fluids*, **12**, I-176 (1969)
32. R. W. Getzinger and L. S. Blair, *Combustion and Flame*, **13**, 271 (1969)
33. W. R. Gilkerson and N. Davidson, *J. Chem. Phys.*, **23**, 687 (1955)
34. G. B. Kistiakowsky and P. H. Kydd, *J. Chem. Phys.*, **25**, 824 (1956)
35. T. Just and H. Gg. Wagner, *Z. Physik. Chem. (Frankfurt)*, **13**, 241 (1957)
36. P. F. Bird, M.S. Thesis, University of New Mexico, Albuquerque, 1961
37. P. F. Bird and G. L. Schott, *J. Quant. Spectry. Radiative Transfer*, **5**, 783 (1965)
38. R. F. Stubbeman and W. C. Gardiner, Jr., *J. Phys. Chem.*, **68**, 3169 (1964); *J. Chem. Phys.*, **40**, 1771 (1964)
39. T. Asaba, W. C. Gardiner, Jr., and R. F. Stubbeman, Tenth Symposium (International) on Combustion (The Combustion Institute, Pittsburgh, 1965) p. 295
40. D. L. Ripley and W. C. Gardiner, Jr., *J. Chem. Phys.*, **44**, 2285 (1966)
41. D. L. Ripley, Dissertation, University of Texas, Austin, 1967
42. W. C. Gardiner, Jr., K. Morinaga, D. L. Ripley, and T. Takeyama, *J. Chem. Phys.*, **48**, 1665 (1968)
43. W. C. Gardiner, Jr., K. Morinaga, D. L. Ripley, and T. Takeyama, *Phys. Fluids*, **12**, I-120 (1969)
44. C. B. Wakefield, Dissertation, University of Texas, Austin, 1969
45. B. F. Walker, Dissertation, University of Texas, Austin, 1970
46. H. Miyama and T. Takeyama, *J. Chem. Phys.*, **41**, 2287 (1965)
47. T. Just and H. Gg. Wagner, *Ber. Bunsenges. Physik. Chem.*, **64**, 501 (1960)
48. J. B. Homer and I. R. Hurle, *Proc. Roy. Soc.*, **A314**, 585 (1970)
49. W. S. Watt and A. L. Myerson, *J. Chem. Phys.*, **51**, 1638 (1969)
50. E. Meyer, H. A. Olschewski, J. Troe, and H. Gg. Wagner, Twelfth Symposium (International) on Combustion (The Combustion Institute, Pittsburgh, 1969) p. 345
51. J. Troe, *Ber. Bunsenges. Physik. Chem.*, **73**, 946 (1969)
52. R. A. Alpher and D. R. White, *Phys. Fluids*, **2**, 153 (1959)
53. D. R. White, *Phys. Fluids*, **4**, 40 (1961)
54. D. R. White and G. E. Moore, Tenth Symposium (International) on Combustion (The Combustion Institute, Pittsburgh, 1965) p. 785
55. D. R. White, Eleventh Symposium (International) on Combustion (The Combustion Institute, Pittsburgh, 1967) p. 147
56. W. G. Browne, D. R. White, and G. R. Smookler, Twelfth Symposium (International) on Combustion (The Combustion Institute, Pittsburgh, 1969) p. 557
57. R. A. Strehlow and A. Cohen, *Phys. Fluids*, **5**, 97 (1962)

58. A. Cohen and J. Larson, Report Number 1386, Ballistic Research Laboratories, Aberdeen Proving Ground, Maryland, 1967
59. V. V. Voevodsky and R. I. Soloukhin, Tenth Symposium (International) on Combustion (The Combustion Institute, Pittsburgh, 1965) p. 279
    R. I. Soloukhin, Shock Waves and Detonations in Gases (Mono Book Corp., Baltimore, 1966) Chapter 4
60. L. S. Blair and R. W. Getzinger, *Combustion and Flame*, **14**, 5 (1970)
61. C. W. von Rosenberg, N. H. Pratt, and K. N. C. Bray, *J. Quant. Spectry. Radiative Transfer*, **10**, 1155 (1970)
62. D. Gutman and G. L. Schott, *J. Chem. Phys.*, **46**, 4576 (1967)
63. D. Gutman, E. A. Hardwidge, F. A. Dougherty, and R. W. Lutz, *J. Chem. Phys.*, **47**, 4400 (1967)
64. D. Gutman, R. W. Lutz, N. Jacobs, E. A. Hardwidge, and G. L. Schott, *J. Chem. Phys.*, **48**, 5689 (1968)
65. G. L. Schott, Twelfth Symposium (International) on Combustion (The Combustion Institute, Pittsburgh, 1969) p. 569
66. T. A. Brabbs, F. E. Belles, and R. S. Brokaw, Thirteenth Symposium (International) on Combustion (The Combustion Institute, Pittsburgh, 1971) p. 129
67. F. E. Belles and T. A. Brabbs, Thirteenth Symposium (International) on Combustion (The Combustion Institute, Pittsburgh, 1971) p. 165
68. A. G. Gaydon, The Spectroscopy of Flames (Chapman and Hall Ltd., London, 1957) Chapter 5
69. P. J. Padley, *Trans. Faraday Soc.*, **56**, 449 (1960)
70. M. Steinberg and W. E. Kaskan, Fifth Symposium (International) on Combustion (Reinhold Press, New York, 1955) p. 664
71. S. Fujimoto, *Bull. Chem. Soc. Japan*, **36**, 1233 (1963)
72. G. B. Skinner and G. H. Ringrose, *J. Chem. Phys.*, **42**, 2190 (1965)
73. F. E. Belles and M. R. Lauver, *J. Chem. Phys.*, **40**, 415 (1964)
74. R. W. Getzinger, L. S. Blair, and D. B. Olson, Proceedings of the Seventh International Shock Tube Symposium, I. I. Glass (Editor)—(University of Toronto Press, 1970) p. 605
75. E. M. Bulewicz, C. G. James, and T. M. Sugden, *Proc. Roy. Soc.*, **A235**, 89 (1956)
76. I. Prigogine and R. Defay, *J. Chem. Phys.*, **15**, 614 (1947)
77. W. E. Kaskan and G. L. Schott, *Combustion and Flame*, **6**, 73 (1962)
78. R. S. Brokaw, Tenth Symposium (International) on Combustion (The Combustion Institute, Pittsburgh, 1965) p. 269
79. C. B. Wakefield, D. L. Ripley, and W. C. Gardiner, Jr., *J. Chem. Phys.*, **50**, 325 (1969)
80. C. J. Jachimowski and W. M. Houghton, *Combustion and Flame*, **15**, 125 (1970)
81. D. L. Baulch, D. D. Drysdale, and A. C. Lloyd, High Temperature Reaction Rate Data (Department of Physical Chemistry, University of Leeds, England, 1968) No. 2
82. A. A. Westenberg and N. deHaas, *J. Chem. Phys.*, **50**, 2512 (1969)
83. N. R. Greiner, *J. Chem. Phys.*, **51**, 5049 (1969)
84. A. Nalbandyan, *Acta Physicochim. U.R.S.S.*, **19**, 483 (1944)
85. D. R. White and R. C. Millikan, *J. Chem. Phys.*, **39**, 2107 (1963)
86. F. E. Belles and M. R. Lauver, Tenth Symposium (International) on Combustion (The Combustion Institute, Pittsburgh, 1965) p. 285
87. D. B. Olson, M.A. Thesis, University of Texas, Austin, 1969
88. T. Just and F. Schmalz, AGARD Conference Proceedings No. 34, Part 2, Paper 19, September, 1968
89. F. Schmalz, DLR FB 71-08, Deutsche Forshungs- und Versuchsanstalt für Luft- und Raumfahrt, Institut für Luftstrahlantriebe, Porz-Wahn, 1971

90. G. Rudinger, *Phys. Fluids*, **4**, 1463 (1961)
91. W. C. Gardiner, Jr. and C. B. Wakefield, *Astronautica Acta*, **15**, 399 (1970)
92. C. W. Hamilton and G. L. Schott, Eleventh Symposium (International) on Combustion (The Combustion Institute, Pittsburgh, 1967) p. 635
93. T. A. Jacobs, R. R. Giedt, and N. Cohen, *J. Chem. Phys.*, **47**, 54 (1967)
94. H. A. Olschewski, J. Troe, and H. Gg. Wagner, Eleventh Symposium (International) on Combustion (The Combustion Institute, Pittsburgh, 1967) p. 155
95. G. Dixon-Lewis, *Proc. Roy. Soc.*, **A317**, 235 (1970)
96. K. H. Eberius, K. Hoyermann, and H. Gg. Wagner, Thirteenth Symposium (International) on Combustion (The Combustion Institute, Pittsburgh, 1971) p. 713
97. G. Dixon-Lewis and A. Williams, *Nature*, **196**, 1309 (1962)
98. C. P. Fenimore and G. W. Jones, Tenth Symposium (International) on Combustion (The Combustion Institute, Pittsburgh, 1965) p. 489
99. M. J. McEwan and L. F. Phillips, *Combustion and Flame*, **11**, 63 (1967)
100. G. Dixon-Lewis, M. M. Sutton, and A. Williams, Tenth Symposium (International) on Combustion (The Combustion Institute, Pittsburgh, 1965) p. 495
101. J. L. J. Rosenfeld and T. M. Sugden, *Combustion and Flame*, **8**, 44 (1964)
102. C. J. Halstead and D. R. Jenkins, *Combustion and Flame*, **14**, 321 (1970)
103. R. R. Baldwin, *Trans. Faraday Soc.*, **52**, 1344 (1956)
104. S. C. Kurzius and M. Boudart, *Combustion and Flame*, **12**, 477 (1968)
105. R. R. Baldwin, Ninth Symposium (International) on Combustion (Academic Press, New York and London, 1963) p. 218
106. R. R. Baldwin and C. T. Brooks, *Trans. Faraday Soc.*, **58**, 1782 (1962)

Chapter 3

Chemical Reaction and Ionization in Flames

F. M. Page

*Department of Chemistry, University of Aston in Birmingham,
Birmingham 4, England*

3.1. THE FLAME

3.1.1. What is a Flame?

Although the concept of a flame is clear to everyone, no completely satisfactory definition has been put forward. A flame results from an exothermic gaseous reaction, but not all such reactions sustain flames. A flame temperature may vary between a few hundred and over five thousand degrees, and the luminosity may be due to the spectra of species present in the hot gases, or to the continuous radiation from hot solid particles. A flame, in short, has characteristics in common with many other phenomena but a particular flame may not show all the characteristics. This is due in large part to the complex nature of a flame which does not correspond to any one system capable of definition in terms of equilibrium, but is rather an open assembly of such systems.

3.1.2. Flame Structure

Let us consider a mixture of two gases, A and B, which may react exothermically to produce C and D. If this reaction is slow at ambient temperatures but has a significant activation energy the mixture will remain relatively unchanged for an appreciable time, i.e. it is in a metastable state. If a fluctuation of sufficient magnitude occurs, possibly as the result of an external stimulus, reaction will occur, and the local temperatures will rise to a point where the thermal reaction proceeds at a significant rate with the resulting production of more heat which will raise the temperature of the surrounding gas enough to enable reaction to take place. The reaction will thus spread through the system until all has reacted, and the system may be divided, rather arbitrarily, into three regions—the *unburnt gas*, the *reaction zone* and the *burnt gas*. The temperature will be ambient in the unburnt gas, but will be high in the burnt gas. In the reaction zone there will be a transition, but it may not be proper to speak of temperature in this zone. The flame, as usually understood, extends from the perturbed region of the unburnt gas through the reaction zone, and into the burnt gas.

3.1.3. Flame Propagation

The simple structure set out in Fig. 3.1 is not static, in that the reaction zone is continuously being generated in the unburnt gas, and

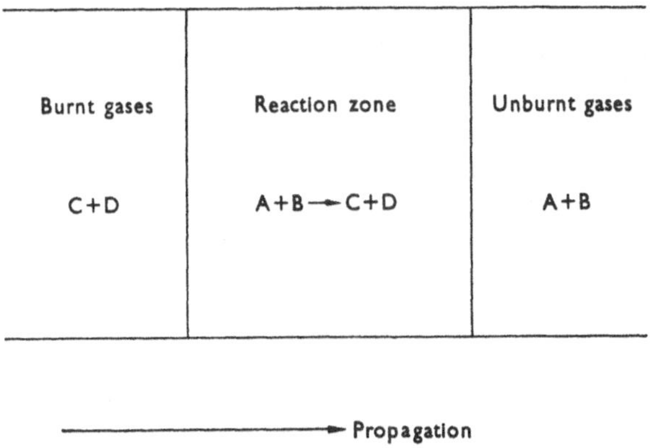

Burnt gases	Reaction zone	Unburnt gases
C+D	A+B → C+D	A+B

————————————————→ Propagation

FIG. 3.1. A schematic flame

is at the same time dying away into burnt gas. It is therefore moving through the system at a rate which corresponds to the rate of generation from the moment of inception to the moment of complete consumption of unburnt gas. If the system is closed, as for example in the cylinder of an internal combustion engine, the flame then ceases. If the system is open with an infinite source of unburnt gas, the flame will be propagated, e.g. down the pipe of a Bunsen burner. The propagation of the reaction zone can then be balanced by the flow, and the flame becomes stationary.

The rate at which the reaction zone is propagated is the rate at which the unburnt gases become perturbed and brought into the reactive state. On the hypothesis that this is achieved by heat alone, as suggested above, this rate of perturbation will be a function of the thermal conductivity of the unburnt gas. The majority of flame propagation reactions proceed, however, through free radical chain mechanisms, usually involving hydrogen atoms, and the perturbation could equally be the result of the diffusion of free radicals into the unburnt gas. The rate of diffusion of the hydrogen atom is far higher than that of any other species, and so is its thermal conductivity; it is not surprising that the velocity of flame propagation may be directly related to the concentration of hydrogen atoms.[1]

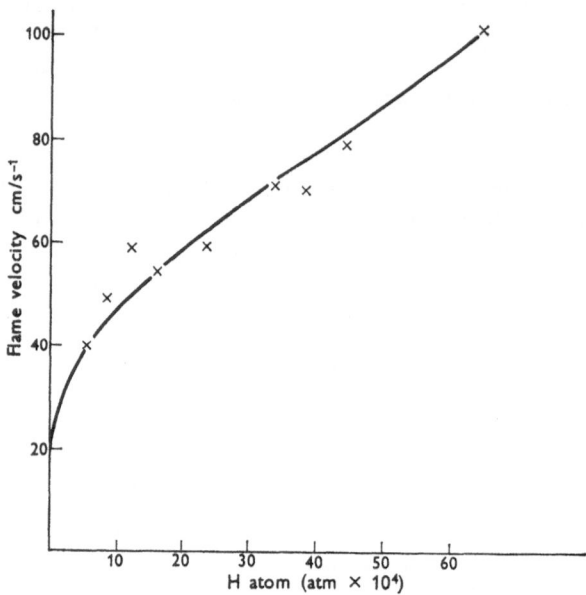

FIG. 3.2. Burning velocity and hydrogen atoms (after Tanford)

3.1.4. Burning Velocity

The rate at which the reaction zone moves into the unburnt gases is known as the *burning velocity* of the unburnt gas mixture. Since the reaction zone may have a considerable area, but, at least at atmospheric pressure, is of negligible thickness, it may be regarded simply as a surface, the boundary between the burnt and unburnt gases, and this surface is known as the *flame front*. The normal burning velocity is then defined as the burning velocity in a direction normal to the flame front. It is a function of the unburnt gas composition and may vary from a few cm s^{-1} to many m s^{-1}, depending on the fuel and oxidant;[2] the magnitude can be related loosely to the kinetics of the combustion reaction. The range of burning velocities observed for a particular fuel/oxidant combination is smaller, but is still large as Fig. 3.3 shows.[3]

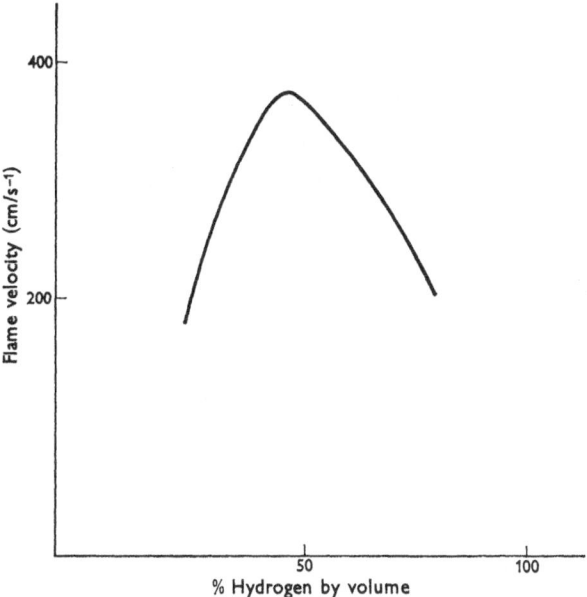

FIG. 3.3. Burning velocity and flame composition

3.1.5. Flame Stabilization

A flame is said to be stabilized when the flame front remains stationary in space. This is achieved when the linear velocity of the unburnt gases is equal and opposite to the normal burning velocity. This criterion implies a very delicate balance between gas flow and composition: however, in practice very broad limits of stability are observed for burner flames (Fig. 3.4). It is true that, at very low unburnt gas flow rates, the flame may *flash back* down the burner tube, while for very high flow, *blow off* may occur. However since the flame is stable over a wide range of intermediate flows, there cannot be any delicacy in the stabilization mechanism.

The unburnt gases flowing down the burner pipe will be normally in the laminar regime—the fully turbulent regime is only reached for very high flow rates—and there will be a parabolic distribution of longitudinal velocities across the pipe. It is thus not possible to balance the flow of unburnt gas against a single burning velocity for a real flame.

The key to this paradox is found in Fig. 3.1 which shows that the burning velocity is a linear function of the hydrogen atom concentration. The walls of the burner, or other stabilizing body, will act as a sink for hydrogen atoms, whether their function is to conduct heat or, more probably, to diffuse ahead of the reaction zone and initiate combustion chains. There will therefore be a concentration gradient

F. M. PAGE

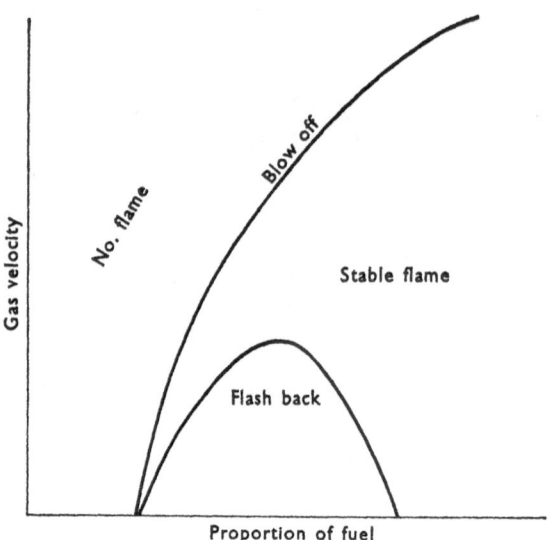

FIG. 3.4. The regions of flame stability

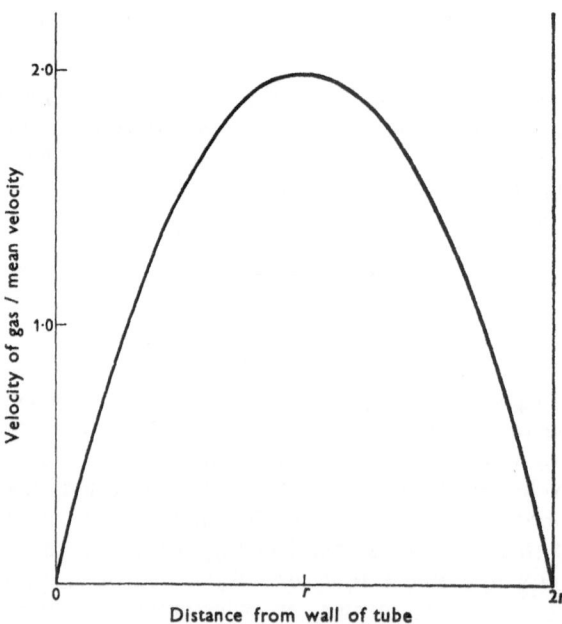

FIG. 3.5. The velocity profile in laminar flow through a tube

across the flame, from zero at the wall to a maximum on the centre line, and this will be reflected in the variation of burning velocity across the flame. An increase in the linear flow of unburnt gas will carry the flame front away from the stabilizing wall, and will thereby decrease the rate of loss of hydrogen atoms. The burning velocity will rise to balance the increased flow rate, and the flame will remain stable though with an altered shape.

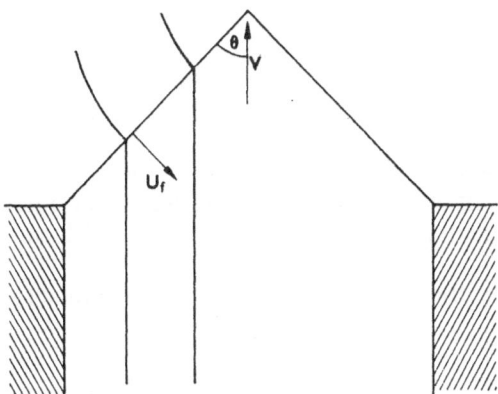

Fig. 3.6. The determination of normal burning velocity

The general shape of a premixed flame stabilized on a pipe approximates to a cone of semi-angle θ. At any point on the flame front, resolution of velocities in the longitudinal directions shows that

$$V = U_f \sin \theta \qquad (3.1)$$

Linear gas velocity = (normal flame velocity) $\times \sin \theta$

This relation may be refined by using the ratio of the areas of the pipe to that of the flame front instead of $\sin \theta$. It is found that the relation holds well within broad limits. When the flow of unburnt gas rises too far the stabilizing mechanism can no longer operate; there is also a clear lower limit to the gas flow at which $\sin \theta = 1$ or $V = U_f$. The flame front then lies straight across the pipe entrance, and any further decrease in flow results in this flame burning back down the pipe. The efficiency of the stabilizing mechanism increases for short distances so there is always a limiting pipe diameter below which flash back can never occur—a fact long utilized in the Stephenson–Davy safety lamp.

3.1.6. Time Scales in Flames

Since the stabilization of a flame is determined by the burning velocity and the unburnt gas flow, and a stabilized flame is stationary in space, it is possible to transform the temporal parameter into a spatial one.

A hydrogen/air premixed flame has a burning velocity of the order of 0·5 m s⁻¹ which the linear gas flow must exceed. A typical value would be 1·0 m s⁻¹. Such a flame would show a slight contraction if burned isothermically. However the burnt gases are actually at a temperature of the order of 2000 K, so there is a sevenfold expansion. Some part of this is taken up by a transverse expansion, but a significant portion produces an acceleration of the burnt gas, with a resultant back pressure across the flame front. The burnt gases pass up the flame with a velocity of the order of 2·0 m s⁻¹; 1 cm height corresponds to 500 μs. Several techniques are available to study flames which have a spatial resolution of 100 μs. The flame photometer in particular can resolve measurements at intervals of 10 μs.

Even with spatially cruder techniques, or slower burning flames, a resolution of 1 ms is common: this determines the time scale of the kinetics which are important in flame gases and which can be studied in flames.

As an example, the following parameters are characteristic of a flame at 1 atm.

In one millisecond

	Molecule	Hydrogen atom	Electron
Diffusion length	0·06 cm	0·15 cm	2·2 cm
Number of collisions made	$3·0 \times 10^6$	$2·2 \times 10^7$	10^9

These collision numbers indicate that a reaction with a second molecule whose partial pressure is less than 0·01 atm will be slow (i.e. have a half life greater than 1 ms) if the reaction has an activation energy greater than 67 kJ mol⁻¹ (or 100 kJ mol⁻¹ for a hydrogen atom). A three-body process will occur about 1000 times less often. It will always be slow and therefore kinetically observable.

3.1.7. Theories of Flame Propagation and Quenching

It is an unfortunate fact that the more complex and challenging a problem is, the less important becomes the detailed chemistry. This is particularly true of flame propagation in hydrocarbon/air mixtures, where the multitude of possible intermediates and mechanisms contrasts with the paucity of observable parameters. Those which can be observed easily—the burning velocity and extinction diameter—are physical rather than chemical in nature. They can be explained by a predominantly physical process, even if chemistry does play a part in that different fuels have different burning velocities under identical stoichiometries. The intrinsic unobservability of the local concentrations and conditions of the chemically distinguishable participants in the propagation of a flame has led to the development of these generalized theories to explain observable phenomena over limited ranges. These theories may be classified by their controlling process as:

(a) thermal,
(b) thermal-diffusion,
(c) radical diffusion,
(d) kinetic.

(a) In pure thermal theory, as developed by Le Chatelier,[4] Nusselt[5] and Bartholomé,[6,7] it was assumed that the unburnt gas reacted when a critical ignition temperature (T_i) was reached, and that the gas was heated by the conduction of heat from the burnt gas at a temperature T_f. The flame velocity will then be given by

$$U_f = \frac{K \cdot \kappa}{C_p} \frac{T_f - T_i}{T_i - T_0} \tag{3.2}$$

Here κ is the thermal conductivity, C_p the heat capacity at constant pressure and K is a constant of proportionality. The greater part of the early work consisted of attempts to predict the values of this constant, and was of only limited success. These attempts did however agree that the flame velocities should be a square-root rather than a linear function of the temperature differentials.

(b) The thermal theory focusses attention on the enthalpy of the flame, which is transported by conduction, and which reaches a maximum in the flame zone. In reality diffusion, which is ignored in the thermal theory, plays an important part in the transport of enthalpy, both into the reaction zone as unburnt gas, and from the reaction zone as hot combustion products. The consequence of including diffusion terms is that the enthalpy is constant across the reaction zone; application of the conservation equations for mass and energy, together with a generalized reaction rate term enabled Semenov[8] to derive an equation for the flame velocity which, for a second order combustion reaction becomes

$$U_f = \sqrt{\frac{2\lambda_f k\alpha \, C_{p(f)}^2}{C_p^3} \left[\frac{T_i}{T_f}\right]^2 \left[\frac{\lambda}{C_p\rho D_v}\right]^2 \left[\frac{RT_f}{\Delta E}\right]^3 \left[\frac{\alpha_1}{\alpha_2}\right]^2 \frac{\exp\left(-\Delta E/RT_v\right)}{(T_f - T_i)^3}} \tag{3.3}$$

The Semenov equation is widely used and reproduces experimental measurements well for a variety of rather slow flames,[9,10] but it does not predict the observed dependence of flame velocity on pressure. It is also inadequate for fast flames, such as those dominated by a branched-chain mechanism involving hydrogen atoms. These flames are much faster than the hydrocarbon flame to which the Semonov equation was originally applied, and have lower overall energies of activation. The rate at which the reaction propagates is therefore not determined by the temperature, or rapidity with which the unburnt gas is brought to reaction temperature, as in the thermal theory, nor by the rate at which materials are brought into the reaction zone, as in the diffusion theories, but rather by the rate at which the active centres—radicals or atoms— can diffuse into the unburnt gas and initiate reaction.

(c) This view was developed by Tanford and Pease[11,12] as a result of their study of the burning velocities of carbon monoxide/oxygen flames. These depend on the presence of hydrogen atoms to propagate the combustion chain. They derived the relation

$$U_f^2 = \sum \frac{k_i D_i P_i}{B_i} \cdot L\theta \qquad (3.4)$$

Here k_i is the rate constant for propagation, D_i the diffusion coefficient, P_i the partial pressure and B_i a constant characteristic of radical recombination, all referring to the ith species of radical. L is the number of gas molecules per unit volume at the temperature of the flame and θ a stoichiometric coefficient to take into account the fact that oxidant and fuel interchange the roles of reaction limiter and diluent at the stoichiometric mixture.

The theory of Tanford and Pease has received a good deal of support from studies of fast-burning flames,[13,14] but, of necessity, the diffusion coefficients and partial pressures were obtained by calculation. The experimental approach described later in section 3.5 has shown that the hydrogen atom concentration in the reaction zone may be an order of magnitude greater than the equilibrium concentration. Recently, diffusion coefficients for a number of simple species have been measured in flames, which will improve the implication of this theory.[14a]

It is scarcely surprising that some deviations from the theory have been observed. One of the most favourable flames, hydrogen/oxygen, was examined by Padley and Sugden.[15] They compared their experimentally-determined hydrogen atom concentrations in the reaction zone with the observed burning velocities. The excellent fit obtained emphasizes the dominance of this mechanism in these flames, rather than the general applicability of the theory.

(d) Attempts to produce a more general theory of flame propagation, based on the appreciation of the importance of radicals, and attempts to include both the reaction kinetics and kinetic theory of gases in the formation, have met with some success. However efforts to produce usable solutions have involved such averaging and smoothing approximations[16] as to vitiate the original intentions. A complete analysis by computer is now feasible since the rate constants for individual steps are nearly all known but these add little to our insight into the propagation of a flame.

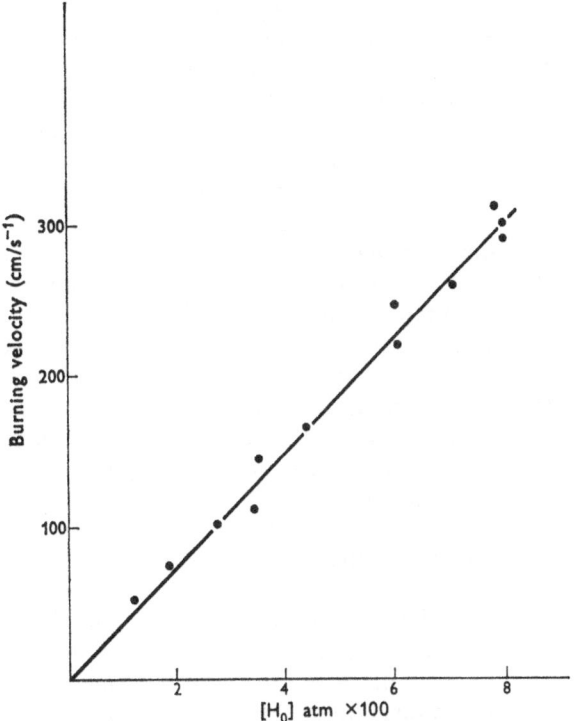

FIG. 3.7. The burning velocity in hydrogen flames, and the experimental hydrogen atom concentration in the reaction zone (after Padley)

3.1.8. Flame Quenching

Allied to the problem of the burning velocity of a flame is that of the *quenching distance*—the minimum separation of two walls which contain the flame for it to propagate rather than be quenched. If a flame is burning on a circular burner, the distance is referred to as the *quenching* or *extinction diameter*.

The action of the walls is both to cool the flame gases and to terminate the propagation chains. The theories of extinction distance have suggested that the extinction distance[17] to a first approximation is inversely proportional to the burning velocity, and also depends somewhat on burner geometry. The major importance of extinction distance to kinetics in flames is, however, indirect: its magnitude dictates the design of burner to be used in flame studies.

3.2. TYPES OF REACTION IN FLAMES

3.2.1. General Considerations

So far, no restriction has been placed on the flame, its chemistry or its physical condition. In the sections which follow, consideration will be limited to premixed flames, in which the fuel, oxidant, diluent and additive are mixed before entering the burner. The flames will be fast-burning and at atmospheric pressure. They will usually be hydrogen/oxygen, though acetylene/oxygen and carbon monoxide/oxygen will also be considered. They will be assumed to be laminar rather than turbulent. They will be used simply as a high-temperature matrix within which chemical reactions may be studied. Some interactions with the flame gases may occur, but usually the detailed chemistry of the flame will not determine the reactions of added species. These reactions are also independent of transport of material within the flame.

The temperature reached in the burnt gases depends somewhat on the unburnt gas temperature and on the design of the burner. It may be as low as 1400 K for a very dilute hydrogen flame or as high as 3300 K for a stoichiometric flame without diluent. Most work has been carried out in the range 2000–2500 K, and almost exactly in the centre of the range, at 2102 K, an energy of 40 kJ mol^{-1} represents a Boltzmann factor of 0·1. Since this temperature is seven times room temperature, energy barriers in flames are seven times less important than at room temperature. The relatively slight importance of the energy terms emphasizes the importance of the entropy of reaction: the dominant reactions are those which lead to many fragments. The chemistry of reactions in flames is therefore the chemistry of the reactions of simple species—atoms, diatomic molecules, a few triatomic species but virtually nothing more complex. Since the species involved in reactions are simple, the associated entropies of activation will be small, i.e. the frequency factors will be normal with gas kinetic cross sections. It follows that the half life of most bimolecular processes will be of the order of microseconds rather than the milliseconds characteristic of the flame flow, and that therefore the flame may be considered to be at or near equilibrium for those processes. A molecule undergoes about 10^{10} collisions s^{-1}; the half life of a process involving that molecule will be less than a millisecond unless the product of the Boltzmann factor for activation and the partial pressure of the reaction partner is less than 10^{-7}. This situation is reached only for trace constituents or for high activation energies.

3.2.2. Time and Concentration Scales

It is useful to examine the time scales in flames a little more closely. The exact scale is a function of the particular mixture being burnt, and of the burner used, but the best burners are those designed to give a cylindrical flame, laminar rather than brush-shaped, with many small cones of reaction. These Meker type burners are composed of a bundle of capillary (hypodermic) tubes. To obviate pressure drop these should have the greatest possible diameter consistent with the extinction distance of the flame mixture used.[18] The burnt gas is hot, so its volume V_B is greater than that of the unburnt gas, V_u. The tubes are spaced out so that the total cross sectional area of the tubes (A_u) is related to that of the burner top (A_B) by the relation

$$[A_B/A_u]^3 = [V_B/V_u]^2 \qquad (3.5)$$

Under these circumstances the flame will be a vertical cylinder whose diameter is that of the burner, and the velocity of the burnt gas V_B will be given by

$$U_B = U_u(V_B/V_u)^{1/3} \qquad (3.6)$$

If the flame is not cylindrical but has cross section A_B, the velocity at that point is

$$U_B = U_u \frac{V_B}{A_B} \cdot \frac{A_u}{V_u} \qquad (3.7)$$

and the calculation of burnt gas velocity requires a measurement of the flame diameter.

The volume ratio is obtained from the conservation of mass

$$V_B = V_u \cdot \frac{T_B}{T_u} \cdot \frac{P_u}{P_B} \qquad (3.8)$$

The pressure drop $(P_u - P_B)$ may contain a term for the head needed to drive the gases through the burner, but is largely aerodynamic; the gases are accelerated across the reaction zone, so that there must be an appropriate pressure differential across the zone. This differential is not large and is often neglected.

There will be a change in volume due to reaction which can be significant for flames with little diluent. For hydrogen/oxygen three volumes of unburnt gas produce two volumes of steam, and V_B must be corrected accordingly; there is a further correction if the flame is so hot that dissociation of the water to OH and H radicals becomes significant. The burnt gas velocity (V_B) is given by

$$V_B = U_u \left(\frac{T_B}{T_u} \cdot \frac{P_u}{P_B} \right)^{1/3} \qquad (3.9)$$

or approximately, if the pressure drop is small, by

$$U_B = (T_B/T_u)^{1/3} \qquad (3.10)$$

The burnt gas velocity is roughly twice that of the unburnt gas.

The usable range of unburnt gas velocities depends on the interaction of several factors. If the velocity is too high, the flame may blow off, or flow within the burner may exceed the Reynolds criterion and become turbulent. If too low, the flame will either flash back, with wide burner tubes, or be extinguished if the tubes are below the extinction diameter. Each factor depends on the burner design, and on the flame composition, but the overall result is not too restrictive, particularly for the favoured fast burning fuels. Hydrogen or acetylene supplied at 1 ft³/min of unburnt gas will sustain a flame 1 cm in diameter over a wide range of composition. The burnt gas velocity in such a flame will be

$$U_B = \frac{28\,060}{0.25\pi} \times \frac{T}{298} \times \frac{1}{60} \approx 2T \text{ cm s}^{-1}$$

Thus 4 cm represents about 1 ms near 2000 K.

The flame is affected by diffusion of the atmosphere in which it is burning, and slowly mixes with it. Excess fuel is consumed and the hot gases are eventually cooled by dilution. The use of a sheath, either of inert gas[19] or, preferably, of a second flame,[20] delays these processes, and under normal operating conditions about 15 cm height, or four milliseconds time, are available for study.

The time resolution which may be achieved depends upon the system of analysis. Resonant circuits and microwave alternation techniques average over about 5 cm of flame and have poor resolutions. Microwave cavity techniques, however, can be used down to 3 mm height; Langmuir–Williams probes are about 1 mm in diameter though the sheath diameter may be significantly larger, while spectroscopic techniques, operating at about $f/12$ can resolve to 0.5 mm. Such photometric measurements, usually of neutral species, have a resolution of 10 microseconds; the electrical techniques used to study ions and electrons are about ten times worse. Nevertheless, a resolution of 0.1 milliseconds over a range of 4 milliseconds is adequate for a great deal of work, and limitations are more likely to be in the accuracy of measurement rather than in the resolution which may be achieved.

3.2.3. Concentration in Flames

Since the majority of flames which have been studied are open to the atmosphere, it is convenient to specify concentrations as partial pressures in atm, i.e. as mol fractions. In the unburnt gas there will be fuel, oxidant and diluent; for example, a hydrogen flame at 2200 K might be fed with 0.400 atm H_2, 0.139 atm O_2 and 0.460 atm of N_2. If any substance is added to the flame, these quantities will be altered—in one experiment to study the effect of chlorine on potassium a portion of the nitrogen was saturated with chloroform vapour to produce a partial pressure of 0.008 atm, while a further portion carried a spray of 0.2 M potassium chloride solution in to the flame, producing a partial pressure of 7×10^{-5} atm of potassium and 0.018 atm of water.[21]

The resulting composition of the unburnt gas will therefore be:

Hydrogen	0.391	Water	0.018
Oxygen	0.136	Chloroform	0.008
Nitrogen	0.447	Potassium chloride	<0.0001

The constituents may conveniently be divided into two groups, those in the first column being denoted as *major* constituents, and those in the second as *minor* constituents. Clearly, in any reaction between a major and a minor constituent the concentration of the major constituent will be effectively constant, and *in any balanced pair of such reactions, the minor constituents will adjust so that the ratio of their concentrations is the same as that of the major constituents.*

In passing through the reaction zone, the composition is altered considerably, and new species make their appearance. Water now becomes a major constituent and oxygen a minor constituent, while the concentration of atoms and radicals, completely negligible at room temperature, becomes significant. The burnt gas composition will approximate to:

Nitrogen	0·447
Water	0·290
Hydrogen	0·101
Oxygen	$4·2 \times 10^{-8}$
Oxygen atoms	$3·1 \times 10^{-7}$
Hydrogen atoms	$18·5 \times 10^{-1}$
Hydroxyl radicals	$5·5 \times 10^{-4}$
Carbon monoxide	0·008
Chlorine atoms	$3·1 \times 10^{-5}$
Hydrogen chloride	0·022
Potassium atoms	4×10^{-5}
Potassium chloride	3×10^{-5}
Potassium hydroxide	2×10^{-6}
Charged species	10^{-9}

The distinction between major and minor species is still valid and has proved to be important in the analysis of kinetics in flame gases. At the present juncture it should be noted that, as well as the major species (partial pressure $\approx 0·1$ atm) and the residues of the additives which are usually fairly inactive kinetically (partial pressures $\approx 0·01$ atm), there is now a group of free radicals and changed species whose partial pressures are 10^{-4} atm or less, but which are active and observable participants in flame kinetics.

3.2.4. Balanced Reactions

It was suggested earlier that if the half life of a reaction was to be less than one millisecond, the product of the Boltzmann factor and the partial pressure of the species involved had to be greater than 10^{-7}. The consideration of time scales indicated that a reaction could only be considered to be balanced, that is, to depart negligibly from the equilibrium distribution of concentrations, if the half life was less than the time resolution of the system, i.e. less than 100 microseconds. The critical value of the product [concentration] × [Boltzmann factor] must therefore be greater than 10^{-6}. Since many of the concentrations are only of this order, it is apparent that concentrations may be determined by the kinetics rather than by the thermodynamics of the processes involved. The complexity of a flame as a reaction system prevents a detailed analysis of the concentration-time dependence of every species present and many minor constituents must be presumed to be at a steady state. However the loose application of steady state criteria to kinetic mechanisms without consideration of their validity can be misleading, and it is necessary to consider generally whether such criteria may ever be applied.

The combustion process in the reaction zone is a branched chain reaction, and the individual steps in such a reaction may be classed as:

(a) initiation,
(b) chain branching,
(c) propagation/balancing,
(d) termination.

The initiation process may be ignored for the present purpose. The propagation and branching steps must be fast since their effect on the kinetics is limited to the reaction zone. The termination step, however, is usually rather slow and will determine the transition from the reaction zone to the high temperature equilibrium state.

If we consider the H_2/O_2 reaction, the propagation and branching steps are:

$$H + O_2 \underset{k_{-1}}{\overset{k_1}{\rightleftarrows}} OH + O \qquad (3.11)$$

$$O + H_2 \underset{k_{-2}}{\overset{k_2}{\rightleftarrows}} OH + H \qquad (3.12)$$

$$OH + H_2 \underset{k_{-3}}{\overset{k_3}{\rightleftarrows}} H + H_2O \qquad (3.13)$$

If molecular oxygen is regarded as a stable molecule, and atomic oxygen as a diradical, the first step is branching, in the sense that one free valence becomes three; if the concentration of O and OH rise sufficiently, the inverse reaction can act as a terminating step in decreasing the number of free valencies. If, on the other hand, molecular oxygen is regarded as a diradical, in keeping with its triplet nature, then the three reactions have, on each side, three, two and one free valence respectively, none of them leads to a change in the number of free valencies, and the effect of the inverse reaction is to redistribute the total number of free valencies (which are wholly as O_2 in the unburnt gas) but not to alter this number in any way. This may be expressed by:

$$\frac{d(\text{radical})}{dt} = 2\frac{d(O_2)}{dt} + 2\frac{d(O_2)}{dt} + \frac{d(H)}{dt} + \frac{d(OH)}{dt}$$

$$= 2(k_{-1}[O][OH] - k_{-1}[H][O_2])$$

$$+ 2(k_1[H][O_2] - k_{-1}[O][OH] - k_2[O][H_2] + k_{-2}[H][OH])$$

$$+ k_{-1}[O][OH] - k_1[H][O_2]$$

$$+ k_2[O][H_2] - k_{-2}[H][OH]$$

$$+ k_3[OH][H_2] - k_{-3}[H][H_2O]$$

$$+ k_1[H][O_2] - k_{-1}[O][OH]$$

$$+ k_2[O][H_2] - k_{-2}[H][OH]$$

$$- k_3[OH][H_2] + k_{-3}[H][H_2O]$$

$$= 0 \qquad (3.14)$$

The number of free valencies will therefore remain constant unless some other reaction can terminate the process. Such a reaction is the three-body process

$$H + H + X \longrightarrow H_2 + X \qquad (3.15)$$

or an analogous process involving OH. This reaction is slow, and in fact has a time scale comparable with that of the flame; the concentration of free radicals in a hydrogen flame decays towards the equilibrium level over the region of observation. This is not necessarily true of all flames: for example, hydrocarbon flames show smaller deviations from equilibrium because of the faster termination steps possible.

The excesses of hydrogen atoms, or any other radical, over the equilibrium values will rapidly reach a common level, since rapid bimolecular exchanges of the propagation steps of the type

$$OH + OH_2 \rightleftharpoons H + H_2O \qquad (3.16)$$

will ensure that

$$\frac{[H]}{[OH]} = \frac{[H]_{eq}}{[OH]_{eq}} = K_3\frac{[H_2]}{[H_2O]} \qquad (3.17)$$

Here the suffix "eq" denotes the equilibrium concentration and K_3 is the equilibrium constant for the reaction. This arises because water and hydrogen are major constituents which are virtually unaffected by the transition to the equilibrium state. Hence

$$\frac{[H]}{[H_{eq}]} = \frac{[OH]}{[OH_{eq}]} = \gamma \qquad (3.18)$$

where γ is termed the *disequilibrium parameter*.

By similar arguments, [O] and $[O_2]$ are above the equilibrium value by a factor of γ^2.

Since all the free radicals derived from the flame gas matrix are buffered to their equilibrium proportions by the reservoir of the major constituents, they can themselves act as buffers to the minor constituents which will distribute themselves over the possible states accordingly. The interactions may however be with the radical excess or against it. For example, a metallic hydroxide may be made by two alternative processes[22]

$$M + OH\,(+\,X) \rightleftharpoons MOH\,(+\,X)$$

$$M + H_2O \rightleftharpoons MOH + H$$

The ratio of [MOH]/[M] has been denoted as ϕ_M in the literature, and in the first reaction is directly proportional to [OH] while in the second reaction it is inversely proportional to [H]. If [H] and [OH] had their equilibrium values, there would be no distinction, but since they are not, we write:

$$\phi_M^1 = K_1[OH] = K_1\gamma[OH]_{eq} \tag{3.19}$$

$$\phi_M^2 = K_2[H_2O]/[H] = K_2[H_2O]/\gamma[H]_{eq} = K_1[OH]_{eq}/\gamma \tag{3.20}$$

since
$$K_1 = \frac{K_2[H_2O]}{[H]_{eq}[OH]_{eq}} \tag{3.21}$$

ϕ_M^1 is therefore greater than ϕ_M^{eq} by the factor γ, while ϕ_M^2 is less than it by a similar factor.

If both processes were to operate at the same time, they would cause a catalysed recombination of hydrogen atoms. This does not occur, and in every case studied balancing occurs by one or other of the processes alone.

Sugden has considered the kinetics which will lead to the dominance of one process.[22] He shows that either reaction is likely to bring the metal and hydroxide into balance within the time resolution normally achieved: the observed values of ϕ do not lag behind the local radical concentration. This balance is achieved by the first reaction if the M—OH bond dissociation energy is low, but by the second if it is high. If both reaction rate constants are normal, the first reaction would be expected to have a small or zero energy of activation, but as a termolecular process would have an efficiency of 10^{-4}. As it involves a collision with a minor constituent OH, it will be further slowed by a factor of $10{:}^3$ the reaction will go once in every 10^7 collisions made by the metal atom. The second reaction will have an efficiency of unity, but involves a specific major constituent, normally present at about 10^{-1} atom, and will have an activation energy at least as great as the heat of reaction. It will therefore be slower than the first reaction unless the Boltzmann factor is greater than 10^{-6}. At a temperature of 1800 K this corresponds to 210 kJ mol^{-1}. The heat of reaction is the difference between the M—OH and H—OH bond energies. Taking the latter to be 515 kJ mol^{-1} at 1800 K, it is apparent that the three-body process will be the faster unless the M—OH bond energy is greater than about 300 kJ mol^{-1}.

3.2.5. Determination of Reaction Order for Radicals

It is frequently necessary to decide on a mechanism under circumstances which are more complex than that described above, and the use of the disequilibrium parameter may be extended beyond the rather simple case of the hydroxides by an application of ideas advanced in the field of electrochemistry by Pourbaix.[23]

If a process is rate limiting, the rate may be expressed by:

$$\text{rate} = kC_1^\alpha C_2^\beta \cdots \text{etc.}$$

where the process is

$$\alpha C_1 + \beta C_2 \longrightarrow \text{products}$$

At the same time, the concentrations C_1, C_2 are related to other species by balanced reactions

$$a_1 + b_1 \rightleftharpoons c_1 + d_1 \cdots$$

If these reactions are truly balanced to within the validity of the steady state approximation, they may be rewritten as linear algebraic equation

$$a_1 + b_1 - c_1 - d_1 \cdots = 0 \qquad (3.22)$$

The whole set of these equations may then be combined with that for the rate limiting process; the latter equation is simplified by reducing the number of independent variables which are related to each other by the balanced equations.

In flame kinetics, this technique may be applied to deduce the possible mechanism from the observed dependence on hydrogen atoms. It has been observed that the intensity of radiation from a flame doped with lithium decreases with height: the concentration of lithium is a function of the hydrogen atom concentration, more lithium hydroxide being formed as [H] falls.

If the reaction is written

$$\text{Li} + \text{OH} + M_1 \rightleftharpoons \text{LiOH} + x\text{H} + M_2 \qquad (3.23)$$

where M_1 and M_2 are unspecified singlet molecular species present in major amounts then

$$\text{Li} + \text{OH} - \text{LiOH} - x\text{H} + \sum M = 0 \qquad (3.24)$$

This may be combined with the flame gas equation balancing H and OH, in the form

$$\text{H} - \text{OH} + \sum M = 0 \qquad (3.25)$$

to give

$$Li - LiOH + (1 - x)H + \sum M = 0 \qquad (3.26)$$

The total number of free valencies must be even, so that

$$x = 1 \text{ (Li runs parallel to H)}$$

or

$$2 - x \text{ is zero or even}$$

The simplest solution to these relations is $x \equiv 2$ (the solution $x = 4$ implies a multibody collision) so that the equation for the reaction is

$$Li + M \longrightarrow LiOH + H$$

or

$$Li + OH + M \longrightarrow LiOH + 2H$$

Again, the latter formulation implies an improbably complex reaction, and the former, with $M \equiv H_2O$ is to be preferred.

This approach was used by Page and Sugden[24] in discussing the ionization of alkali metals in flames. They observed that the electron concentration in a hydrogen flame doped with lithium was independent of hydrogen atom concentration. However when the additive was sodium the square of electron concentration was inversely proportional to the hydrogen atom concentration.

The possible species involved in ionization are A, A^+, OH^-, e^- and AOH. The Pourbaix equations relating these are

$$A - A^+ - e^- + xH - \sum M = 0 \qquad (3.27)$$

$$OH^- - e^- + yH - \sum M = 0 \qquad (3.28)$$

$$A - AOH + zH - \sum M = 0 \qquad (3.29)$$

These correspond to hypothetical equilibrium constants which may be combined with the equations for charge neutrality and for conservation of A to yield an equation connecting the electron concentration with the total amount of A. This equation simplifies under limiting conditions to give the following expressions for the order of the electron concentration with respect to hydrogen atom concentration. In all cases, the degree of ionization is small ($A \ll A$)

(1) OH^-, AOH negligible
$$\text{Order} = x/2$$

(2) OH^- negligible, AOH dominant
$$\text{Order} = (x - z)/2$$

(3) OH^- dominant, AOH negligible
$$\text{Order} = (x + y)/2$$

(4) OH^-, AOH dominant
$$\text{Order} = (x - z)/2$$

Experimentally lithium, which would fit case 2 or 4, showed an order zero, while sodium, whose hydroxide is unstable, fits case 1 or 3 with an order of $-\frac{1}{2}$. Hence $z = -1$, and LiOH is made by the reaction

$$Li + H_2O \rightleftharpoons LiOH + H$$

As before, the total number of free valencies must be even for each Pourbaix equation, so that x must be even and y odd. The possible solutions were

$$\begin{aligned} x &= -2 & y &= 1 \\ x &= 0 & y &= -1 \\ x &= 2 & y &= -3 \end{aligned} \qquad (3.30)$$

corresponding to the ionization processes

$$A + H_2 \longrightarrow A^+ + e^- + H + H$$
$$A + X \longrightarrow A^+ + e^- + X \qquad (3.31)$$
$$A + H + H \longrightarrow A^+ + e^- + H_2$$

Page and Sugden considered only cases 3 and 4 and concluded that the only acceptable solution was $x = 0$, $y = -1$ on the grounds that the other reaction schemes involved a quarternary collision at some stage. Furthermore, they argued that neither of the simple processes corresponding to the solution were acceptable on grounds of energy transfer, and favoured linear combinations of these simple processes:

$$A + OH \rightleftharpoons A^+ + OH^-$$
$$OH^- + H_2 \rightleftharpoons H_2O + H + e$$

Some doubt has been cast on this conclusion by later mass spectrometric studies; these have consistently failed to demonstrate the presence of OH^- in flames in quantities sufficient to justify the approximation used.[25] If the alternative (case 1) approximation is used, then the solution is that $x = -1$, which contradicts the basic requirement that x be even. Despite the internal consistency of the results obtained, the explanation of the order with respect to hydrogen atoms must be sought elsewhere than as a consequence of a balanced reaction mechanism. The work must therefore be viewed as an ingenious application of kinetics rather than as a positive contribution to the understanding of ionization. It paved the way to the later work on competitive ionization processes in the presence of phosphorus or the halogens which is described in section 3.7.6.

3.2.6. Kinetic Description of a Flame

It is now possible to expand the schematic diagram of the flame given earlier by the inclusion of a time scale, and by a more detailed subdivision of the various zones. The unburnt gas is heated and the

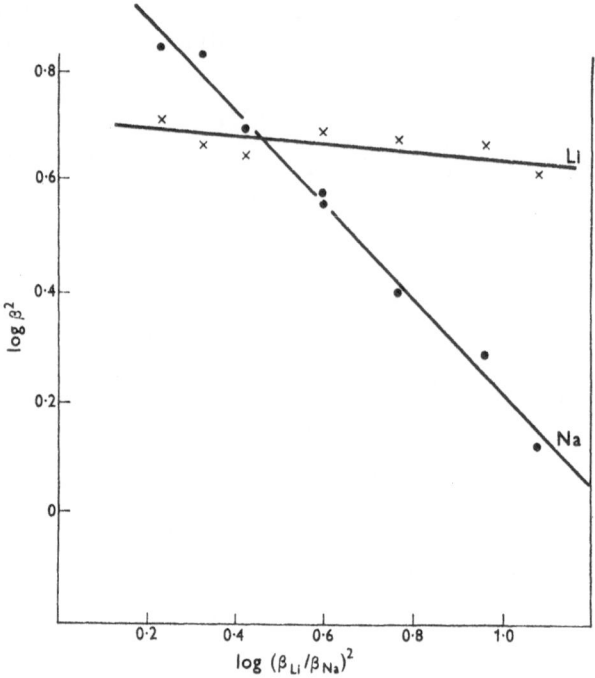

FIG. 3.8. The order of reaction with respect to hydrogen atoms for
alkali metal ionisation

radical concentration raised by the up-stream diffusion of hydrogen
atoms and thermal conduction from the reaction zone, and enters the
reaction zone at about 700 K. The rapid chemical reaction which
ensues results in a high concentration of radicals, close to that of the
oxidant in the unburnt gas. These radicals are often excited. No
temperatures can be assigned to this region, as the energy is not properly
partitioned. Physical equilibration rapidly ensues and the burnt gases
leave this region with a temperature of about 1500 K but with both the
number and distribution of radicals far from equilibrium. In the next
stage, chemical equilibration of the distribution takes place through the
rapid balancing reactions, but the total radical concentration remains
high; the temperature increases only slightly. A further region may be
described in which the radical concentration falls rapidly towards the
final equilibrium concentration, with a consequent release of heat; this
raises the temperature by about 500 K so that it approaches the final
flame temperature. There then follows an extended region where the
radical concentration falls only slightly, and the temperature, through a
combination of radiative and convective heat loss and diffusion cooling
balanced by recombination heating and combustion of excess fuel,
remains practically constant. Finally, the effects of indrawn air become

dominant, and the flame degenerates to hot gas mixing rapidly with the surrounding air and cooling as it does so.

3.3. TECHNIQUES OF MEASUREMENT

3.3.1. Introduction

The study of the kinetics of a reaction demands a technique for determining the concentrations of relevant species. Since the techniques used in the field of flame kinetics differ markedly from those in other fields, it is of interest to survey briefly the more important methods. A flame, being an open system, is not easy to sample and quench in the classical manner, though Fristrom et al.[26,27] have devised aerodynamic probes and have discussed the problems inherent in their use. Much greater use has been made of the high temperatures in a flame in determining the composition by measuring the concentrations of excited or ionized species. The earliest studies[28,29] measured only the d.c. conductivity of the flame gas either between two electrodes or between one electrode and the burner,[30] a technique commonly used in gas chromatography. These were little more than qualitative surveys, although a study of the Hall effect in flames[31] did show that the negative charge carriers were electrons. Later work showed increasing sophistication and a steady improvement in the specificity of analysis and in the time resolution which could be achieved. The technique may be considered in three groups—spectroscopic and electrical methods and mass spectroscopy.

3.3.2. Spectroscopic Methods

There is a voluminous literature concerned with the study of flame spectra, but the application of spectroscopy to the study of flame kinetics followed the introduction of flame photometry as a general analytical tool. The chief interest before this was in the spectra of the flames, which could serve to demonstrate the presence of intermediates in the combustion process.[32] These were in general detected by the emission spectra of excited species and therefore were not necessarily indicative of the concentrations of ground state species. The difficulties of constructing burners which were sufficiently large and uniform to allow the study of absorption spectra prohibited a measurement of the species in their ground states, until the development of the multiple pass technique.[33]

The emission spectra of substances added to flames, e.g. sodium atoms added in the form of a spray of sodium chloride solution, was a much more profitable study.[34–37] In the first place, the amount of additive could readily be varied by altering the concentration of the solution sprayed, and in the second place, it was possible, by comparing the intensity of first and second resonance lines, to demonstrate that the

7

additive was truly thermally excited. Alternatively, one resonance line could be examined at different temperatures.[38]

The intensity of the resonance radiation of wavelength λ emitted from a body of gas at a temperature T, and thickness l is

$$I = \frac{8\pi ahc}{\lambda^3} \exp\left(-hc/2kT\right) \int [1 - \exp\left(-k_\nu l\right)] \qquad (3.32)$$

where k_ν is the absorption coefficient. For low concentration of emitting species ($k_\nu l \ll 1$) the integral becomes $(\pi e^2/mc)N_0 fl$ where N_0 is the number of emitting species in the ground state, and f is the transition probability for the line. Then

$$I = \left(\frac{8\pi^2 ahe^2}{m}\right)\left(\frac{1}{\lambda^3}\right)(Nfl)(\exp\left[-hc/\lambda kT\right]) \qquad (3.33)$$

The general form of k_ν is complicated, and involves a detailed consideration of the effects of the various broadening factors. At high values of (Nfl) the dominant term is the Lorenz collisional broadening which James and Sugden[39] showed to lead to a square root dependence of I upon (Nfl). The detailed theoretical analysis of the shape of the log I–log (Nfl) graphs, including the change from a slope of 1 to a slope of $\frac{1}{2}$ was made by van der Held,[40] and such graphs are referred to as van der Held plots, a term often extended, incorrectly, to any plot showing a transition from linear to square root dependence on concentration as the concentration is raised, whatever the reason for such transition. An excellent review of the process of excitation and de-excitation in flames has recently been published.[40a]

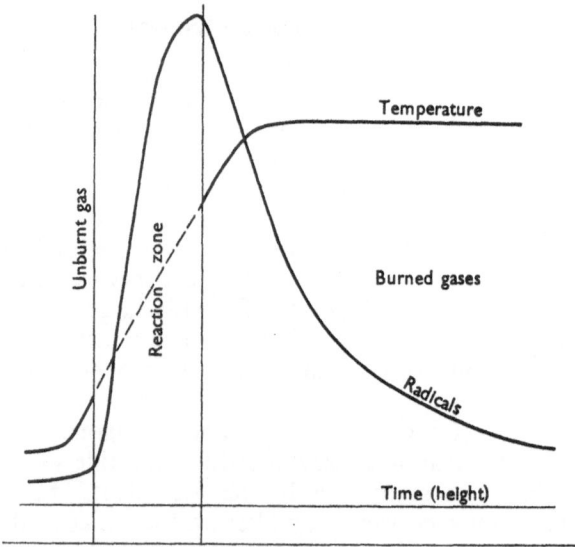

FIG. 3.9. Profiles in a flame

The intensity of radiation is usually recorded as a photocell current, while the concentration of ground state atoms in the flame is related to the concentration of salt in the sprayed solution by an atomizer delivery factor. Both this factor and the photocell sensitivity can be determined experimentally, though they are difficult to measure. The van der Held plot is absolute, and if the corresponding logarithmic plot of photocell current against concentration of sprayed solution is superimposed upon the theoretical van der Held plot, both the calibrations may be made without the necessity of measuring the quantities individually.

3.3.3. Spectrometer Calibration

The measurements of the intensity of radiation are usually made with a photomultiplier and a monochromator. Some early measurements used optical filters to isolate an atomic line, but this limited the studies to

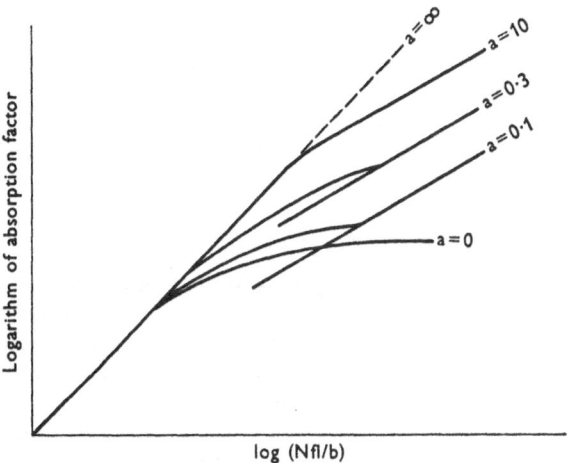

FIG. 3.10. The van der Held plot

effectively monochromatic flames, and the extension to many line, or band spectra demanded the much better resolution provided by a prism or grafing monochromator. It is customary to improve the signal to noise ratio by chopping the light beam from the flame, in synchrony with a separate light bulb/photocell circuit, and feeding the two chopped signals through a phase-sensitive detector to a multi-range amplifier and meter.

The spectral response of the photocell and monochromator varies with wavelength, and must be properly calibrated if measurements at different wavelengths are to be compared. This is conveniently done by determining the emission characteristics of a tungsten lamp over the spectral range at various temperatures and estimating the intensity of light passing through the monochromator from the known emissivity of tungsten, and the monochromator slit function.

3.3.4. Thermal Excitation

If the measurements of the intensity of radiation from excited species are to be related to the concentration of ground state species, it is necessary to demonstrate that the species are in thermal equilibrium so that equation (3.33) may be applied. This may be done by three methods. In the first place, the determination of flame temperature by the line reversal technique demands that there be thermal equilibration, and the concordance of a temperature so determined with that determined by other methods, or from lines known to be equilibrated is evidence for equilibration. Secondly the comparison of measured intensities of the first and higher resonance lines should be in accord with (3.33), a comparison which may be extended to a pair of different atoms if compound formation is not important. Thirdly, and most commonly, (3.33) may be transformed into

$$\log_e I = \log_e N + \log_e \frac{8\pi^2 \alpha h e^2 fl}{m\lambda^3} - \frac{hc}{\lambda kT} \tag{3.34}$$

so that a plot of $\log I$ against $1/T$ should yield a straight line whose slope, the excitation energy, corresponds to the wavelength of measurement.[38]

3.3.5. Further Spectroscopic Techniques

Spectroscopic studies are not limited to the resonance lines of metals and to electronic excitation. The emission of electronically-excited polyatomic species such as $CuOH$[41] have been widely studied, as have the infra-red emission from such species and the rotational and vibrational structure in the electronic region. Atomic absorption spectra have been used in the conventional manner of flame photometry to determine the ground state population in the linear region of the van der Held plot; Sugden and James[38] used a two flame technique in the square root region which predated the more conventional methods of atomic absorption spectroscopy. In this region the intensity is proportional to the square root of (Nfl) so that if the intensities from two flames are measured separately, and then together, with the radiation from one flame passing through the other, the change in intensity will be

$$\Delta I = I_1 + I_2 - I_{12} \tag{3.35}$$

Since $I = k(Nfl)^{1/2}$, this may be transferred to

$$\frac{\Delta I}{N_2^{1/2}} = a - bN^{1/2} \tag{3.36}$$

Some spectra observed in flames are continuous rather than line spectra. Such spectra are amenable to the same analysis as the line

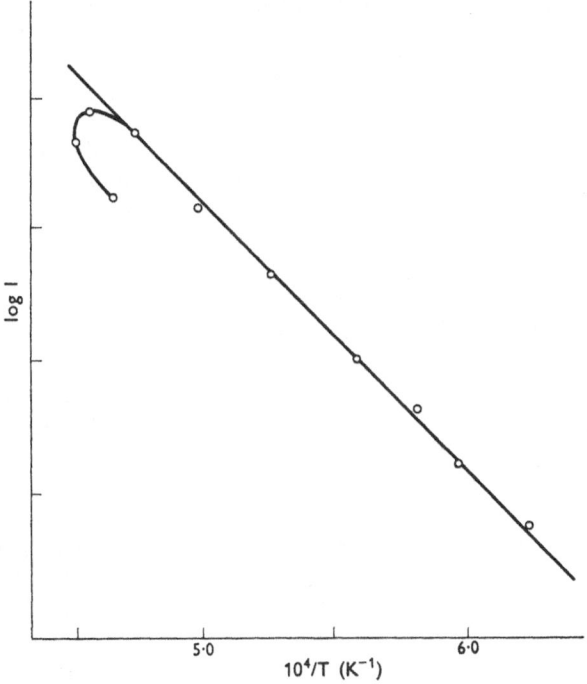

FIG. 3.11. The intensity of NaD radiation from a series of flames

spectra, but do not give absolute results. Nevertheless they have been used in the study of flame processes, notably the alkali metal — OH, the H—OH and the NO—O recombination continua.

3.3.6. Microwave Techniques

The study of ionization in flames demands an analytical technique for the determination of the concentration of ions. There are many difficulties, both experimental and in interpretation, associated with the measurement of the d.c. conductivity of a flame and some specificity in analysis is desirable.

It is customary to determine the electron concentration by a.c. conductivity, the positive ion concentration by diffusion limited d.c. conductivity, and to apply mass spectrometry to a more specific analysis both of positive and negative ions.

The a.c. conductivity may be determined from the absorption coefficient for microwaves, or from the alternation in the resonance response of a tuned circuit either in the megacycle region (tank circuit) or in the microwave region (resonant cavity). The attenuation (β) (in dB cm^{-1}) of microwave radiation of frequency ω by a flame containing

n ions of mass m in unit volume is given by

$$\beta = \frac{17 \cdot 36\pi n e^2}{mc}\left(\frac{\omega_c}{\omega^2 + \omega_c^2}\right) \tag{3.37}$$

The factor $17 \cdot 36$ arises from the use of decibel units and the magnetic permeability has been taken to be 1. Strictly β is a summation over all ionic species. However the mass of the ion in the denominator means that only electrons contribute significantly unless ω is much less than ω_c, the collision frequency.

The application of (3.37) to derive the electron concentration for the measured attenuation requires a knowledge of the electron collision frequency. Belcher and Sugden,[42] who compared the attenuation at various values of ω, found this to be $8 \cdot 8 \times 10^{10}$ s^{-1}. Later work[45,54] showed that this was too low by a factor of 3, and that the electron concentrations measured by them, and subsequent workers,[60] were therefore too small.[43,44] These low electron concentrations were ascribed to the presence of OH$^-$ ions,[46-48] and the use of the later and higher values of the collision theory obviates the necessity of invoking OH$^-$ as a significant ionic constituent. Page, Soundy and Williams[49] have examined the temperature dependence of the collision frequency, and shown that the collision frequency is related to the temperature and composition by

$$\omega_c = [H_2O]T^{-3/2} \tag{3.38}$$

Since they worked in fuel-rich flames at fairly low temperatures, the water concentration will, to a first order, be proportional to the temperature, so that the collision frequency was proportional to $T^{-1/2}$. These workers argued that the collision cross-section for water was far larger than for any other species at thermal electron energies, but did not carry out a sufficiently wide range of experiments to prove this. Bulewicz and Padley,[50] using the allied cyclotron resonance technique examined a wide variety of fuels and deduced composition dependent electron molecule cross-sections which increased as the amount of hydrogen in the fuel, and hence water in the burnt gases, increased.

The microwave resonant cavity, first applied to flames by Sugden and Thrush,[51] offered few advantages over the direct measurement of attenuation. Since the theoretical equation for power loss was not soluble in real conditions, the apparatus had to be calibrated with known electron concentrations. The development of the loop coupled E_{010} cavity by Horsfield[52] and Pennycook[53] and its application to the kinetics of flame ionization by Sugden, Padley and Jensen,[55,56] changed the situation. This cavity is not quantized in the vertical dimension, and can therefore be used to study narrow sections of flame: resolutions as high as 2 mm have been achieved. It was still necessary to calibrate the apparatus, usually with caesium assuming complete ionization, but the kinetics of alkali metal ionization could be followed successfully.

It is not essential to work in the microwave region, and Sugden and Smith,[47] Knewstubb,[57] Williams[58] and Borgers[59] have used tank circuits (capacitor-inductances) operating at megacycle frequencies in various forms. Only Knewstubb was able to approach the kinetics of ionization. Since in the megacycle region $\omega \ll \omega_e$ the dominance of the electron is lost, and the technique can in principle give information about heavy ions as well. The very poor height resolution, and the field distortion in the more highly conducting flames have however restricted the application of this method despite its sensitivity.

3.3.7. Electrostatic Probe Techniques

The measurement of the positive ion concentration in flames is most effectively carried out by diffusion-limited negatively biased electrostatic probes, frequently known as Langmuir probes.[61,62]

The general current voltage relation characteristic of a small spherical probe with large counterelectrodes immersed in a neutral plasma is shown in Fig. 3.12.

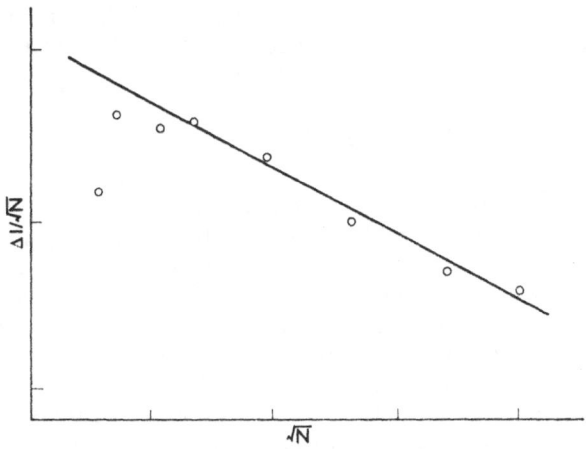

FIG 3.12. The James-Sugden two-flame technique.

At both limits, there is a relation between the concentration of ions of appropriate sign and the limiting current, and the measurement of probe characteristics has been applied widely in plasma studies. In deriving the basic relation it is assumed that the probe is cold, or rather that it is non-emitting. This is valid in discharge plasmas where the gases are at ambient temperatures, even though the ions may have "temperatures" of tens of thousands of degrees. It rapidly ceases to be valid in flames where heat transfer to the probe rapidly raises it to temperatures at which thermionic emission may occur. Most of the early studies[63] of ionization in flames using small probes up to the work of Kinbara and Nakamura[64,65] took no special precautions to avoid this source of error. Calcote *et al.*[66-70] introduced the watercooled probe and Soundy and Williams[71,72] designed the low duty cycle rotating probe which has been the basis for most of the more recent kinetic studies in flames. In this technique, the probe, properly insulated, is swept through the flame at the end of a rotating arm, and the current is recorded on an oscilloscope, with the time base triggered just before the probe enters the flame. As a variant, the stabbing probe which is driven into and out of the flame by a crank has also been used.

The probe current has been calculated by Su and Lam[73,74] to follow the relation

$$N_0 = \frac{1}{4\pi kT} \times \frac{1}{(4r_p\mu^2)^{2/3}} \left(\frac{I^2}{\phi_p}\right)^{2/3} \qquad (3.39)$$

While the current may be calculated either for positive or for negative ions, the chief use has been for positive ions, whose temperatures are accepted to be close to the flame gases.[163,167] Some doubt exists about the temperatures of the electrons which von Engels argues should be significantly higher,[75,76] as found by Calcote. Williams[77] has disputed this and points out that leakage currents can cause the same numerical observation as high temperatures. Because of this uncertainty, it has been found better to determine electron concentrations by the resonant cavity methods.

3.3.8. Mass Spectrometry

The mass spectrometer has proved to be a most powerful tool for the study of flames, both in its conventional form, with a source unit to ionize flame gas species, and in sourceless form, to study the natural flame ionization. The field has recently been surveyed by Matthews,[78] from the early work by Eltenton[79] on radicals in methane flames, and Foner and Hudson,[80] to the later studies on flame composition of Fristrom[81] and Klein and Herron.[82] The majority of this work was confined to low-pressure flames, where radical recombination during sampling was reduced. Similar problems faced the students of natural flame ionization, and Calcote and Reuter[83] used a radiofrequency mass spectrometer which is small enough to use at a relatively high pressure to study acetylene/oxygen flames at a few millimetres pressure. Deckers and van Tiggelen[84-87] also used a low-pressure flame to study a variety of fuels and oxidants, but the chief information on the mechanism and kinetics of ion reactions in flames at atmospheric pressure comes from the work of Sugden, Knewstubb and others using a triply pumped mass spectrometer.[88-90] No ionizing source was needed in these experiments, the flame burning against a short conical probe, usually gold plated to reduce the catalytic activity, at the apex of which was a small sampling hole. Ions passed through this hole into the first pumping chamber of the mass spectrometer, and considerable trouble was taken in the design of the pumping chambers and ion lens system to reduce as far as possible the chance of reactions within the instrument. The effects of the stagnant boundary at the sample hole could be diagnosed by using sample holes of different diameter.

Too many positive ions have been observed for there to be any point in cataloguing them. Broadly speaking, the number of ions and their abundance increases with increasing fuel and with increasing unsaturation in the fuel, although oxygen-containing ions do not behave so simply. Negative ions are much rarer than positive ions except either in the outermost, cold, parts of the flame where attachment may proceed readily,[91-94] or if a highly electronegative element such as a halogen is present. One paper by Green and Sugden[95] describes a study of the negative ions in the reactive zone of a hydrogen flame containing traces of acetylene. A large number of carbon-containing species were found which had a short life, and which depended to a very high order on the amount of acetylene added.

3.4. The determination of flame composition

3.4.1. Introduction

Although a flame is ideally an inert matrix for the study of kinetics at high temperatures and some reactions are indeed known for which the flame is in this sense ideal, in the great majority of the reactions studied in flames is some interaction between the reactants and the flame, and it is therefore necessary to specify the local flame composition before a meaningful study of kinetics can be achieved. Even before these local compositions can be specified, the local temperature and time scale must be established, and, in the present section, the methods whereby these local parameters are established will be briefly considered.

3.4.2. Temperature and Time

Although many methods have been used to establish the temperature of a flame, the most widely accepted technique is the line reversal method. The flame is heavily doped with an element, usually sodium, which has a conveniently placed and easily excited resonance line. This resonance line is then viewed by spectroscope against the continuous background of a lamp, whose operating temperature is adjusted until the resonance lines disappear. If the flame is hotter than the lamp, the lines appear in emission (bright) while if the lamp is the hotter, the lines appear in absorption (dark). When the lines are not visible, the lamp and flame are believed to be at the same temperature. The temperature of the lamp filament is then measured independently with a pyrometer. The technique is discussed in detail, and the necessary precautions are outlined by Wolfhard and Gaydon[96] or Lawton and Weinberg.[97]

The time scale in the flame is obtained from the linear flow rate, which in turn is calculated from the area of cross section of the flame, and its volume flow rate. The volume flow rate is calculated from the preburned flow rate, the composition of the burnt gases and their temperature. The composition need be known only roughly; it is rare for the minor constituents to contribute to a change in the number of moles of gas passing in one second.

$$\frac{dt}{dx} = \text{time in seconds corresponding to 1 metre}$$

$$= \frac{\text{cubic metres of unburnt gas supplied per second}}{\text{area of flame in square metres}}$$

$$\times \text{(moles of burnt gas derived from one mol of unburnt gas)}$$

$$\times \frac{\text{temperature of unburnt gas}}{\text{temperature of burnt gas}}$$

The mole ratio is calculated for the actual composition of the unburnt and burnt gases.

3.4.3. Flame Photometry of the Alkali Metals

The measurement of the concentration of the minor species in the burnt gases—H atoms, OH radicals, etc—can be achieved in several ways. The majority of the studies of flame kinetics have been carried out in flames for which the minor constituent composition has been determined by a photometric technique, and the first and most important of these is the sodium/lithium comparison technique of Bulewicz, James and Sugden.[98]

The intensity of light emitted from a flame containing free atoms is given by

$$I = KN_0 lf\lambda^{-3} \exp(-he/2kT) \tag{3.40}$$

where λ is the wavelength of the light (assumed to be resonance radiation), N_0 the number of atoms in the ground state and f is the transition probability. This relation is discussed in detail by Mitchell and Zemansky[99] or Mavrodineau[100] and the corresponding relation for the integrated intensity of a molecular band by Penner.[101]

Equation (3.40) was used to describe the emission from flames doped with alkali metals by James and Sugden[39] and they showed that, while it accounted well for the emission from sodium, the emission from a flame containing lithium was lower than expected because of the formation of significant amounts of undissociated lithium hydroxide.

If equimolar solutions of two elements are sprayed into a flame, and the first does not form compounds (e.g. Na), while the second does (e.g. Li), then the partial pressures of the elements in all forms in the flame will be proportional to the strength of the solutions and will be the same, but the total partial pressure of the second element (X_{20}) is made up of free metal (X_2) and the hydroxide (X_2OH).

$$[X_{20}] = [X_2] + [X_2OH]$$
$$= [X_2](1 + \phi_2)$$

where
$$\phi_2 = [X_2OH]/[X_2]$$

Since $[X_1] = [X_{10}] = [X_{20}]$ we have
$$[X_1] = [X_2](1 + \phi)$$

or
$$[X_1]/[X_2] = 1 + \phi$$

The ratio of the intensities, I_1/I_2 may be written down from (3.40):
$$I_1/I_2 = (N_1/N_2)(f_1/f_2)(\lambda_2/\lambda_1)^3 \exp(-hc/kT\{1/\lambda_1 - 1/\lambda_2\})$$

Since the wavelengths and transition probabilities of the two resonance lines are known, the ratio of the line intensities leads to the ratio of the partial pressures of free X_1 and X_2, and hence to the value of ϕ.

If $K\phi$ is the equilibrium constant for the reaction
$$X_2OH \rightleftharpoons X_2 + OH$$
$$K\phi = [OH]/\phi$$

The value of [OH] may be calculated from the flame composition and a plot of log $[OH]/\phi$ against $1/T$ should yield a straight line whose slope is the dissociation energy of X_2OH. In fact, a curve results, which tends to a limiting slope (Fig. 3.13). The curvature is a function of height in the flame, becoming less as the flame is ascended.

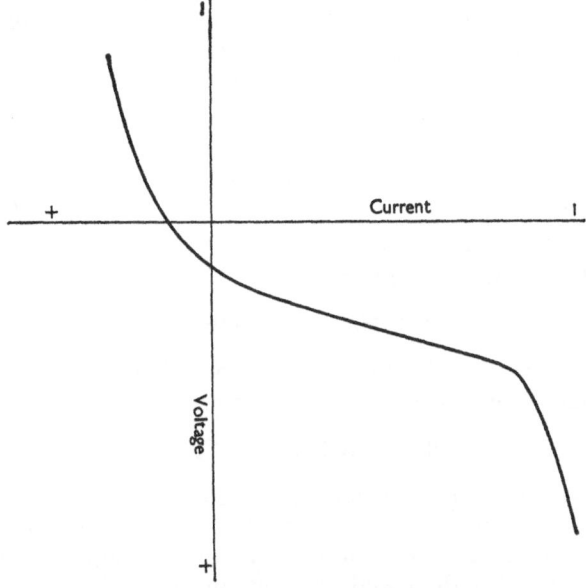

FIG. 3.13. The probe characteristic

These observations were accounted for by assuming that lithium hydroxide was formed in a fast, balanced, reaction between lithium atoms and water molecules

$$Li + H_2O \rightleftharpoons LiOH + H$$

$$K_1 = \frac{[LiOH][H]}{[Li][H_2O]} \tag{3.41}$$

$$\phi = K_1[H_2O]/[H]$$

or

$$[H] = K_1[H_2O]/\phi \tag{3.42}$$

If the hydrogen atoms are formed in the reaction zone, in excess of the equilibrium amount, and recombine in slow three-body processes, the excess will decrease at greater heights in the flame, and at higher temperatures, yielding precisely the shape of curves found by James and Sugden.

The observation can be used, alternatively, to calculate the partial pressure of hydrogen atoms from the observed value of ϕ by equation (3.42). In order to do this accurately, precise values for the equilibrium constant K_1 (equation (3.42)) are needed, and these in turn require an exact knowledge of the dissociation energy of LiOH. Since this is effectively determined by a measurement of ϕ, the process is in some measure cyclic, but many workers have made the measurements, among the most reliable being Cotton and Jenkins,[102] who used an absorption technique. The measurements have recently been reviewed by Zeegers and Alkemade[103] who recommend the weighted average value for D_0° of 438 ± 4 kJ mol^{-1} ($104 \cdot 5 \pm 1$ kcal mol^{-1}). Combining this with Cotton and Jenkins value for K_1, at 2370 K leads to

$$K_1 = 1 \cdot 6 \times 10^9 \exp(-53\,220/T)\,\text{atm}$$

3.4.4. The Sodium/Chlorine Method

The Na/Li method for determining the concentration of hydrogen atoms described in the previous paragraph depended on the establishment of the rapid, balanced reaction

$$Li + H_2O \rightleftharpoons LiOH + H$$

When a halogen is added to a flame containing an alkali metal, some alkali halide is formed by the analogous reaction

$$Na + HCl \rightleftharpoons NaCl + H$$

and a study of this reaction can also be used to establish the hydrogen atom concentration. Although any alkali metal and halogen may in principle be used, the formation of hydroxide complicates the system for all metals other than sodium. The approximation of assuming all halogen to be present as the halogen acid is not valid for bromine and iodine, and the extreme stability of HF renders fluorine of little practical use. Consequently only the pair Na/Cl has been used.

As the amount of halogen added to the flame increases, the amount of free alkali metal is reduced, and therefore the intensity of resonance radiation emitted falls as shown in Fig. 3.14.

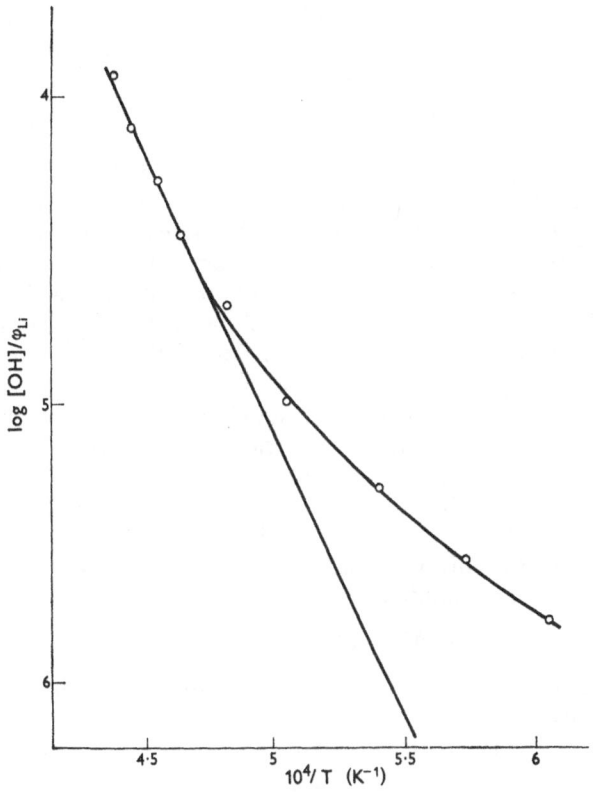

FIG. 3.14. The photometric determination of the dissociation constant for lithium hydroxide (after James)

If

$$K_y = \frac{[NaY][H]}{[Na][HY]}$$

Then

$$I_0/I = \text{total Na/free Na}$$

$$= ([NaY] + [Na])/[Na]$$

$$= \frac{[NaY]}{[Na]} + 1$$

Hence

$$[H] = K_v[HY][Na]/[NaY]$$
$$= K_v[HY] \cdot I/(I_0 - I)$$

This method was shown by Bulewicz, James and Sugden to lead to H atom concentrations in close agreement with those found by the Na/Li method. It has been studied in detail by Phillips and McEwan.[104,105] It requires careful attention to the design of the burner system but can be of use in special circumstances. The halogen is readily lost if the system contains rubber,[106] an effect first reported, though not explained, by Collins and Polkinhorne.[107]

3.4.5. Radical Concentrations in Flames

It was suggested by Arthur[108] that premixed flames at atmospheric pressure might have concentrations of radicals, particularly of hydrogen atoms, which were in excess of those calculated on the basis of full thermodynamic equilibrium at the temperature of the flame. The work of Bulewicz, James and Sugden[98] showed this suggestion to be correct, and demonstrated the pattern of such excess concentration. At high temperatures, the measured concentrations were indistinguishable from the calculated values, but at progressively lower temperatures the two deviated more and more, temperature having little effect on the actual H atom concentration below 2000 K. This is particularly true of the more fuel-rich systems (right hand points in a group).

The H atom concentration in a flame decreased markedly with height, through the recombination process

$$H + H + M \longrightarrow H_2 + M$$

The use of these measurements to study the kinetics of this reaction is discussed below.

It is believed that the pairs of reactions

$$H_2O + H \rightleftharpoons H_2 + OH \qquad\qquad K_\alpha$$
$$OH + H \rightleftharpoons H_2 + O \qquad\qquad K_\beta$$

are close to equilibrium, the known data suggesting that they should be in balance after 1 μs. It is therefore possible, at all times of significance, to write

$$[OH] = K_\alpha[H_2O][H]/[H_2]$$
$$[O] = K_\beta[OH][H]/[H_2]$$
$$= K_\alpha K_\beta[H_2O][H]^2/[H_2]$$

In all but the most extreme flames, H_2O and H_2 will be major constituents of the flame gases, whose ratio will not depend to any significant extent on whether or not the radicals are equilibrated, though it will depend on the fuel/oxidant ratio. Determination of the local H atom concentration in a premixed flame therefore also determines the OH and O concentration; conversely, the dependence of any effect on the H atom concentration is indistinguishable from a dependence on the OH concentration, unless a comparison can be made between flames of different composition.

3.4.6. Relative Methods of Measuring Radical Concentrations

Either the Na/Li or the Na/Cl method may be used to determine the hydrogen atom concentration in a flame, as well as the variants discussed below. However these techniques are not very easy to apply: particularly in the study of kinetics in flames, there is a call for a method which will rapidly trace the relative H atom concentration in one flame. Such a method is provided by the copper flame band method developed by Bulewicz and Sugden. They showed[109] that the intensity of the band system around 428 nm, due to CuH $(A^1\Sigma^+-X^1\Sigma^+)$ and the broad system in the green, ascribed by them to CuOH,[41] depends upon height in the flame in the same way as the concentration of H atoms; however the intensity of the Cu resonance doublet at 325 nm was not affected by the composition of the flame. Sugden[110] had demonstrated mathematically that if a compound AB can be balanced against free A by the two processes

$$A + HB \rightleftarrows AB + H \qquad (3.43)$$

$$A + B + M \rightleftarrows AB + M \qquad (3.44)$$

it would follow the first process (3.43) if the A—B bond energy were greater than 300 kJ mol^{-1}, but the second if the bond energy were less. LiOH and NaCl follow the former process. CuOH and CuH, with low bond energies, follow the latter so that the equilibrium constants may be written down

$$\frac{[CuH]}{[Cu][H]} = K_H \qquad \frac{[CuOH]}{[Cu][OH]} = K_{OH}$$

If the temperature in a given flame is sensibly uniform, and no dilution occurs through diffusion, or entrainment of air, the concentration of copper atoms, which represent the major form of copper in the flame, will be constant, so that

$$[CuH] = K_H[Cu][H] \qquad [CuOH] = K_{OH}[Cu][OH]$$

or

$$I_{CuH} = \alpha[H] \qquad I_{CuOH} = \beta[OH]$$

The relative intensities of the CuH or CuOH bands are therefore a direct measure of the relative concentration of H or OH respectively at the points of measurement within any one flame.

3.4.7. Other Methods of Determining Radical Concentrations

It is not necessary to use spectrophotometry to determine the concentration of H atoms: other techniques, such as the determination of the conductivity can be used. Page and Sugden[24] used the microwave attenuation of flames containing sodium and lithium to obtain the values of ϕ_{Li} at different heights, and hence the hydrogen atom concentrations. These methods, however, offer only the advantage of convenience. They are in general less precise and have a poorer height resolution than the photometric method.

Certain continua have been observed in flames which are due to the direct recombination of radicals and atoms. The best known of these are the alkali metal–hydroxyl recombination continuum, which extends throughout the visible range. This continuum is obscured by the very intense resonance lines of Na and Li. It accounts for the majority of the visible radiation from a K laden flame, where the first and second resonance doublets are at the limits of visibility. James and Sugden[110] showed that the integrated intensity of this continuum was proportional to both the [OH], derived from their measurements of [H] atoms, and to the concentration of free alkali metal, and that the intensity could therefore be used as a relative measure of [OH] concentration. The natural continuum arising from the reaction $H + OH \rightarrow H_2O + h\nu$ has also been used, as has the $H + Cl$ continuum.[113] The direct measurement of the intensity of the OH (306 nm) band can be used to determine the OH profile in a flame;[114] however Hollander has pointed out that there is considerable overexcitation and the method is unreliable near the reaction zone.

Oxygen atoms show two recombination continua which have been used to monitor their concentration. When nitric oxide is added to a flame, the greenish continuum observed was identified with that ascribed by Gaydon[115] to the reaction

$$NO + O \longrightarrow NO_2 + h\nu$$

Unfortunately, the addition of NO to a flame induces a catalytic recombination of hydrogen atoms by a complex series of reactions,[116] and the method therefore has limited value.

Another spectrum which has been suggested to be of value in monitoring O atoms is the "iodine flame bands." Phillips and Sugden[117] showed that the (4, 0) band of the IO system is excited in flames, and its intensity is proportional to the local O atom concentration.

3.4.8. Chemiluminescent Peaks

Padley and Sugden[118] noted that, in sodium-doped flames burning at low temperatures where there is expected to be a great excess of hydrogen atoms near the reaction zone, there was an intense emission from the Na D lines at the reaction zone, but that the main body of the flame only developed the characteristic golden yellow some distance downstream. They ascribed this to the chemiluminescent excitation of sodium by the recombination process

$$Na + H + H \longrightarrow Na^* + H_2$$

This process rapidly becomes less important as the excess H atoms fall. Since the recombination of H atoms provides much of the heat in the flame, the temperature of the bulk gases rises as the H atoms decay, and so the normal thermal excitation also rises, as shown in Fig. 3.15.

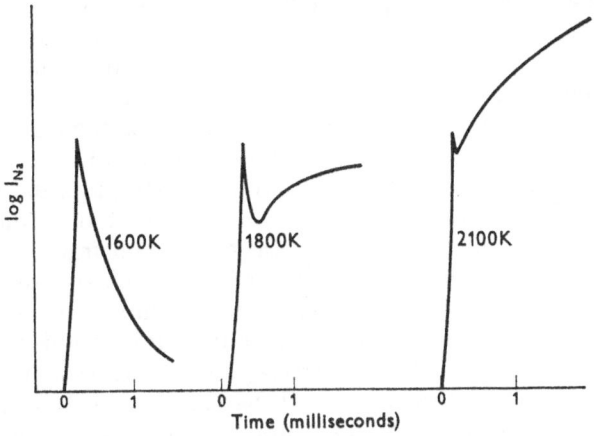

FIG. 3.15. Chemiluminescent peaks (after Padley)

Padley and Sugden[119] extended these observations to other elements. They showed that while sodium catalysed the H + H recombination, lead favoured the H + OH recombination and silver was intermediate.

The study of these peaks is of particular value in determining the H atom concentration just after the reaction zones in cool flames, where there is insufficient intensity of radiation to use the more conventional methods.[120–122]

3.4.9. Bulk Composition and Radical Concentration

The concentrations of the radicals [H], [OH] and [O] are interrelated by the equations of section 3.4.5. In any one flame they rise or fall together since the concentrations of the bulk constituents [H_2] and [H_2O] do not alter. If two flames are compared with the same flame temperatures but different unburnt composition, the ratios of the radical concentrations will differ since the bulk constituents will be present in differing amounts. Page[123] used this fact to determine equilibrium constants by varying the flame composition under iso-thermal conditions. James and Sugden[110] made this a very powerful diagnostic technique by setting up a series of flames. In any one family the $N_2:O_2$ ratio was held constant while the $H_2:O_2$ ratio varied; different families had different $N_2:O_2$ ratios. The sets were not iso-thermal, but showed a very characteristic pattern when the [H], [OH], the [H]^2 or [O] concentrations were plotted as a function of temperature.

3.5. Kinetics in Flames

3.5.1. Introduction

The reactions which occur in flames, and of which the kinetics have been studied in flames, may be grouped in various ways. We shall group together those reactions which are concerned with additives or impurities, and separate those reactions which are a function of the flame gases alone. We shall be only minimally concerned with the flame gas reactions as such for two reasons. Firstly, the interaction of additives with the flame gases, and the ionization of the additive are kinetically more interesting and illustrative. Here the flame plays the part of an inert matrix enabling experiments to be carried out which would otherwise be inaccessible. Secondly, the detailed chemistry of the flame gas reactions is exceedingly complicated, and each fuel oxidant combination should be considered separately. Although much effort has been directed to studying flame gas reactions, this complexity, and the aerodynamic features of flame propagation, have rendered progress slow. Much more informative work has been carried out under pre-flame oxidation conditions. As Ashmore[124] has pointed out, "In the reaction zone many of the elementary reactions are of the same kind as those that occur at low temperatures, though their relative importance changes with temperature".

3.5.2. Balanced Reactions

In an earlier section, some consideration was given to the criteria for balancing a reversible pair of reactions. One of the simplifying effects of working with high-temperature flames (1800 K) is that the multiplex reaction mechanisms proposed to account for oxidation kinetics around 400 K are reduced to a few, balanced, bimolecular exchanges, which form the propagation steps of the chain reaction at low temperature.

In the case of the hydrogen/oxygen flame, these are:

$$H + O_2 \rightleftharpoons OH + O$$
$$O + H_2 \rightleftharpoons OH + H$$
$$OH + H_2 \rightleftharpoons H_2O + H$$

All these reactions are balanced, in the sense that the instantaneous concentration product differs from the true equilibrium constant by less than about 10%. Since the departure from equilibrium is slight, kinetic studies are not possible: the ratios of radical concentrations are close to their equilibrium values, so it is not possible to separate the effects due to one radical from those due to another.

Similar sets of reactions may readily be written down for carbon monoxide/oxygen, acetylene/oxygen, hydrogen/chlorine, methane/oxygen, or ammonia/oxygen flames, and are subject to the same comments. It is worthy of note that while these balanced reactions dictate the distribution of unpaired spins in a given system, changes in the total number of spins result only from recombination or termination. While the distribution and recombination rates are characteristic of any one system, addition of a few percent of one fuel to the major fuel in a flame can result in a rapid catalytic recombination of the radicals. This explanation has been invoked in detail to explain the effect of acetylene or nitric oxide on excess hydrogen atoms in hydrogen flames.

Flame studies *per se* have added little to our knowledge of the rates of those balanced reactions. Their kinetics have been well studied at lower temperatures, and they will therefore not be considered further.

3.5.3. Termination Reactions

There are three types of termination reactions which have been studied in flames:

$$R_1 + R_2 \longrightarrow R_1R_2 + h\nu \quad \text{radiation}$$
$$R_1 + R_2 + X \longrightarrow R_1R_2 + X^* \quad \text{chemiluminescence}$$
$$R_1 + R_2 + M \longrightarrow R_1R_2 + M \quad \text{third body}$$

If two simple radicals R_1, R_2 collide, they will have sufficient energy to redissociate the newly-formed R_1R_2 molecule unless part of the energy of the newly-formed bond can be dissipated. Radiation stabilization can occur by the emission of light; the probability is slight, but increases rapidly with the complexity of R_1 and R_2. A third body may be present at the collision and carry away the excess energy, either as kinetic energy (general third body M) or as specific electronic excitation of an added material (chemiluminescence). The kinetics of many of these processes have been examined either directly or by following the rate of radical decay, usually by photometry. The most widely studied process is undoubtedly recombination in hydrogen flames with either a flame gas molecule or an added metal atom as third body.

3.5.4. Recombination of Hydrogen Atoms

The determination of hydrogen atom concentrations by Bulewicz, James and Sugden[98] showed that in cold, fuel-rich flames, the hydrogen atoms were in excess of the equilibrium concentration. The second order recombination coefficient defined as $-1/[H]^2 . d[H]/dt$ was 2×10^{-14} cm^3 molecule^{-1} s^{-1}. Page and Sugden,[24] using an ionization rather than photometric techniques, found a rather smaller value, 0.7×10^{-15} cm^3 molecule^{-1} s^{-1} at 2400 K. The concentrations in the flame were probably too close to equilibrium for much weight to be attached to this figure. The first detailed study of recombination was due to Bulewicz and Sugden[125] who used the CuH method to determine [H] at various heights in a series of flames in the temperature range

1600–2400 K. The second order rate constant k' determined experimentally from the observed kinetic law

$$1/[H] - 1/[H]_0 = k't \qquad (3.45)$$

was analysed in terms of the two processes

$$H + H + X \longrightarrow H_2 + X \qquad k_2 \qquad (3.46)$$

$$H + OH + X \longrightarrow H_2O + X \qquad k_3 \qquad (3.47)$$

where X is a third body, most probably a bulk constituent (H_2, H_2O or N_2).

The concentrations of [H] and [OH] are closely linked by the balanced reaction

$$H + H_2O \rightleftharpoons H_2 + OH \qquad K_1 = \frac{[H]_2[OH]}{[H_2O][H]}$$

It is therefore not possible to separate [H] and [OH] and the basis equation becomes

$$\frac{-d[radicals]}{dt} = \frac{-d([H] + [OH])}{dt} = \frac{-d[H]}{dt}(1 + K_1[H_2O]/[H_2])$$

$$= (2k_2[H]^2 + k_3[H][OH]) + k_3[H][OH]$$

$$= 2(k_2 + k_3K_1[H_2O]/[H_2])[H]^2 \qquad (3.48)$$

or

$$k' = 2(k_2 + k_3K_1[H_2O]/[H_2])/(1 + K_1[H_2O]/[H_2]) \qquad (3.49)$$

which differs from the original formulation only in the factor of 2.

The experimental value of k' was therefore multiplied by $(1 + K_1[H_2O]/[H_2])$ calculated from the data of Gaydon and Wolfhard[96] and plotted against $K_1[H_2O]/[H_2]$ to evaluate k_2 and k_3 (Fig. 3.16).

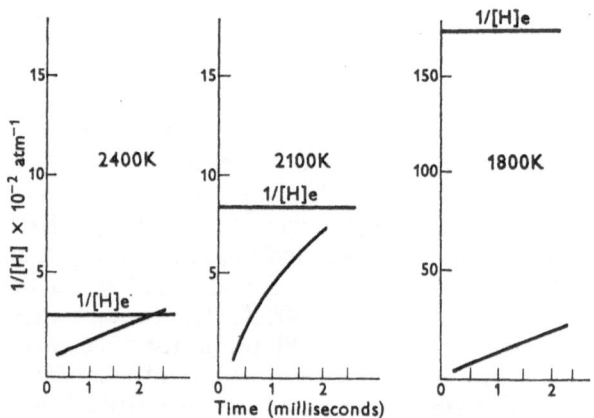

FIG. 3.16. Recombination in carbon monoxide flames (after Zeegers)

The rate constants k_2, k_3 are, of course, composite and could in principle be analysed into the various parts by considering an isothermal set of flames

$$k_2 = k_{2x}[X][Z] + k_{2y}[Y] + k_{3z}[Z] + \cdots \qquad (3.50)$$

where X, Y and Z represent H_2, H_2O, N_2 etc.

It was not possible to do this partly because of an indistinguishability between, for example, the processes

$$H + H + H_2O \longrightarrow H_2 + H_2O$$

$$H + OH + H_2 \longrightarrow H_2O + H_2$$

but more cogently because the recombination would be expected to be dominated by $[H_2O]$, which is practically constant in an isothermal set of flames. Bulewicz and Sugden therefore considered only water as the third body. The ratio of k_3 to k_2 will then be independent of composition, and was found to be 24 ± 5 over the whole temperature range while $[H_2]$ varied by sixfold and $[N_2]$ by fourfold. Their assumptions are thus justified since H_2 cannot contribute more than 8% or N_2 14% to the rate. The fact that k_3 is greater than k_2 was attributed to a combination of the greater collision cross section for OH, and to the greater possibility of "sticky collisions".

The actual value of $k_3^{H_2O}$, the rate constant for the reaction

$$H + OH + H_2O \longrightarrow H_2O + H_2O$$

was only slightly dependent on temperature varying from $1 \cdot 5 \times 10^{-30}$ cm^6 molecule^{-2} s^{-1} at 1650 K to $1 \cdot 8 \times 10^{-30}$ cm^6 molecule^{-2} s^{-1} at 2290 K.

The same reaction was studied by Padley and Sugden,[118] who used two approaches. In the first, the chemiluminescent excitation of the sodium D lines, which occurs through analogous three-body processes, was assumed to be dependent on radical concentrations which depended in turn on the molecular three-body recombination processes (the excitation process only contributes insignificantly to the total recombination rate). Only one numerical result is given: $k_3^{H_2O} = 8 \times 10^{-32}$ cm^6 molecule^{-2} s^{-1} at 1880 K.

These workers also considered the rate of heat release by radical recombination. Assuming that this was the sole cause of the increase in temperature after the reaction zone, they evaluated the radical recombination rate. Again it was assumed that H_2O was the only effective third body and that the recombination reaction was

$$H + OH + H_2O \longrightarrow H_2O + H_2O$$

The agreement with Bulewicz and Sugden was close except at the lowest temperature.

Dixon-Lewis, Sutton and Williams[126,127] estimated hydrogen atoms by deuterium exchange as well as by chemiluminescence. They found an overall recombination coefficient of $5 \cdot 1 \times 10^{-12}$ cm^3 molecule^{-1} s^{-1} at 1072 K. A further analysis assuming that nitrogen and water were equally efficient third bodies (on the basis of a single experiment with added water), gave (cm^6 molecule^{-2} s^{-1}):

$$k_2^{N_2} = k_2^{H_2O} = 4 \cdot 7 \times 10^{-33} \quad k_2^{H_2} = 3 \cdot 0 \times 10^{-32}$$
$$k_3^{N_2} = k_3^{H_2O} = 3 \cdot 0 \times 10^{-31}$$

Rosenfeld and Sugden[128] also analysed the chemiluminescence from lead at 406 nm in a series of very cold flames at atmospheric pressure. They succeeded in obtaining individual values for the coefficients averaged over the temperature range 1350–1450 K (cm^6 molecule^{-2} s^{-1}):

$$k_2^{N_2} = 2 \times 10^{-33} \qquad k_3^{N_2} = 1 \cdot 3 \times 10^{-31}$$
$$k_2^{H_2} = 4 \cdot 8 \times 10^{-32} \qquad k_3^{H_2O} = 2 \cdot 8 \times 10^{-31}$$
$$k_2^{H_2O} + 2K_1 k_3^{H_2} = 2 \cdot 0 \times 10^{-32}$$
$$K_1 = 0 \cdot 021$$

If H$_2$O and N$_2$ were equally efficient in either reaction so that

$$k_3^{H_2O}/k_2^{H_2O} = k_3^{N_2}/k_2^{N_2}$$

the figures give

$$k_2^{H_2O} = 4 \cdot 3 \times 10^{-33} \qquad k_3^{H_2} = 3 \cdot 7 \times 10^{-31}$$

However Sugden is quoted[129] as saying $k_3^{H_2} < 1 \times 10^{-31}$. Zeegers[130] and Zeegers and Alkemade[131] have also examined this problem, but found that at temperatures around 2300 K the small range of composition which is accessible to experiment vitiated a detailed analysis. Arguing from their experimental data that the dominant factor was $k_3^{H_2}$, they found

$$k_3^{H_2} \leq 9 \times 10^{-31} \text{ cm}^6 \text{ molecule}^{-2} \text{ s}^{-1}$$
$$2k_3^{N_2} + k_3^{H_2O} = 1 \cdot 5 \times 10^{-31} \text{ cm}^6 \text{ molecule}^{-2} \text{ s}^{-1}$$

McAndrew and Wheeler[132] in a propane/air flame at 2080 K examined the effects of CO$_2$ and SO$_2$ as third bodies as well, finding

$$k_3^{H_2O} = 4 \cdot 1 \times 10^{-31} \qquad k_3^{N_2} = 6 \times 10^{-32}$$
$$k_3^{CO_2} = 3 \cdot 2 \times 10^{-31} \qquad k_3^{SO_2} = 1 \cdot 1 \times 10^{-24}$$

In hydrocarbon, and more particularly in dry carbon monoxide flames, the termination reactions are:

$$CO + O + X \longrightarrow CO_2 + X \qquad k_4 \qquad (3.51)$$
$$O + O + X \longrightarrow O_2 + X \qquad k_5 \qquad (3.52)$$

These were also examined by Zeegers who concluded that in fuel-rich, dry carbon monoxide flames the oxygen atom concentration was too low for k_5 to contribute. He determined the recombination rate from the temperature profile during the approach to equilibrium, and after allowing for radiation losses obtained second order recombination coefficient of 2·7 and 2·2 × 10^{-15} cm³ molecule^{-1} s^{-1}. Zeegers attributed the difference to a different concentration of molecular oxygen in the two flames and re-examined the work of Hollander[133] to obtain other second order coefficients.

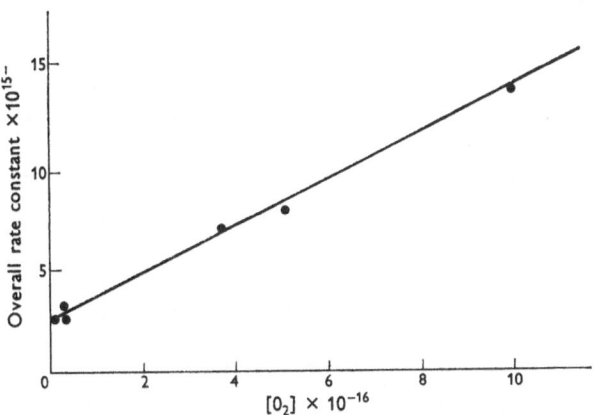

FIG 3.17. Hydrogen atom recombination in flames.

The linear dependence of the second order coefficient on $[O_2]$ shown in Fig. 3.17 confirms the dominant importance of oxygen. The close agreement of a flame in which argon was substituted for nitrogen as diluent indicates that nitrogen is not important. If carbon monoxide had any effect the line would be curved, so that it was deduced that the intercept corresponded to carbon dioxide acting as a third body, and the rate constants were:

$$k_4^{O_2} = 1\cdot2 \times 10^{-31}\,\text{cm}^6\,\text{molecule}^{-2}\,\text{s}^{-1}$$

$$k_4^{CO_2} = 2\cdot4 \times 10^{-33}\,\text{cm}^6\,\text{molecule}^{-2}\,\text{s}^{-2}$$

These figures are adequate to explain the observed second order recombination rates in acetylene flames without invoking CO_2 as a third body in hydrogen atoms recombination as suggested by McAndrew and Wheeler; Zeegers suggests that the close agreement of their second-order rate constant in a propane flame with his values for acetylene flames is an indication that his recombination mechanism holds equally for hydrocarbon flames in general.

The recombination rates in hydrogen flames were also examined by Halstead and Jenkins[134,135] in a careful and extensive series of studies. They used not only nitrogen but also helium, argon, steam and carbon dioxide as diluent. Only a limited range of temperature was covered and the results were presented as averages over the temperature range, in view of the known slow variation of rate with temperature. The results presented in Table 3.1 include those for carbon dioxide, but the authors point out that the effects of carbon monoxide and carbon dioxide are inseparable and in view of the work of Zeegers on the CO–O recombination, attempt no more than to present a global CO_2 recombination coefficient.

TABLE 3.1

Recombination coefficients

$(cm^6 \ molecule^{-2} \ s^{-1} \times 10^{32})$ at 1900 K

X	k_2^x	k_3^x
H_2	(0·20 assumed)	*
H_2O	*	7·5
N_2	0·53	0·9
He	0·5	2·2
Ar	0·5	0·9
CO_2/CO	1·5	1·0

$* \ k_2^{H_2O} + K_1 k_3^{H_2} = 1·0 \times 10^{-32} \ cm^6 \ molecule^{-2} \ s^{-1}$

$K_1 = 0·0845$

3.5.5. Excitation Recombination

The intensity of radiation from a cool hydrogen flame doped with sodium shows a marked peak at the bottom of the flame (Fig. 3.15) which Padley and Sugden[118] attributed to chemiluminescent excitation by the process

$$H + H + Na \longrightarrow H_2 + Na^*$$

the bond energy of the nascent hydrogen molecule providing the excitation energy for the sodium atom. In view of the trace amounts of sodium added, this process does not contribute significantly to the total recombination but rather is buffered by the major processes.

In their analysis, Padley and Sugden examined the peak intensity of chemiluminescence as a function of composition. They showed that, for sodium, both of the processes

$$Na + H + H \longrightarrow Na^* + H_2$$

$$Na + OH + H \longrightarrow Na^* + H_2O$$

were operative.

The number of quanta of sodium D radiation emitted per cubic centimetre per second, n, will be given by

$$n(1 + \phi\tau) = q_2 z_2 [Na][H]^2 + q_3 z_3 [Na][H][OH] \qquad (3.53)$$

where $(1 + \phi\tau)$ is a Stern–Volmer quenching correction in which τ is the mean radiative lifetime of sodium atoms in the excited state, and ϕ is the number of quenching collisions made per second. The q's are the probabilities of formation of an excited sodium atom at a collision and the z's the collision frequencies for the ternary processes.

If the meter reading of the spectrophotometer is I, and γ a constant of proportionality

$$n = \gamma I \qquad (3.54)$$

and we may write

$$Ir(1 + \phi\tau) = C_1[H]^2 + C_2[H][OH] \qquad (3.55)$$

$$= \left\{ C_1 + C_2 K_1 \frac{[H_2O]}{[H_2]} \right\} [H]^2$$

$$= C_f[H]^2 \qquad (3.56)$$

This may be rearranged to give

$$\frac{I_0}{[H]_o^2} = \frac{C_f}{r(1 + \phi\tau)} \qquad (3.57)$$

where subscript o indicates the value at the peak of chemiluminescence and $[H]_o$ was determined by the CuH technique.

As seen in Fig. 3.18 this function is composition dependent, tending to a limiting line in very hydrogen-rich flames. Padley and Sugden argue that ϕ would be expected to depend on the major constituents and would show a smooth composition dependence, so that the dependence of Fig. 3.18 was to be ascribed entirely to the dependence of C_f; and that

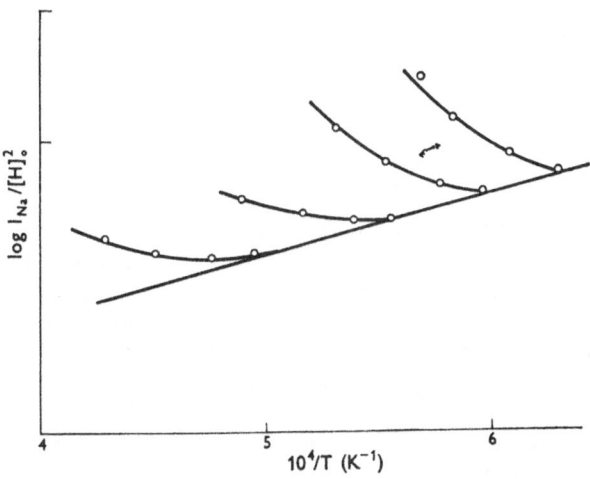

FIG. 3.18. The decay of chemiluminescence.

the limiting value in fuel-rich flames was that due to the process

$$Na + H + H \longrightarrow Na^* + H_2$$

The composition C_f may be separated to yield the ratio C_2/C_3 which was found to be $0\cdot4 \pm 0\cdot1$ over the range of flames examined and from the limiting value (C_2) and the ratio, the individual values of zq are calculated, knowing the concentration of sodium atoms

$$q_1z_2 = 2 \times 10^{-32} \text{ cm}^6 \text{ molecule}^{-2} \text{ s}^{-1}$$

$$q_3z_3 = 6 \times 10^{-32} \text{ cm}^6 \text{ molecule}^{-2} \text{ s}^{-1}$$

This approach was developed by Zeegers and Alkemade,[131] who examined chemiluminescence of OH (306·4 nm) and K (404·5 nm). They made careful correction for self absorption and for background radiation from the process $K + OH \rightarrow KOH + h\nu$. The only process found to be important for OH was

$$H + OH + OH \longrightarrow H_2O + OH^*$$

for which $q_3z_3 = (7 \pm 1) \times 10^{-33} \text{ cm}^6 \text{ molecule}^{-2} \text{ s}^{-1}$. Substitution of argon for nitrogen as diluent had no significant effect on the rate of radiation, nor did a tenfold variation in the ratio $[H_2]/[H_2O]$.

Zeegers also recalculated the value found by Kaskan[136] and quoted it as $(10 \pm 3) \times 10^{-33} \text{ cm}^6 \text{ mol}^{-2} \text{ cm}^{-1}$. Similar results were found for potassium when

$$q_3z_3 = 3\cdot5 \times 10^{-33} \text{ cm}^6 \text{ molecule}^{-2} \text{ s}^{-1}$$

Padley and Sugden[118] also examined other elements, reporting the chemiluminescent excitation of Tl (535·0 nm), Fe (371·9 nm), Ag (338·2 nm) and Pb (368·3 nm) resonance line. These are illustrated in Fig. 3.19, together with the functions $[H]_o^2$ and $[H]_o[OH]_o$. It is obvious that the thallium excitation follows the function $[H]_o^2$, while lead follows $[H]_o[OH]_o$ equally closely. This implies that the excitation reactions are

$$Tl + H + H \longrightarrow Tl^* + H \qquad C_2 \gg C_3$$

$$Pb + H + OH \longrightarrow Pb^* + H_2O \qquad C_2 \ll C_3$$

No kinetic data were, however, reported.

A further type of chemiluminescence was reported by Reid and Sugden,[137] who showed that the emission of the alkaline earth hydroxide bands was not thermal. They suggested that in addition to the normal high bond energy mechanism of hydroxide formation (cf. lithium)

$$Li + H_2O \rightleftharpoons LiOH + H$$

a mechanism of formation from the oxide also operated, yielding excited hydroxides

$$BaO + H_2 \rightleftharpoons BaOH^* + H$$

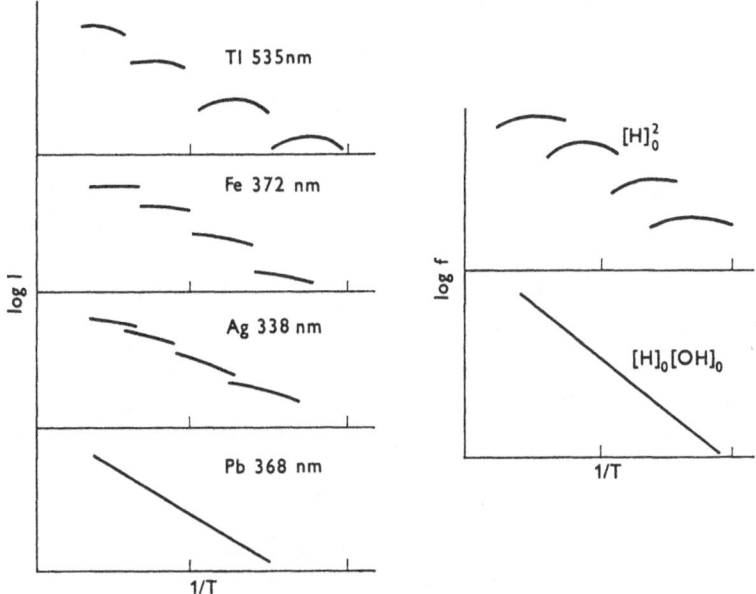

FIG 3.19. Chemiluminescence with various metals.

Again, the processes were too complex to be analysed in detail, and no numerical data emerged.

3.5.6. Radiative Recombination

Most flame photometric studies reveal the presence of a faint background continuum due to radical recombination, and the continuum may be enhanced by additives—the pale lilac colour of the potassium flame test in elementary analysis is due in large measure to the K—OH recombination continuum. While not *strictly* a problem in chemical kinetics, it is interesting to survey some of the results obtained from the study of continua.

One of the more widely studied continua is that of the carbon monoxide/oxygen flame. The intensity of this continuum (J) is closely related to the concentrations of CO and atomic oxygen:

$$J = a[CO][O] \qquad (3.58)$$

Kaskan[136] determined the value of a to be $1 \cdot 5 \times 10^9$ quanta s^{-1} cm^{-3} $Å^{-1}$ ($1 \cdot 5 \times 10^{25}$ qs^{-1} m^{-4}) at $434 \cdot 5$ nm.

He pointed out that the direct recombination of ground state CO and O was spin forbidden if the product was ground state CO_2, and therefore proposed that an excited CO_2 was formed in a fast reaction, some part of which could radiate

$$CO(^1\Sigma) + O(^3P) \longrightarrow CO_2^*$$
$$CO_2^* + M \longrightarrow CO_2 + M \text{ radiationless}$$
$$CO_2^* \longrightarrow CO_2 + h\nu$$

Zeegers supported Kaskan but found a rather higher absolute radiation. He derived the coefficients a_λ and c_λ in the expanded expression

$$J_\lambda = a_\lambda[CO][O] + c_\lambda \exp(-h\nu/kT)[CO_2]$$

and showed that they represented the continuous background in a series of carbon monoxide and acetylene flames reasonably well.

Zeegers found little or no radiation from a clean hydrogen flame, in contrast to Padley[112] who observed a continuous emission ascribed to the process

$$H + OH \longrightarrow H_2O + h\nu$$

Analogous continua were reported by Phillips and Sugden[113] in hydrogen flames to which halogens were added:

$$H + Cl \longrightarrow HCl + h\nu$$
$$H + Br \longrightarrow HBr + h\nu$$
$$H + I \longrightarrow HI + h\nu$$

Another continuum which has received considerable attention is the nitric oxide continuum. This was first studied in flames by James and Sugden[111] who found that small amounts of nitric oxide added to a flame produced a greenish continuum ascribed to

$$NO + O \longrightarrow NO_2 + h\nu$$

This process was studied in great detail by Bulewicz and Sugden,[115] with a view to establishing a method of measuring the concentration of oxygen atoms in a flame. They succeeded in demonstrating the complexity of the reaction scheme which actually catalyses the recombination of hydrogen atoms and perturbs the concentration which it had been hoped to measure.

The alkali metal–hydroxyl continuum was one of the first to be studied in detail, and Jones and Sugden[110] demonstrated that, in any flame containing alkali metals, there was a background continuum due to processes of the type

$$Na + OH \longrightarrow NaOH$$

They demonstrated the occurrence of this continuum with all alkali metals, with a very similar intensity, for equal concentrations of free alkali atoms, and a similar dependence on flame composition. They suggested that this measurement could be used to monitor the OH concentration in flames.

3.6. THE KINETICS OF IONIZATION

3.6.1. Introduction

The rate of ion formation in flames has been studied by many techniques and shown to be a complex phenomenon which must take account of thermal and chemiionization, of charge transfer and ion recombination. The ionization may be dominated by an added substance of low ionization potential, or it may be the "natural" ionization of a hydrocarbon flame. This natural ionization is of widespread importance, and it has been suggested[138] that even in so-called "pure" hydrogen flames, the background ionization is due to traces of hydrocarbon. Other experiments with ultra-pure gases have yielded strange, though unconfirmed, results.

3.6.2. Ionization in Hydrocarbon Flames

The reaction zone of a hydrocarbon flame contains a large number ($\approx 10^{11}$ cm^{-3}) of ion pairs. These have been studied extensively by mass spectrometry and a partial assignment of the masses made. The particular fuel used has some, but not a large, effect on the ions produced. Fuels which produce sooty flames, such as acetylene or benzene give richer spectra and higher yields of ions.[139-141] As a rough rule, one carbon atom in a million produces an ion. The ions observed have been divided into secondary ions and flame ions experimentally by varying the size of the inlet pinhole. The walls of the pinhole will be covered by a layer of cold, stagnant gas and many secondary ions will be formed in this layer. If the pinhole is small, the boundary layer will fill it, and all ions will be secondary ions. If the pinhole is enlarged, the boundary layer will not seal it completely, and a proportion of the flame gases will penetrate. Ions which are relatively insensitive to pinhole size are likely to be formed in the boundary layer, while those whose intensity depends strongly on the orifice are likely to be true flame ions.

When the count of flame ions was followed as a function of distance from the reaction zone, it was found that the majority of hydrocarbon ions rose and fell together, but that mass 19 (H_3O^+) rose slightly less rapidly and fell very much more slowly, so that after a few millimetres it was effectively the only ion. The earliest ion to be formed was found to be mass 29, supposed to be CHO^+ which reached its maximum value close to the maximum rate of increase of H_3O^+.[142]

The dominant flame ion, H_3O^+, appears rapidly in the reaction zone, and in such large amounts that it is probably produced by a proton exchange reaction between a precursor ion and water. Calcote,[67,143] on

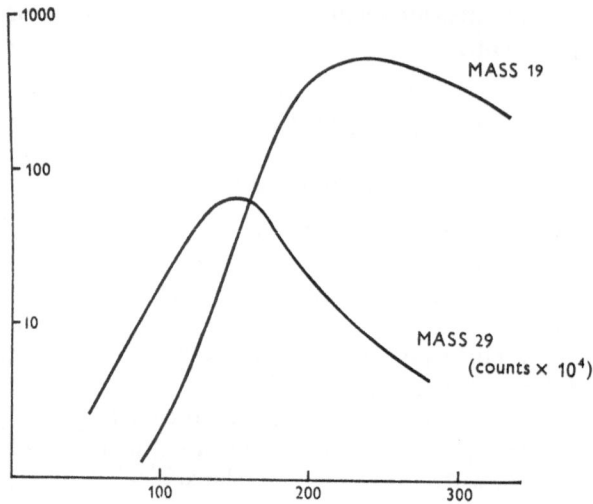

FIG 3.20. Rates of reaction with H_3O^+.

thermochemical considerations, postulated that the initial reaction was

$$CH + O \longrightarrow CHO^+ + e$$

This reaction was at one time contested but now is generally accepted as the most probable chemical reaction for ion production.

The ion CHO^+ satisfies the general requirement in that mass 29 appears ahead of mass 18, and the kinetics of H_3O^+ formation may be set down as follows.

$$CH + O \longrightarrow CHO^+ + e \qquad k_0$$
$$CHO^+ + H_2O \longrightarrow CO + H_3O^+ \qquad k_2$$
$$H_3O^+ + e \longrightarrow H_2O + H \qquad k_1$$

Applying the steady state criteria to CHO^+

$$[CHO^+] = \frac{k_0[CH][O]}{k_2[H_2O]}$$

At the maximum of the H_3O^+ curve, it is the dominant positive ion, so that $[H_3O^+] = [e]$ and $d[H_3O^+]/dt = 0$

$$[H_3O^+]_{max} = \left(\frac{k_0[CH][O]}{k_1}\right)^{1/2}$$

The value of k_1, discussed below, is about 5×10^{-7} cm³ molecules⁻¹ s⁻¹, and $[H_3O^+]_{max}$ of the order of 10^{11} molecules cm⁻³. Hence

$$k_0[CH][O] = 5 \times 10^{15} \text{ cm}^{-3} \text{ s}^{-1}$$

At this point, $[CHO^+]/[H_3O^+] = 2 \times 10^{-5}$ so that $[CHO^+]$ is of the order of 2×10^6 molecules cm^{-3} and $[H_2O]$ of the order 10^{17}. Therefore

$$k_2 = 2 \cdot 5 \times 10^{-8} \text{ cm}^3 \text{ molecules}^{-1} \text{ s}^{-3}$$

The concentration of CH is such that it may just be seen in low-pressure flames by multiple-path absorption. A fair estimate, by comparison with the limiting observable concentration of sodium is therefore 10^{13} molecules cm^{-3}. $[O]$ is known to be 4×10^{14} molecules cm^{-3} so that

$$k_0 = 1 \cdot 2 \times 10^{-12} \text{ cm}^3 \text{ molecules}^{-1} \text{ s}^{-1}$$

Subsequently Green and Sugden[138] refined their observations while Peeters and van Tiggelen[145] using a probe technique but a similar analysis of the ion-forming mechanism found closely comparable values (all in cm^3 molecules^{-1} s^{-1}).

	Green and Sugden	Peeters and van Tiggelen
k_0	3×10^{-12}	5×10^{-12}
k_1	$2 \cdot 3 \times 10^{-7}$	—
k_2	7×10^{-9}	6–8×10^{-9}

Calcote[146] has discussed these processes in detail, and has advanced somewhat different figures for k_0 and k_2, of 1×10^{-11} and 1×10^{-8} cm^3 molecules^{-1} s^{-1}. It is doubtful if the experimental accuracy is sufficient for a debate on the exact value, since Green and Sugden's value for k_0 is itself only an estimate, and the Aerochem[154] selected values of 5×10^{-12} and 1×10^{-8} are acceptable. All workers are agreed that the temperature dependence is slight, and in the present uncertainty of experimental precision should be ignored.

The recombination reaction (k_1) has been studied by many workers. Apart from the work of Green and Sugden, there has been a long history of measurements of ion recombination rates in flames which mass spectrometry has demonstrated to be due to this reaction. Among the early work by Wilson[147] may be found a recombination rate which differs little from the currently accepted figure. King[69,149] was the first to attempt measurements in recent times, but his work in low-pressure flames was shown by Calcote[150] to require correcting for ambipolar diffusion. When such correction was made, the recombination rate showed no pressure dependence.

pressure (mm Hg)	$\alpha \times 10^7$
33	$1 \cdot 6$
66	$2 \cdot 4$
520	$1 \cdot 6$
760	$2 \cdot 0$

It is sensibly independent of fuel in so far as methane ($2 \cdot 5 \times 10^{-7}$), propane ($2 \cdot 9 \times 10^{-7}$) and acetylene ($2 \cdot 8 \times 10^{-7}$) flames show the same value and shows little or no dependence on temperature. This reaction was also studied by Padley and Kelly[157] who quote the following results as well:

Semenov and Sokolik	$2 \cdot 2 \times 10^{-7}$
Calcote, Curzius and Miller	$2 \cdot 4 \times 10^{-7}$
Green and Sugden	$2 \cdot 2 \times 10^{-7}$
Wortberg	$1 \cdot 1 \times 10^{-7}$
Hayhurst and Telford	$3 \cdot 0 \times 10^{-7}$

3.6.3. Reaction of CHO⁺

The great stability of the product CO makes the ion CHO^+ an efficient proton transfer agent, and the probable precursor of most of the ions observed. Nevertheless, only one rate constant for a process other than transfer to water has been discussed. Hurle, Sugden and Nutt,[151] studying the ionization of nitric oxide in CO/O_2 and H_2/O_2 flames with traces of hydrocarbon showed that the process

$$CHO^+ + NO \longrightarrow NO^+ + CHO$$

had a rate constant of $1 \cdot 2 \times 10^{-10}$ cm³ molecules⁻¹ s⁻¹.

This is a charge exchange rather than proton exchange reaction so that the rate constant is not comparable with that for the reaction with water.

3.6.4. Ionization in Other Flames

Since it appears probable that the ionization processes in hydrogen flames are attributable to traces of hydrocarbon, the formation of CHO^+ will explain the natural ionization in almost every flame system which has been studied. The only notable exceptions are the dry carbon monoxide or cyanogen flames, and there has been very little qualitative work on these systems.

The cyclotron resonance studies of Bulewicz and Padley[152] indicate that the ionization in a low-pressure cyanogen/oxygen flame is in excess of the thermal level though this contradicts some early work in the field of gas chromatography. They found a dependence of electron combination on the square of the pressure, which was held to indicate that ions were produced in a termolecular process. Several possible reactions were postulated but no kinetic data were presented. Work on carbon monoxide flames[153] has indicated the widespread occurrence of CHO^+ even in supposedly dry flames.

3.6.5. Recombination Reactions

The most common and most important recombination reaction is that already discussed—the recombination at H_3O^+ with electrons. In the presence of a sufficiently high concentration of electron acceptors, the electron concentration may be relatively small and the recombination between ions correspondingly important. Calcote and Kurzius[154] have made some preliminary studies of this problem, and measured the recombination at H_3O^+ with OH^- and Cl^-

$$H_3O^+ + Cl^- \longrightarrow H_2O + HCl \quad k = 3 \times 10^{-8} \text{ cm}^3 \text{ molecule}^{-1} \text{ s}^{-1}$$

$$H_3O^+ + OH^- \longrightarrow H_2O + H_2O \quad k = 8 \times 10^{-8} \text{ cm}^3 \text{ molecule}^{-1} \text{ s}^{-1}$$

These are of the same order as the rate constant for recombination with electrons, the somewhat higher value for OH^- being as expected.

Other ionic recombination reactions have been postulated without any measurements, for example in cyanogen flames the process

$$NO^+ + CN^- \longrightarrow CO + N_2$$

may compete with the well-established reaction

$$NO^+ + e \longrightarrow N + O$$

studied, among others, by Bascombe, Jenkins, Schiff and Sugden[155] who found $k = 5 \times 10^{-7} \text{ cm}^3 \text{ molecule}^{-1} \text{ s}^{-1}$.

3.6.6. Charge Transfer Reactions of H_3O^+

The addition of metallic salts to a flame markedly affects the rate of recombination observed. This was demonstrated dramatically by Knewstubb.[156] He showed that lead was virtually unionized in a hydrogen flame, but was apparently as able to provide electrons as sodium in an acetylene flame. His explanation was that the natural ionization in the reaction zone of the acetylene flame was much greater than in the hydrogen flame, and that the lead transferred an electron to the natural positive ion. Since this was polyatomic the three-body recombination rate for electrons with lead atoms would be much lower than for dissociative recombination with the polyatomic natural ion, as that ionization would persist for some distance downstream.

The identification of the natural ion as H_3O^+ opened the way to the determination of the rates of electron transfer, and Hayhurst and Telford[158] were able to determine the rates of transfer from sodium, thallium and lead by mass spectrometry. They found little or no dependence on temperature in the range 1800–2500 K and the means of their results in this temperature range were:

$$H_3O^+ + Na \quad 10 \cdot 8 \times 10^{-9} \text{ cm}^3 \text{ molecule}^{-1} \text{ s}^{-1}$$

$$H_3O^+ + Tl \quad 7 \cdot 1 \times 10^{-9} \text{ cm}^3 \text{ molecule}^{-1} \text{ s}^{-1}$$

$$H_3O^+ + Pb \quad 4 \cdot 2 \times 10^{-9} \text{ cm}^3 \text{ molecule}^{-1} \text{ s}^{-1}$$

The first of these was also studied by Hayhurst and Sugden[159,160] who found a similar (1×10^{-8}) value by the same technique. The rates

8

were obtained by a measurement of the increased rate of disappearance of H_3O^+ in the presence of the metal added. Padley and Sugden[161] estimated the recombination rate in the presence of lead as 3×10^{-9} which compares well with those more sophisticated results above.

The rates for these reactions increase with decreasing ionization potential, as pointed out by Hayhurst and Telford.[162] The electron affinity of a metal species may be regarded as the ionization potential

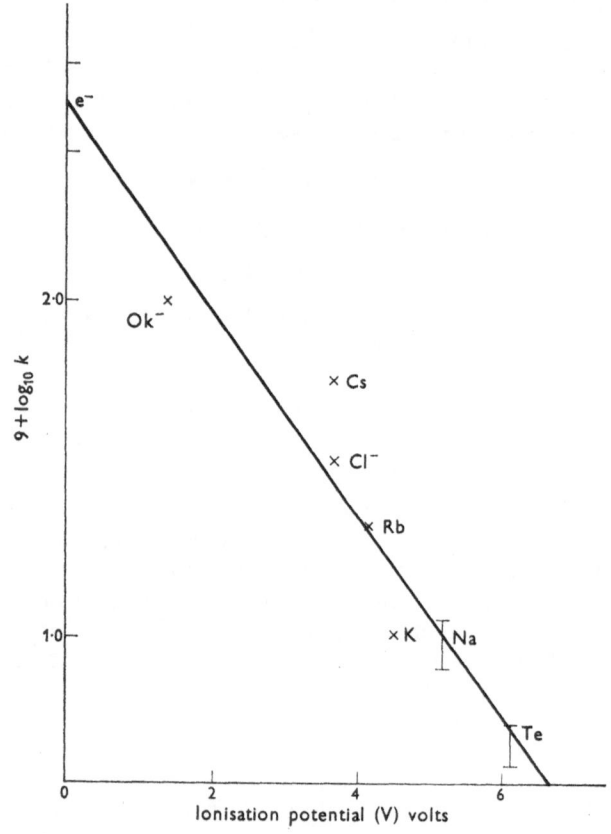

FIG. 3.21. The field between two colliding ions.

of the negative ion and it will be seen from Fig. 3.21 that there is a good logarithmic relation between the rate of the charge transfer reaction

$$H_3O^+ + X \longrightarrow X^+ + H_2O + H$$

and the ionization potential of X; this extends even to the electron, i.e. to zero ionization potential.

It may be seen from Fig. 3.21 that

$$k_x = k_0 \exp(-0.53 V_x)$$

where k_x is the rate for reactant x, whose ionization potential is V_x. This implies that there is an energy barrier to the transfer of an electron even if the process is energetically favourable but the empirical coefficient is far too small for this to be the whole explanation, and the very slight dependence on temperature indicates that the rate is not energy controlled. The concordance of the rate constants for such diverse reaction systems suggests strongly that one common process is operating, and that the process is the transfer of an electron from X to H_3O^+, irrespective of the charge or nature of X. Such a process will show a barrier (Fig. 3.22) which will be of the order of V_x, but quantum mechanical

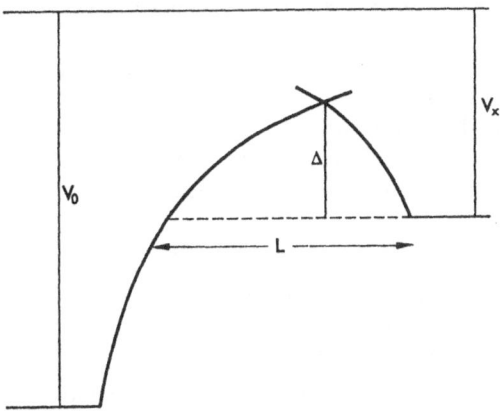

FIG. 3.22. The rates of ionization of alkali metals (after Hollander).

tunnelling may dominate the transfer process, as in similar reactions in solution. In order to test this hypothesis we will use Fowler's[174] formulation for the mean transmission coefficient through an abrupt boundary with a square potential hill

$$d = \left(\frac{2\pi kTH_0}{H^2}\right)^{1/2} \exp\left(-2Kl(H-H_0)^{1/2}\right)$$

where H is the height of the barrier and H_0 the highest occupied electron level of X, both reckoned relative to the lowest unoccupied level of H_3O^+; l is the width of the barrier and $k^2 = 8\pi^2 m/h^2$.

The preexponential term is of the order of unity and will in any case depend markedly on the precise form of barrier considered. Attention may therefore be focussed on the exponential term. Gamov's modification for a triangular barrier is

$$d = \exp\left(-\tfrac{4}{3}kl(H-H_0)^{1/2}\right)$$
$$= \exp\left(-0.68 \, L \, \Delta^{1/2}\right)$$

TABLE 3.2

X	$\log_{10} k + 9$	r	V_x	$-\log_{10} d$	$9 + \log_{10} k/2$
e	2·30	1·54	0	—	2·30
OH⁻	1·90	2 59	1·43	0·25	2·15
Cl⁻	1·48	2·59	3·70	0·77	2·25
Na	1·00	2·49	5·12	1·06	2·06
Tl	0·74	2·98	6·08	1·51	2·25
Pb	0·60	3·01	7·38	1·85	2·44
					2·24 ± 0·12

where L is the barrier width in Å and Δ the barrier height in electron volts, reckoned from X.

Semi-empirical calculation, using coulombic or exponential coulombic potentials agree that Δ is of the order of $V_x/2$ and L of the order $2rV_x/(V_x + V_0)$ where r is the internuclear separation

Hence

$$d = \exp\left(-0\cdot97rV_x^{3/2}/V_x + V_0\right)$$

Calculating r from the crystal radii of X⁺ and the dimensions of the water molecule, and taking V_o to be 6·3 eV yields Table 3.2.

The constancy of the final column is considerably better than the quoted experimental uncertainties in the various papers, which amount to ±0·2 (±50%) and supports strongly the view that all these reactions proceed by a similar mechanism.

Additional data may be found in the paper by Soundy and Williams,[163] who obtained indirect values for the rate of the charge transfer reaction during their study of the atomic ion-electron recombination process. Their analysis does not allow for compound formation, of particular

TABLE 3.3

$$X + H_3O^+ \longrightarrow X^+ + H_2O + H$$

X	$9 + \log_{10} k$	ϕ	r	V_x	$-\log_{10} d$	$9 + \log_{10} k/2$
In	1·05	0·5	2·54	5·76	1·23	2·23
Pb	0·48	0	3·01	7·38	1·86	2·34
Mn	0·41	0	2·62	7·41	1·62	2·03
Cr	0·38	5	2·35	6·74	1·33	2·49
Li	1·40	10	2·22	5·36	0·99	2·39
Mg	0·02	1	2·36	7·61	1·50	1·52
Co	−0·04	0	2·52	7·81	1·65	1·61
Ni	−0·18	0	2·51	7·61	1·60	1·42
Cu	−0·32	0	2·50	7·68	1·60	1·28
Zn	−0·53	0	2·44	9·36	1·87	1·35

importance with In, Li and Mg, and also omits to take CsOH into account in the calibration.[164] The values have been corrected to take this into account. These values will depend on the flames composition and temperature, but are of the correct magnitude.

In view of the uncertainties in the corrections to be applied to the calibration, and of the effects of compound formation, the agreement must be held to be satisfactory.

3.6.7. Recombination of Electrons and Metal Ions

The metal ions produced by the charge exchange reactions discussed in the previous section will recombine with electrons until an equilibrium state has been reached. For elements of low ionization potential, such as the alkali metals, the contribution of the charge exchange reaction is lost in the dominant process of thermal ionization, and serves only to hasten the attaining of the equilibrium level. However for elements of higher ionization potential, the equilibrium level is so low that metal ions can only be produced by the charge exchange process. This point was advanced by Knewstubb and Sugden[156] to explain the high level of ionization produced by lead in an acetylene flame compared to that produced in an otherwise identical hydrogen flame. The recombination of the monatomic metal ion and electron must proceed through either a three-body or a radiative capture process; both will be much slower that the dissociative recombination of the polyatomic H_3O^+.

The limited spatial resolution of available apparatus limited the study of the recombination rates to the case of lead, for which a recombination coefficient of 3×10^{-9} cm^3 molecules^{-1} s^{-1} was given by Padley and Sugden[161] from microwave studies and 6×10^{-9} by Hayhurst and Sugden[160] as a result of direct mass spectrometric observations. Jensen and Padley[165] used a microwave cavity to study lead and obtained the closely similar value of 6.5×10^{-9}; for manganese and chromium they found 7.5×10^{-9} and 26×10^{-9}, respectively. The latter figure (chromium) was slightly dependent on composition, probably because of the presence of solid particles. They considered the possibility that the ion was $CrOH^+$ and not Cr^+ but showed that this was quantitatively improbable, even though Schofield and Sugden[166] had shown that, for the alkaline earths, such hydroxy ions were of dominant importance.

Soundy and Williams,[163] using their rotating electrostatic probe, made a very extensive study of recombination rates, showing that the actual ion current measured was the result of the charge transfer process already discussed, and the subsequent ion-electron recombination process (Table 3.4).

Kelly and Padley[167] subsequently added gallium (7.9×10^{-9}), indium (6.1×10^{-9}) and thallium (6.4×10^{-9}) to the list.

Jensen and Padley also studied the alkali metals near to equilibrium. By assuming that the rate quotients could be calculated from Saha's

TABLE 3.4
Recombination rates for $M^+ + e^- \longrightarrow M$

M	Rate $\times 10^9$ cm^3 molecules^{-1} s^{-1}	M	Rate $\times 10^9$ cm^3 molecules^{-1} s^{-1}
In	3	Mg	37
Pb	9	Co	60
Mn	24	Ni	79
Cr	18	Cu	94
Li	9	Zn	100

equation,[177] they used the full expression for forward and back ionization processes to obtain both the rate of ionization and of recombination. Hayhurst and Sugden[160] used similar arguments to analyse their mass spectrometric results (Table 3.5).

Jensen and Padley[169] also showed that there was a slight temperature dependence of the rates, expressible either as a low negative energy of activation (about -20 kJ mol^{-1}), or as a $T^{-1.5 \pm 0.2}$ power dependence. Qualitatively, this is in accord with Thompson's theory of three-body recombination, which predicts a $T^{-3.5}$ dependence. It is however doubtful if a theory devised for heavy ions can properly be used for electrons. At flame temperature the critical radius within which recombination can occur ($2e^2/3kT$) has the value of 5·1 nm, or 1 % of the mean free path. This leads to a very small probability of recombination, but the low mass and high collision frequency of the electron more than offsets this; the predicted rate (6×10^{-7} at 2200 K) is considerably above the rates observed. These rates would be correct if the negative ion had the mass of a hydrogen atom, or if the critical radius were 1·5 nm.

Thompson's expression for the rate of recombination may be written as[168]

$$\alpha = \pi r^2 (\bar{u}_+^2 + \bar{u}_-^2)(2w - w^2)$$

TABLE 3.5

Metal	Rate $\times 10^9$ cm^3 molecules^{-1} s^{-1}	
	Hayhurst and Sugden	Jensen and Padley
Li	—	9·0
Na	6·4	8·5
K	3·5	6·5
Rh	—	5·5
Cs	4·5	8·5

where r is the critical radius, $\bar{u}$ the r.m.s. velocity and w a probability of collision which, for small values of (r/λ), becomes $8r/\lambda$ where λ is the mean free path.

This may also be written as

$$\alpha = \frac{Z}{n}(2w - w^2) \times \frac{Q \; ion}{Q \; neutral}$$

The Q's are the electron collision cross sections, Z is the electron collision frequency and n the particle density. The Thompson probability factor is now defined in terms of the collision cross section, rather than the electrostatic energy. Taking Bulewicz and Padley's[170] value of 40×10^{-16} cm² for the mean electron collision cross section, the probability factor become 4.4×10^{-3}, and

$$\alpha = \frac{3.0 \times 10^{11}}{2.7 \times 10^{18}} \times 4.4 \times 10^{-3} \frac{Qi}{Qn} = 4.85 \times 10^{-10} \frac{Q \; ion}{Q \; neutral}$$

The factor $Q \; ion/Q \; neutral$ will be of the order of 10, and will be likely to show some specificity. It therefore appears that a simple three-body collisional process provides an adequate explanation for the recombination coefficient found for about half the elements studied. Some recombinations proceed much more rapidly, possibly because of the electronic structures of the elements concerned (all transitional metals). Alternatively the ions concerned may not be the simple metal ions, but hydroxides which react as ions, as in the case of the alkaline earths.

3.7. Kinetics of Ionization of Alkali Metals

3.7.1. Introduction

The early work of Page and Sugden[24] on the order of the ionization process with regard to radicals has already been referred to, as suggesting that the mechanism of ionization of alkali metals was

$$A + OH \rightleftharpoons A^+ + OH^-$$

rather than the direct

$$A + M \longrightarrow A^+ + e^- + M$$

The first direct evidence that the ionization of alkali metals involved a kinetically slow step came almost simultaneously from Knewstubb and Sugden.[171] They showed that the change in slope from 1.0 to 0.5 in a logarithmic conductivity concentration plot occurred at a much higher concentration of alkali metal than is expected and that this discrepancy decreased as time (height in flame) was increased. Furthermore, the discrepancy was much reduced in the presence of a few percent of acetylene in the fuel. Their analysis showed that these results were consistent with the slow ionization of sodium and lithium according to the mechanism proposed by Page and Sugden. The height resolution in all this early work left much to be desired, and little progress was made until an adequate resolution had been achieved.

3.7.2. Ionization in Hydrogen Flames

The first steps towards a more precise study of the kinetics of ionization of alkali metals were taken by Padley and Sugden[161] who used the Horsfield–Pennycook TM 010 microwave cavity to study the ionization of sodium. They showed that the process followed a simple unimolecular rate law. They were able to reject the process:

$$Na + OH \longrightarrow Na^+ + OH^-$$

in favour of

$$Na + M \longrightarrow Na^+ + e + M$$

by the dependence of the observed unimolecular rate constant on the concentration of OH: the rejected process requires that there be a direct dependence, but none was found. The favoured process would be expected to have an activation energy equal to the ionization potential, and to be very slow. Padley and Sugden reported an activation energy of 285 ± 29 kJ mol^{-1}, much lower than expected, and a normal cross section. In discussing this anomaly, Smith[172] suggested that the reaction was really

$$Na + H_2O \longrightarrow Na^+ . H_2O + e$$

The hydration energy of the sodium ion stabilizes the system by about 210 kJ mol^{-1}.

Schofield and Sugden[130] extended this work to potassium. This again posed the dilemma of an abnormally high cross section, with the activation energy equal to the ionization potential, or a normal cross section and abnormally low activation energy. In an endeavour to test the validity of Smith's suggestion about hydrated ions, Hayhurst and Sugden[120,173] searched for such ions with a mass spectrometer. They were successful in detecting them, and in showing that they were true flame ions, though present only in small amounts. They evaluated the ΔE_0^0 for the reaction by statistical mechanics, finding the energies of Table 3.6.

These are fully in accord with elementary electrostatic calculations using a point dipole–model. Polarization of the water molecule could

TABLE 3.6

Element	ΔE_0^0 (kJ mol^{-1})
Li	234
Na	163
K	121

also contribute an equal amount to the energy. Early in these studies, it was noticed that the addition of acetylene in small amounts hastened the onset of equilibrium; it was suggested that the rapid chemiionization leading to the formation of H_3O^+ in the reaction zone, followed by charge transfer to sodium atoms, could bring this about. Schofield[144] pointed out that while this mechanism could operate where the expected level of ionization was only a few percent, it was inadequate where a large proportion of the alkali metal was to be ionized. His studies, where 10–70% of the metal was ionized at equilibrium, showed that acetylene had only a limited effect in catalysing equilibration.

Hayhurst and Sugden,[120] in their mass spectrometric studies, were able to follow the formation of sodium ions by both processes

$$Na + H_3O^+ \longrightarrow Na^+ + H_2O + H$$
$$Na + M \longrightarrow Na^+ + M + e$$

In both cases, they argued that the Na^+ could be hydrated ($Na^+.H_2O$).

They examined the effect of acetylene on these processes, and observed that the rate of ionization of sodium in the reaction zone was far greater than could be accounted for by charge transfer, the count of Na^+ being considerably greater than that of the H_3O^+ from which it was supposedly formed. Furthermore, in the presence of acetylene the sodium count overshot the final value and fell towards it in contrast to the continual increase in absence of added acetylene. It is evident that, while sodium ions are certainly produced both by charge transfer and by thermal ionization, a third process also occurs which is confined to the reaction zone and is dependent on acetylene. This process may be ionization by "hot" electrons, as suggested by Cozens and von Engels:[75]

$$Na + e \longrightarrow Na^+ + 2e$$

3.7.3. Ionization in Carbon Monoxide Flames

Simultaneously with the work just described, the Dutch school of Alkemade and his co-workers were examining rates of recombination and ionization in acetylene and carbon monoxide flames, using optical and R.F. techniques. This work is of particular significance in that the carbon monoxide flames were dry, so that mechanisms which depend on H, OH or H_2O do not operate.

The first experiments of this school were directed to the understanding of ionization interference in flame photometry. Alkemade[175,176] showed that the short-fall in resonance radiation intensity from alkali metals through ionization, and the suppression of such ionization, could be reconciled with the Saha[177] relation only if an excess of electrons over alkali ions were present in acetylene flames, in amounts which decreased with height in the flame. In developing the study Borgers[178,179] used the R.F. coil technique to measure electron concentration. He at first supported this conclusion for the clean acetylene flame, but then found that the electron level in a clean carbon monoxide flame was less than 10^8 cm^{-3}. Hollander,[180] repeating the experiments of Alkemade in carbon monoxide flames, found that the electron level required to reconcile his observations on the emission of K resonance radiation with Saha was $1 \cdot 5 \times 10^{11}$ at 0·4 cm height, falling to 3×10^9 cm^{-3} at 5 cm, while for Cs, the values were 2×10^{11} and 1×10^9, respectively.

It was obvious from this that the assumption of excess electrons was incorrect, both because K and Cs should have shown the same excess, and because the excess of electrons over alkali ions could not exceed 10^8 cm^{-3}, the density in a clean flame. Hollander was therefore forced to the conclusion that the assumption of equilibrium ionization was invalid, and reinterpreted the previous work as the result of slow ionization.[181] If the alkali metal X ionizes according to the opposing reactions

$$X \underset{k_{-1}}{\overset{k_1}{\rightleftharpoons}} X^+ + e$$

Then

$$\frac{dX}{dt} = -k_1[X] + k_1[X^+][e]$$

If the total alkali metal is [X$_0$], and charge balance is maintained

$$[X_0] = [X] + [X^+]$$

$$[X^+] = [e]$$

$$\frac{d[X]}{dt} = -k_1[X] + k_{-1}[X_0 - X^2]$$

Writing $y = X_0 - X = X^+$

$$-\frac{dy}{dt} = k_1 y + k_{-1} y^2 - k_1[X_0]$$

Writing $K = k_1/k_{-1}$

$$-dy/dt = k_1\{y + y^2/K - X_0\}$$

Putting $\alpha^2 = 1 + 4[X_0]/K$

$$-k_1 t = \frac{1}{\alpha} \log\left(\frac{2y/K + 1 - \alpha}{2y/K + 1 + \alpha}\right) + C$$

If significant amounts of the metal X are present as compounds, then K must everywhere be replaced by $K(1 + \phi)$ where

$$K(1 + \phi) = ([X_0] - [X^+])/[X]$$

The boundary conditions are that there is no ionization at zero time, i.e. $y = 0$ at $t = 0$

Hence,

$$-k_1 t = (1/\alpha) \log \left\{ \frac{1 + 2y/K(\alpha + 1)}{1 - 2y/K(\alpha - 1)} \right\}$$

At equilibrium $y = \frac{1}{2}K(\alpha - 1)$. While for small degrees of ionization (t only small)

$$e^{-k_1 t} = 4y\alpha/K(\alpha^2 - 1) = \alpha y/[X_0]$$

or

$$\alpha k = -\frac{d \ln y}{dt}$$

Hollander[182] was able to apply this relation to the ionization of the alkali metals Na, K and Cs in a series of carbon monoxide/oxygen flames (Fig. 3.23).

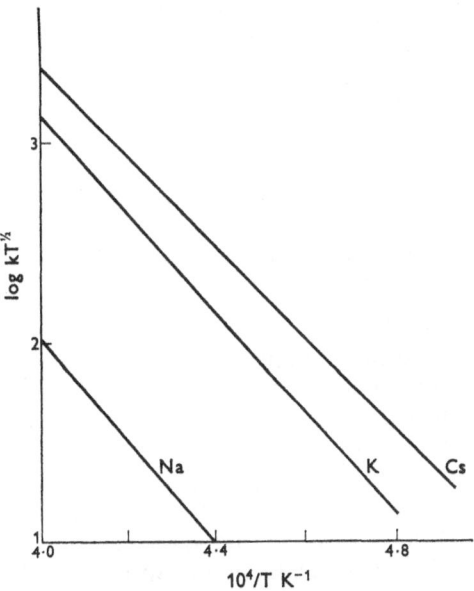

FIG. 3.23. Superequilibrium ionization.

The straight lines in this figure have been drawn with slopes corresponding to the ionization potentials of the elements, and they fit the measured points very closely.

This result, that the activation energy for the ionization process is close to the ionization potential of the element, contradicts the work described earlier. However subsequent studies by Jensen and Padley,[165,169] and by Kelly and Padley[167] have confirmed and extended these observations. The data collected in Table 3.7 show clearly, as pointed out by

F. M. PAGE

TABLE 3.7

	K_i (s^{-1})	Method	E (kJ/mol)	Ip (kJ/mol)	Q (10^{-16} cm^2)
Na	51 55	cavity probe	63 ± 21	496	2·7 × 10^4
K	1500 1700 1600	optical cavity probe	414 ± 17	419	2·3 × 10^4
Pb	2900 2700 2500	optical cavity probe	406 ± 17	404	2·0 × 10^4
Cs	17 000 11 500 12 000	optical cavity probe	385 ± 25	375	2·3 × 10^4
Tl	0·19 0·29	cavity probe	578 ± 17	590	2·3 × 10^4
Ga	0·36	probe	565 ± 33	578	1·4 × 10^4
In	0·42 0·87	cavity probe	548 ± 21	557	1·3 × 10^{-4}

Data obtained in a $H_2/O_2/N_2$ flame at 2440 K
Reference Kelly and Padley[167]

Hollander, Kalff and Alkemade,[182] that the collision cross sections for ionization are similar to those found from fluorescence quenching data.

The agreement between different workers is satisfactory and is strong evidence for the reality of these fast ionization reactions. The observation cannot be accepted without some reservations, however, because of small areas of doubt. In the first place, the observation by Hayhurst and Sugden of ultra-rapid ionization in the reaction zone suggests that ionization is not a simple process. Secondly, the most recent and accurate work of Kelly and Padley depends on a calibration which is admitted to be marginally in doubt. Finally, recent work has shown that the ionization of an alkali salt sprayed into a flame is not uniform, but has rather the nature of a group of meteor trails derived from the evaporation of individual crystallites.[148] Despite these slight objections, the weight of evidence does favour ionization cross sections which are significantly greater than gas kinetic cross sections, which Hollander, Kalff and Alkemade attribute to ionization from the excited levels of the alkali atom.

3.7.4. The Theory of Hollander

In order to explain the very large collision cross sections which he had deduced, Hollander suggested that ionization occurred from the multiple degenerate upper levels of the alkali metal atom. If the energy transferred at a collision comprises rotational and translational contributions, as well as vibrational, the energy distribution may be taken to be continuous and the rate of ionization (R_i) will be given by

$$R_i = NQ_i(8kT/\pi\mu)^{1/2} \exp(-E_i/kT) \qquad (3.59)$$

where Q_i is the cross section for ionization and E_i the ionization energy.

If ionization occurs from excited and metastable states, as suggested by Benson[183] to account for the ionization of the inert gases in shock tubes, the expression must be replaced by a suitably weighted summation over these states.

$$\begin{aligned} R_{ij} &= N_j Q_{ij}(8kT/\pi\mu)^{1/2} \exp(-E_{ij}/kT) \\ &= Z_{ij}n_j \exp(-E_{ij}/kT) \end{aligned} \qquad (3.60)$$

where Z_{ij} is the collision frequency

$$R_i = \sum R_{ij}$$

and

$$n_j = (g_j/Q_e)N_0 \exp(-E_j/kT) \qquad (3.61)$$

where Q_e is the electronic, E_j = energy of level j, E_{ij} = ionization energy of level j and g_j = statistical weight of level j. Hence

$$R_i = \sum Z_{ij}N_0(g_j/Q_e) \exp(-E_j/kT) \exp(-E_{ij}/kT) \qquad (3.62)$$

Provided that Z_{ij} does not depend greatly on i and j, and remembering that $E_{ij} + E_j = E_i$

$$R_i = ZN_0 \exp(-E_1/kT) \cdot \sum g_j/Q_e \qquad (3.63)$$

This differs from the original equation (3.59) by the term $\sum g_j/Q_e$. Since there are an infinite number of excited levels, the term may be made as large as desired, and Hollander proposed an upper limit, by arguing that for all levels within kT of the continuum, every collision would lead to ionization, so that the summation need only be carried up to this limit.

While recognizing the limitation of this model Hollander showed that the calculated rates of ionization of Na, K and Cs were 107, 1600 and 2780 s^{-1} at 2500 K. These are in very good agreement with both his own measurements and those of Padley and co-workers (Table 3.7) though there is a discrepancy in the case of caesium. Hollander was quick to point out that the assumption of a Boltzmann distribution was dubious; unless ionization was equilibrated, the upper levels, fed to a noticeable extent by ion-electron recombination would be underpopulated. There is, moreover, a certain arbitrariness about the limit of $E_{ij} = kT$—raising or lowering this by $\pm\frac{1}{2}kT$ would raise or lower the calculated rates by a factor of about 3.

3.7.5. Distributions in an Ionizing Flame

Since the transition from an unionized flame to an ionized flame is kinetically limited, it is not valid to apply the Boltzmann distribution to the species undergoing ionization, except in the limiting state of equilibrium ionization. Under such limiting conditions, a given level will be populated and depopulated in relation to any other given level at equal rates, according to the principle of microscopic reversibility. If one of the processes of population is restricted by any means the total population will be altered, since the depopulation process is unaffected; it will be reduced by the ratio:

$$\frac{\text{total rate of population by remaining processes}}{\text{total rate of population if restriction were lifted}}$$

Any state for which ionization is a valid depopulating process will be thus affected if the ionizing system is so far from equilibrium that population by recombination is unimportant. All other states associated with this state will also be affected since the population of this state is lowered, which applies a restriction to their population from this state.

Experimental evidence for or against depopulation is not easy to find, but it has long been established that the second resonance doublets for sodium (330·5 nm) and potassium (405 nm) show identical flame reversal temperatures to the first resonance doublets. Since both second resonance doublets lie within 0·75 volt ($\approx 4\,kT$) of the ionization limit, it would appear that the depopulation is slight at this distance from the ionization limit. It is true that the conditions in the flame approximate more closely to equilibrium than to the kinetically limited state, but such evidence as there is supports the application of the Boltzmann distribution to the lower levels.

No such evidence exists for the higher levels, and the possibility of a significant depopulation must be considered.[192,193]

If the energy distribution is continuous and Maxwellian, the chance that a molecule will gain or lose energy between E and $E + dE$ will be $\exp(-E/kT)\,dE$ and if there are n_k molecules in the kth state (energy E_k to $E_k + dE$) the number of molecules which reach some other state defined by E_j in a given time t is

$$Ztn_k \exp(-E/kT)\,dE$$
$$= (ZtN_0/Q_e).\exp(-E_k/kT).\exp(-E/kT)\,dE$$

If E_j is fixed, and E_k defined by

$$E_k = E_j - E$$

the total number of molecules reaching state j from lower states in time t is

$$R\!\uparrow = \int_0^{E_j} (ZtN_0/Q_e)\exp(-E_j/kT)\,dE$$
$$= (ZtN_0/Q_e)E_j \exp(-E_j/kT)$$

The number of molecules reaching state j from above will be equal to the number of molecules which leave state j having gained an energy E to $E + dE$, integrated over all values of E

$$R\!\uparrow = Zt \int_0^\alpha n_j \exp\left(-E/kT\right) dE$$
$$= (ZtN_0/Q_e)g_j \exp\left(-E_j/kT\right)kT$$

The ratio of these rates is therefore

$$R\!\uparrow/R\!\downarrow = g_j kT/E_j$$

Reverting to the level j of Hollander's theory this is populated and depopulated at equal rates from other levels, by the principle of microscopic reversibility. The rates of population may be grouped into:

R_b population from below

R_a population from higher atomic levels

R_i population from the ionization continuum

At equilibrium:

$$R_b = ZE_j \exp\left(-E_j/kT\right)$$
$$R_a + R_i = (g_j kT/E_j) . R_b = Zg_j kT \exp\left(-E_j/kT\right)$$
$$R_i = (Zn_j/kT) \exp\left(-E_{ij}/kT\right)$$

and the corresponding depopulation rates are the same. Under kinetic conditions, the depopulation rate for the ionization process is given by the same expression but the population rate from the continuum is zero. Hence

$$n_j/n_{je} = \frac{R_b + R_a}{R_b + R_a + R_i}$$

$$n_j/n_{je} = \frac{1 + R_a/R_b}{1 + \dfrac{R_a + R_i}{R_b}} = \frac{1 + g_j kT/E_j - R_i/R_b}{1 + g_j kT/E_j}$$

$$= 1 - \frac{R_i E_j}{(E_j + g_j kT)R_b}$$

Since the lowest value of E_j is $\approx 1 - 2\,kT$, the term $g_j kT$ may be neglected. This is equivalent to saying that the rate of population of the state from above is not important.

Hence

$$n_j/n_{je} = 1 - (R_i/R_b)$$

The rate of ionization is then

$$R_i = (Zn_j/kT) \exp(-E_{ij}/kT)$$
$$= Z(1 - R_i/R_b)(n_{je}/kT) \exp(-E_{ij}/kT)$$

Hence

$$R_i[1 + \{Zn_{je}/R_b kT\} \exp(-E_{ij}/kT)] = (Zn_{je}/kT) \exp(-E_{ij}/kT)$$

$$R_i' = \frac{Zn_{je} \exp(-E_{ij}/kT)}{kT + \{Zn_{je}/R_b\} \exp(-E_{ij}/kT)}$$

$$n_{je} \exp(-E_{ij}/kT) = (N_0 g_j/Q_e) \exp(-E_i/kT)$$

so that

$$R_i = (n_0/Q_e) \exp(-E_i/kT) \frac{g_j}{kT + (g_j/E_j) \exp(-E_{ij}/kT)}$$

Since E_{ij} is small the last term may be expanded as

$$\frac{g_j}{kT + (g_j/E_j)(1 - E_{ij}/kT)} \quad \text{or} \quad \frac{g_j E_j kT}{E_j + g_j kT}$$

Extending this result, following Hollander, to the summation over all levels, the factor $\sum g_j$ is replaced by

$$\sum g_j kT/(E_j + g_j kT)$$

Again there is no difficulty in carrying out the summation over enough energy levels to account for any observed collision frequency, but the contribution of any level is smaller by a factor of $kT/E_j k$, and the summation must be extended over proportionately more levels. For most levels, $g_j < E_j/kT$, so that the second term in the denominator may be neglected, also E_j is effectively constant, and equal to the ionization potential, since most levels are less than kT from this limit.

The effects of kinetic depopulation of the upper levels are therefore to decrease the importance of any one level, and also to introduce on additional temperature-dependent term. This is equivalent to a diminution of the activation energy by kT, and it is noteworthy that many of the activation energies reported by Kelly and Padley do fall slightly below the ionization potential, though those reported by Hayhurst and Telford do not.[194]

The weight of evidence thus favours the view that the ionization of simple species in flames is a straightforward collisional process, with an activation energy approximately equal to the ionization potential, the process being rapid through the participation of excited normal frequency factor and lowered energy of activation by a process involving the formation of complex ions. There still remains a residue of uncertainty, which is associated with doubts about the homogeneity of the flame and the processes occurring up to the onset of ionization.

3.7.6. Super-equilibrium Ionization

In 1953, Smith,[184] examining the effect of gaseous chlorine on the ionization of the alkali metals using a resonant circuit, showed that, over certain ranges, the frequency shift produced when sodium or lithium was sprayed into the flame was in the opposite sense to that expected, whereas for most conditions, and for the other alkali metals, the behaviour was as expected. This anomalous effect with sodium and lithium corresponded to an increase in free electron concentration being produced by the addition of an electron acceptor. Page, using a direct microwave attenuation technique, subsequently showed that there was indeed an increase in the free electron concentration for small addition of halogen, but that it passed through a maximum, and large additions of halogen resulted in the expected fall. It was shown that the effect depended on the temperature and on the particular combination of alkali metal and halogen being studied.

In 1963, Karman and Guiffrida[185] announced a similar effect which they had observed in flame ionization detectors for gas chromatography, when an alkali metal and phosphorus-containing compound were present in the flame simultaneously, and Page and Woolley[186] showed that this effect, too, was the result of an increase in the electron level above that to be expected in the absence of acceptor.

In discussing the halogen phenomena, Padley, Page and Sugden[187] put forward the following explanation:

In a system at equilibrium, the electron concentration [E] from an alkali metal at a concentration [X] will fall, in the presence of an acceptor Y, because of the formation of Y^- ions, and of XY, and the ratio $[e]/[e_0]$ will fall from a value $1 \cdot 0$ as [Y] increases along curve A (Fig. 3.24). It is known, however, that the hydrogen atoms are not at equilibrium, and that at any point a disequilibrium parameter $\gamma = [H]/[H_e]$ may be defined.

Negative ions of Y may be formed from HY by two processes

$$HY + e \rightleftharpoons Y^- + H$$
$$HY + H + e \rightleftharpoons Y^- + H_2$$

If the system is truly at equilibrium, either process may represent it but since the hydrogen atoms are not an equilibrium, the processes are not equivalent, the first leading to a ratio $[e]/[Y^-]$ proportional to [H] and therefore to γ, while the second leads to a proportionality to λ^{-1}. A simultaneous operation of both processes is equivalent to the catalysis of the recombination reaction

$$H + H \longrightarrow H_2$$

It is known that ionization in the absence of halogen follows process 2, and the ratio $[e]/[e_0]$ would be expected to follow curve B (Fig. 3.24)

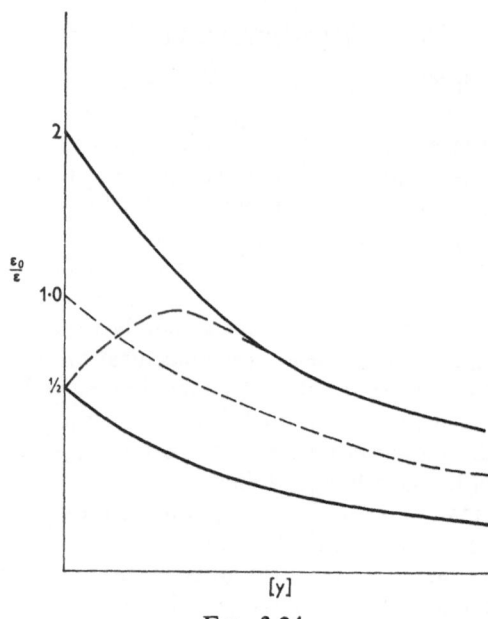

FIG. 3.24

Padley, Page and Sugden postulated that the addition of Y caused a shift to process 1, which would cause $[e]/[e_0]$ to follow curve C. In actual fact, the ratio would start on curve B and cross over to curve C, producing the dotted experimental curves D or E, depending on the efficiency of transfer.

Sugden and Hayhurst[188] later showed that the analysis could be made even more general. They considered the seven chemical processes

(1) $A + X \underset{k_1}{\overset{k_1}{\rightleftharpoons}} A^+ + e + X$ K_1

(2) $A + B \underset{k_2}{\overset{k_2}{\rightleftharpoons}} A^+ + B^-$ K_2

(3) $B^- + H \underset{k_3}{\overset{k_3}{\rightleftharpoons}} HB + e$ K_3

(4) $HB + H \underset{k_4}{\overset{k_4}{\rightleftharpoons}} H_2 + B$ K_4

(5) $AOH + H \underset{k_5}{\overset{k_5}{\rightleftharpoons}} A + H_2O$ K_5

(6) $AB + H \underset{k_6}{\overset{k_6}{\rightleftharpoons}} A + HB$ K_6

(7) $H_2O + H \underset{k_7}{\overset{k_7}{\rightleftharpoons}} H_2 + OH$ K_7

and the three conservation equations

(8) $[A_0] = [A] + [AOH] + [AB]$

(9) $[B_0] = [B] + [HB] + [AB]$

(10) $[A^+] = [e] + [B^-] + [OH^-]$

In the actual analysis, [AB] was omitted in equation 9 and [OH$^-$] in equation 10, as being small.

The rates of direct ionization (process 1) and indirect ionization (processes 2 and 3) were related by defining the ratios

$$R_2 = k_{-2}/k_{-1}[X] \qquad R_3 = k_{-3}/k_{-1}[X]$$

also

$$K_1 = k_1/k_{-1}$$

Attention was confined to processes 1–3, as the others were all considered, from flame photometric evidence, to be fast and balanced at all times.

Certain distribution parameters were defined and related to elementary equilibrium constants.

$$\eta = [B^-]/[e]; \qquad \eta_e = [B]_e/K_\eta; \qquad \eta/\eta_e = \gamma^{-1}$$

$$\phi = [AOH]/[A]; \qquad \phi_e = [OH]_e/K_\phi; \qquad \phi/\phi_e = \gamma^{-1}$$

$$\theta = [HB]/[B]; \qquad \theta_e = [H]_e/K_\theta; \qquad \theta/\theta_e = \gamma^{-1}$$

$$\zeta = [AB]/[A]; \qquad \zeta_e = [B]_e/K_\zeta; \qquad \zeta/\zeta_e = \gamma^{-1}$$

where γ is the disequilibrium parameter [H]/[H$_e$]. Processes 1–3 are assumed to have brought ions and electrons to a steady state.

$$d[A^+]/dt = k_1[A][X] - k_{-1}[A^+][e][X] + k_2[A][B] - k_{-2}[A^+][B^-] = 0$$

$$d[B^-]/dt = k_2[A][B] - k_{-2}[A^+][B^-] - k_3[H][B^-] + k_{-3}[HB][e] = 0$$

$$d[e]/dt = k_1[A][X] - k_{-1}[A^+][e][X] + k_3[H][B^-] - k_{-3}[HB][e]$$
$$= 0$$

Taking the negative ions first,

$$\{k_2[A] + k_{-3}[e]\theta\}[B] = \{k_{-2}[A^+] + k_3[H]\}[B^-]$$

$$\frac{[B^-]}{[B]} = \frac{k_2[A] + k_{-3}[e]\theta}{k_{-2}[A^+] + k_3[H]} = \frac{K_2R_2[A] + R_3[e]\theta}{R_2[A^+] + K_3R_3[H]}$$

For electrons

$$k_1[A][X] - k_{-1}[A^+][e][X] = k_{-3}[e][B]\theta - R_3K_3[H][B^-]$$

$$K'[A] - [A^+][e] = R_3[e][B]\theta - R_3K_3[H][B^-]$$

$$= R_3[e][B]\theta = K_3[H][B]\left\{\frac{K_2R_1[A] + R_3[e]\theta}{R_2[A^+] + K_3R_3[H]}\right\}$$

$$= R_3[B]\left\{\frac{R_2[A^+][e]\theta - K_2K_3R_2[H][A]}{R_2[A^+] + K_3R_3[H]}\right\}$$

Noting that:

$$[A^+] = [e] + [B^-] = [e](1 + \eta)$$
$$[B_0] = [B] + [HB] = [B](1 + \theta)$$
$$[A_e] = [A] + [AOH] + [AB] = [A](1 + \phi + \zeta)$$

the relation becomes

$$K'[A] - [e]^2(1 + \eta) = \frac{R_2R_3[B_0]}{(1 + \theta)}\left\{\frac{[e]^2\theta(1 + \eta) - K_2K_3[H][A]}{R_2[e](1 + \eta) + K_3R_3[H]}\right\}$$

This is a cubic in $[e]$, which is not easy to use in analysing experimental results. However Hayhurst and Sugden showed that, at least for the halogens, the term $R_2[e](1 + \eta)$ in the denominator of the R.H.S. can be neglected if R_3 were sufficiently large; the equation now simplifies to a linear equation in $[e]^2$. In doing this, it is useful to relate the real equilibrium constants K_2 and K_3 to the elementary equilibrium constants K_η etc.

$$K_2 = [A^+][B^-]/[A][B] = [A^+][e]\eta/[A][B]$$
$$= K'\eta/[B] = K'/K_\eta$$
$$K_3 = [HB][e]/[H][B^-] = [HB]/[H]\eta$$
$$= [B]\theta/[H]\eta$$
$$= \theta_e K_\eta/[H]_e$$

The relation then becomes

$$[e]^2(1 + \eta)\left\{\frac{R_2[B_0]}{(1 + \theta)}\frac{\theta[H]_e}{\theta_e K_\eta[H]} + 1\right\} = [A]K' + \frac{R_2[B_0]}{(1 + \theta)}\frac{K'[A]}{K_\eta}$$

Hence

$$[e]^2 = \frac{K'[A]}{(1 + \eta)}\left\{\frac{1 + R_2[B_0]/K_\eta(1 + \theta)}{1 + R_2[B_0]/K_\eta(1 + \theta)\gamma^2}\right\}$$

Inserting the value for $A = [A_0]/(1 + \phi + \zeta)$

and writing $X = R_2/K_\eta(1 + \theta)$

$$[e]^2 = \frac{K'[A_0](1 + X[B_0])}{(1 + \eta)(1 + \theta + \zeta)(1 + X[B_0]\gamma^{-2})}$$

If $[e_0]$ is the electron concentration when $[B_0] = 0$ and $E = [e]/[e_0]$

$$E^2 = \frac{(1 + X[B_0])(1 + \phi)}{(1 + X[B_0]\gamma^{-2})(1 + \theta + \zeta)(1 + \eta)}$$

The corresponding expression for $P = [A^+]/[A_0^+]$ is

$$P^2 = \frac{(1 + X[B_0])(1 + \phi)(1 + \eta)}{(1 + X[B_0]\gamma^{-2})(1 + \theta + \zeta)}$$

On differentiating these expressions, the initial slope of the E^2 or P^2 graph against $[B_0]$ may be determined

$$\frac{dE^2}{d[B_0]_{[B_0]=0}} = \frac{R_2(\gamma^2 - 1) - \gamma^2(1 + \theta)(1 + K_\eta/K_\zeta)}{K_\eta\gamma^2(1 + \theta)}$$

$$\frac{dP^2}{d[B_0]_{[B_0]=0}} = \frac{R_2(\gamma^2 - 1) + \gamma^2(1 + \theta)(1 - K_\eta/K_\zeta)}{K_\eta\gamma^2(1 + \theta)}$$

The conditions for a maximum in the graph may also be evaluated with the results

$$E^2_{max} = \frac{R_2(\gamma^2 - 1)}{\gamma^2(1 + \theta)} \frac{L}{(1 + X[B_m]\gamma^{-2})^2} \frac{1}{(1 + K_\eta/K'_\zeta + 2[B_m]/K_\eta K'_\zeta)}$$

where $[B_m]$, $E'_{\zeta max}$ are the values at the maximum and $K'_\zeta = K_\zeta(1 + \phi)$

$$P^2_{max} = \frac{R_2(\gamma^2 - 1)}{\gamma^2(1 + \theta)} \left[\frac{(1 + \eta_{max})}{(1 + X[B_0]\gamma^{-2})}\right]^2 \frac{1}{(K_\eta/K'_\zeta - 1)}$$

Since $\gamma > 1$, a maximum in P will only be observed if $K_\eta > K'_\zeta$: this condition which will not, in general, be satisfied.

These relations have been exploited by Page and Woolley[186] and by Woodfield[189] for the case of phosphorus acting as acceptor, though without definite kinetic results, and by Hayhurst and Sugden who studied the halogens as acceptors. The detailed data needed to calculate the subsidiary functions were available, and the negative ion, B^- was clearly identified. These authors worked in the temperature region 1800–2300, the measured values of γ falling between 3 and 17. Kinetic limitation of the ionization of the alkali metal was avoided by the addition of 0·8% of acetylene so that the steady state treatment was valid.

They used both the initial slope and the conditions at the maximum in the electron concentration to evaluate R_2, and found that the initial slopes were much less sensitive to errors in γ.

Typically, for sodium and the halogens the following values of R_2 were obtained.

TABLE 3.8
Value of R_2 for Sodium

Halogen	Flame temperature		(K)
	1800	2000	2270
Cl	0·030	0·20	0·32
Br	0·017	0·15	0·50
I	0·018	0·13	0·33

Potassium and rubidium gave almost identical results, while no maxima were observed with caesium. Lithium ionized too little to give useful results. The values of R_2 were combined with the known recombination rates to give the values of k_{-2}, the rate constant for the process $A + B \rightarrow A^+ + B^-$ at 1800 K (Table 3.9).

TABLE 3.9
Rate constant in $cm^3\ s^{-1}$ for the process
$A + B \longrightarrow A^+ + B^-$ at 1800 K

A	B	K_{exp}	K_{theor}
	Cl	6×10^{-10}	2×10^{-9}
Na	Br	$3·5 \times 10^{-10}$	—
	I	4×10^{-10}	—
	Cl	$1·6 \times 10^{-10}$	—
K	Br	$2·3 \times 10^{-10}$	6×10^{-11}
	I	$2·5 \times 10^{-9}$	2×10^{-9}

The rate constant k of an exothermic reaction, with an internuclear spacing of r_e in the transition complex may be written, following Magee, as

$$k = 2\ Kr_e^2 \sqrt{\frac{2\pi kT}{\mu_{AB}}}$$

Here k is a transmission coefficient and μ_{AB} the reduced mass of the colliding system. The spacing r_e is equated to $e^2/(I_A - E_B)$ where e is the electronic charge and I_A is the ionization potential of A and E_B is the electron affinity of B. The transmission coefficient k is related to ρ, the probability of crossing over from one potential curve to another by

$$K = 2\rho(1 - \rho)$$

and ρ in turn may be related to the energy separation of the curves, and their slopes, by the Landau–Zener formula

$$\rho = \exp\left(4\pi^2\varepsilon^2/h\nu(S_I - S_{II})\right)$$

The theoretical values of Table 3.9 were calculated in this way, and the agreement with experiment is impressive, in view of the criticism of the simple Landau–Zener formula by Bates[190] and by Coulson.[191]

3.8. PERSPECTIVES IN FLAME KINETICS

3.8.1. Introduction

The study of reaction kinetics in flames has always been severely limited in scope. It has been concerned primarily with the reactions of trace additions to flames, and only secondarily with the flame reactions as such. The temperatures are too high for the detailed chemistry to be studied conveniently, and too low for the detail to be irrelevant. A great deal of work was done during the past two decades under the impetus given firstly by rocketry and later by magnetohydrodynamics but the immediate questions have been answered and the necessary numerical data, as has been shown for the hydrogen–oxygen reaction, are more conveniently obtained by other techniques. Flame and shock-tube results for the hydrogen-oxygen reaction are compared in Chapter 2 of this book.

The impact of the computer on complex reaction kinetics has directed attention to the individual steps in the mechanism which are rarely high-lighted in flames. Nevertheless, there remains much work to be done—the hyperionization reactions are a virtually unexplored field, while the contribution of transport processes, of heat or mass, to flame kinetics is only beginning to be assessed in real terms.

An increasing amount of work is devoted to processes in two-phase systems—and a reasonably complete account of the transformation of a droplet or particle, whether of fuel or additive, into the final state of uniformly dispersed vapour is as yet far away. In these studies many of the techniques used in flame kinetics are applied to follow physical rather than chemical changes.

3.8.2. Turbulent Flames

One of the most recent areas to be attacked is the turbulent regime. Turbulent flames are of dominant importance, and are typically emitters of audible noise. Since Reynolds' criterion for turbulence involves the characteristic dimension, it is found that almost all large industrial flames are turbulent. In a turbulent system it is not possible to describe the instantaneous local functions in continuous terms, but only to talk of mean values over a range of space or time.

If a probe is measuring the instantaneous value of a concentration, the signal may be regarded as a mean signal with a statistically random component superimposed on it. The random component is associated with the turbulent fluctuations in the flame which are constantly dying away and being reborn. If the signal from the probe is compared with the value which it had τ seconds earlier, there will be a relation between the two values since the fluctuations have a finite duration in space and time. The signals are said to correlate, and a correlation coefficient R may be defined

$$ R = \frac{\overline{S_1 S_2}}{[\langle \overline{S_1^{-2}} \rangle \langle \overline{S_2^{-2}} \rangle]^{1/2}} $$

where S_1 and S_2 are the two signals.

A measurement of R may be achieved by splitting the signal and passing one part through a time delay unit before recombining the signals in an analogue device. If the random component is small, the values of S will be close to each other and R will be close to 1; even if the random component is large the correlation coefficient may still be close to 1 if the time delay is less than the lifetime of the turbulent fluctuation at the probe.

At zero time delay, the two parts of the signal correlate exactly, but after some long delay there is no coherence at all, and the correlation coefficient reaches zero. Some recent results obtained by Roberts and Williams which illustrate this for an electrostatic probe following ionization in a turbulent diffusion flame of hydrogen containing 1 % of acetylene. The time delay required for the correlation coefficient to reach a value of e^{-1} is known as the "e-folding distance" and, transformed into spatial rather than temporal co-ordinates through a knowledge of the flow field in the flame, represents the scale of macroscopic turbulence, or size of the smallest significant eddies. The integral under the curve is also a time, since R is dimensionless, and is a measure of the microscopic turbulence in the flame.

In the flame without chemical change, the loss of correlation will be due solely to turbulent decay, but if chemical change also affects the measurement, the loss of correlation will be enhanced and when loss by chemical change is comparable to loss by turbulent decay, the e-folding distance will be significantly altered. The effect of kinetics where the decays of ionization by recombination in the presence of sodium or chromium are slow, and the correlation curves indistinguishable, but the rapid recombination of H_3O^+ and electrons dominates the correlation curve.

3.8.3. Auto and Cross Correlation

The technique just described, in which a signal is correlated with itself after a time delay is known as *auto correlation*. It is equally possible to correlate two independent signals, and such technique is known as *cross correlation*. In general, there will be no correlation between two independent signals, and R will be everywhere zero. If, however, the two signals refer to the same element of volume, there will be a correlation. In particular if one total signal is made up of signals from elements along one path, and a second total signal from elements along a path at right angles to the first, and the two paths have one element of volume in common, the cross-correlated signal will derive only from that element, the signals from all other elements in either path averaging to zero. The correlation coefficient will be much less than 1, since the relevant product $S_1 S_2$ will always be much less than the product of the RMS signals, but will be finite, and affords a method of measuring local concentrations in clearly defined elementary regions without disturbing the flame. Furthermore, if the cross beams are displaced spatially (and hence temporarily) with respect to each other, and a suitable time delay incorporated, a correlation may still be observed if the element observed in the first beam has reached the second beam in the relevant time. Again the correlation will be affected both by turbulence and by kinetics, and both processes may therefore be studied by this technique.

It is apparent that the situation is ripe for a further step, and that the experimental tools are available to gather the necessary information whereon to base an understanding of kinetics in turbulent flames, and even if it is unlikely that a change in classical kinetic thought will result, the extension of ideas which will be needed to establish physical and chemical processes going on in a statistically variable medium as a normal and proper vehicle for the application of kinetic principles will ensure the keen interest and attention of all who study kinetics in flames.

REFERENCE

1. E. Pungor, Flame Photometry Theory (Van Nostrand, London, 1967) p. 66
2. B. Lewis and G. von Elbe, Combustion, Flames and Explosions of Gases (Academic Press, New York, 1959)
3. K. Wohl, N. M. Kapp and C. Gazley, 3rd Symp. on Combustion (Williams and Wilkins, Baltimore, 1969) p. 3
4. E. Mallard and H. Le Chatelier, *Ann. Min.*, **4**, 274 (1883)
5. W. Nusselt, *Z. Ver. Deutsche. Ing.*, **59**, 872 (1915)
6. E. Bartholmé and C. Herrmann, *Z. Electrochem.*, **54**, 165 (1950)
7. K. Bechert, *Z. Naturforsch.*, **3A**, 584 (1968)
8. N. N. Semenov, *Prog. Phys. Sci. USSR*, **24**, 433 (1960)
9. G. L. Dugger, *J. Am. Chem. Soc.*, **72**, 5271 (1950)
10. P. L. Walker and C. C. Wright, *J. Am. Chem. Soc.*, **75**, 750 (1953)

11. C. Tanford and R. N. Pease, *J. Chem. Phys.*, **15**, 431 (1947)
12. C. Tanford and R. N. Pease, *J. Chem. Phys.*, **15**, 861 (1947)
13. P. Wagner and G. L. Dugger, *J. Am. Chem. Soc.*, **77**, 227 (1955)
14. W. H. Clingman, R. S. Brokaw and R. N. Pease, 4th Symp. on Combustion (Williams and Wilkins, Baltimore, 1953) p. 310
14a. A. F. Ashton and A. N. Hayhurst, *Trans. Faraday Soc.*, **67**, 2348 (1971)
15. P. J. Padley and T. M. Sugden, 7th Symp. on Combustion (Butterworths, London, 1959) p. 235
16. A. van Tiggelen and J. Deckers, 6th Symp. on Combustion (Reinhold, New York, 1957) p. 61
17. A. E. Potter and A. L. Berlad, 6th Symp. on Combustion (Reinhold, New York, 1957) p. 27
18. E. R. Miller, Ph.D. Thesis, University of Aston (1968)
19. H. Smith, Ph.D. Thesis, University of Cambridge (1952)
20. C. G. James, Ph.D. Thesis, University of Cambridge (1953)
21. F. M. Page, Ph.D. Thesis, University of Cambridge (1955)
22. T. M. Sugden, *Trans. Faraday Soc.*, **52**, 1465 (1965)
23. E. Pourbaix, Thermodynamics of Dilute Aqueous Solutions (Edward Arnold, London, 1949)
24. F. M. Page and T. M. Sugden, *Trans. Faraday Soc.*, **53**, 1092 (1957)
25. P. F. Knewstubb, 9th Symp. on Combustion (Academic Press, New York, 1963) p. 635
26. R. M. Fristrom, W. H. Avery, and R. C. Grunfelder, 7th Symp. on Combustion (Butterworths, London, 1959) p. 658
27. R. M. Fristrom and A. A. Westenberg, 8th Symp. on Combustion (Williams and Wilkins, Baltimore 1962) p. 438
28. S. Arrhenius, *Ann. Phys. Chem.*, **42**, 30 (1891)
29. H. A. Wilson, *Mod. Phys.*, **3**, 156 (1931)
30. E. Andrew, D. W. E. Axford and T. M. Sugden, *Trans. Faraday Soc.*, **44**, 427 (1948)
31. E. Marx, Handbuch der Radeologie, **4**, (1927)
32. A. G. Gaydon, Spectroscopy and Combustion Theory (Chapman & Hall, London, 1948)
33. A. G. Gaydon, G. N. Spokes and J. van Suchtelen, *Proc. Roy. Soc.*, **256A**, 323 (1960)
34. P. Lenard, *Ann. der physik*, 730 (1931)
35. G. Kirchoff and R. Bunsen, *Phil. Mag.*, **20**, 89 (1860)
36. J. Janssen, *Comptes rendus*, **71**, 626 (1870)
37. H. Lundegardh, *Arkiv Kemi. Mineral Geol.*, **10**, No. 1 (1928)
38. C. G. James and T. M. Sugden, *Proc. Roy. Soc.*, **227A**, 312 (1955)
39. C. G. James and T. M. Sugden, *Nature*, **175**, 252 (1955)
40. E. F. M. van der Held and J. H. Herreman, *Physica*, **3**, 31 (1936)
40a. C. Th. J. Alkemade and P. J. Th. Zeegens, Quantitative analysis of atoms and molecules, ed. J. D. Winefordner (J. Wiley and Sons, 1971) p. 3
41. E. M. Bulewicz and T. M. Sugden, *Trans. Faraday Soc.*, **52**, 1481 (1956)
42. H. Belcher and T. M. Sugden, *Proc. Roy. Soc.*, **201A**, 17 (1950)
43. T. M. Sugden, 10th Symp. on Combustion (Combustion Institute, Pittsburg, 1965) p. 636
44. J. Kay and F. M. Page, *Trans. Faraday Soc.*, **62**, 3081 (1966)
45. J. Schneider and F. W. Hofman, *Phys. Rev.*, **116**, 244 (1959)
46. H. Belcher and T. M. Sugden, *Proc. Roy. Soc.*, **202A**, 17 (1950)
47. H. Smith and T. M. Sugden, *Proc. Roy. Soc.*, **211A**, 31 (1952)
48. F. M. Page, *Disc. Faraday Soc.*, **19**, 87 (1955)
49. F. M. Page, R. G. Soundy and H. Williams, Unpublished work (1963)
50. E. M. Bulewicz and P. J. Padley, 9th Symp. on Combustion (Academic Press, New York, 1963) p. 638

51. T. M. Sugden and B. A. Thrush, *Nature*, **168**, 703 (1953)
52. A. Horsfield, Ph.D. Thesis, University of Cambridge (1957)
53. A. A. J. Pennycook, private communication (1958)
54. E. M. Bulewicz and P. J. Padley, *J. Chem. Phys.*, **35**, 1590 (1961)
55. P. J. Padley and T. M. Sugden, 8th Symp. on Combustion (Williams and Wilkins, Baltimore, 1962) p. 164
56. D. E. Jensen and P. J. Padley, 11th Symp. on Combustion (Combustion Institute, Pittsburg, 1967) p. 351
57. P. F. Knewstubb and T. M. Sugden, *Trans. Faraday Soc.*, **54**, 372 (1958)
58. H. Williams, 8th Symp. on Combustion (Williams and Wilkins, Baltimore 1962) p. 179
59. A. J. Borgers, 10th Symp. on Combustion (Combustion Institute, Pittsburg, 1965) p. 627
60. K. E. Shuler and J. Weber, *J. Chem. Phys.*, **22**, 491 (1954)
61. H. F. Calcote and I. R. King, 5th Symp. on Combustion (Reinhold, New York, 1955) p. 423
62. H. F. Calcote, Progress in Astronautics, **12**, 107 (1962)
63. T. Heumann, *Specrochim. Acta*, **1**, 293 (1940)
64. T. Kinbara, J. Nakamura and H. Ikegami, 7th Symp. on Combustion (Butterworths, London, 1959) p. 263
65. L. Rolla and G. Piccardi, *Atti. Accad. Lincei. VI*, **2**, 29 (1925)
66. I. R. King and H. F. Calcote, *J. Chem. Phys.*, **23**, 2203 (1955)
67. H. F. Calcote, 8th Symp. on Combustion (Williams and Wilkins, Baltimore, 1962) p. 184
68. H. F. Calcote, 9th Symp. on Combustion (Academic Press, New York, 1963) p. 622
69. I. R. King, *J. Chem. Phys.*, **27**, 817 (1957)
70. G. Wortberg, 10th Symp. on Combustion (Combustion Institute, Pittsburg, 1965) p. 651
71. R. G. Soundy and H. Williams, AGARD. P & E panel 26th meeting, p. 165, Pisa (1965)
72. B. E. L. Travers and H. Williams, 10th Symp. on Combustion (Combustion Institute, Pittsburg, 1965)
73. C. H. Su and S. H. Lam, Physics of Fluids, **6**, 1479 (1963)
74. G. J. Suchulz and S. C. Brown, *Phys. Rev.*, **88**, 1942 (1955)
75. J. R. Cozens and A. von Engels, *Nature*, **202**, 480 (1964)
76. J. R. Cozens, 10th Symp. on Combustion (Combustion Institute, Pittsburg, 1965) p. 670
77. H. Williams, 10th Symp. on Combustion (Combustion Institute, Pittsburg, 1965) p. 669
78. C. J. Matthews, *Rev. Pure Appl. Chem.*, **18**, 311 (1968)
79. G. C. Eltenton, *J. Chem. Phys.*, **15**, 474 (1947)
80. S. N. Foner and R. L. Hudson, *J. Chem. Phys.*, **21**, 1608 (1953)
81. R. M. Fristrom 9th Symp. on Combustion (Academic Press, New York, 1963) p. 560
82. F. S. Klein and J. T. Herron, *J. Chem. Phys.*, **41**, 1285 (1969)
83. H. F. Calcote and J. L. Reuter, *J. Chem. Phys.*, **38**, 310 (1963)
84. J. Deckers and A. van Tiggelen, *Combustion and Flame*, **1**, 281 (1957)
85. J. Deckers and A. van Tiggelen, *Nature*, **182**, 863 (1958)
86. J. Deckers and A. van Tiggelen, 7th Symp. on Combustion (Butterworths, London, 1959) p. 254
87. S. de Jaegere, J. Deckers and A. van Tiggelen, 8th Symp. on Combustion (Williams and Wilkins, Baltimore, 1962) p. 155
88. P. F. Knewstubb and T. M. Sugden, *Nature*, **181**, 1261 (1958)
89. P. F. Knewstubb and T. M. Sugden, 7th Symp. on Combustion (Butterworths, London, 1959) p. 247

90. P. F. Knewstubb and T. M. Sugden, *Proc. Roy. Soc.*, **A255**, 520 (1960)
91. P. F. Knewstubb and T. M. Sugden, *Nature*, **196**, 1311 (1962)
92. W. J. Miller and H. F. Calcote, *J. Chem. Phys.*, **41**, 4001 (1964)
93. H. F. Calcote, S. C. Kurzius and W. J. Miller, 10th Symp. on Combustion (Combustion Institute, Pittsburg, 1965) p. 605
94. A. Feugier and A. van Tiggelen, 10th Symp. on Combustion (Combustion Institute, Pittsburg, 1965) p. 621
95. J. A. Green and T. M. Sugden, AGARD P & E panel 26th Meeting, p. 130, Pisa (1965)
96. A. G. Gaydon and H. G. Wolfhard, Flames (Chapman & Hall, London 1960)
97. J. Lawton and F. J. Weinberg, Electrical Aspects of Combustion (Clarendon Press, Oxford, 1969)
98. E. M. Bulewicz, C. G. James and T. M. Sugden, *Proc. Roy. Soc.*, **235A**, 89 (1956)
99. A. C. G. Mitchell and M. W. Zemansky, Resonance Radiation and Excited Atoms (C.U.P., Cambridge, 1961)
100. R. Mavrodineau, Flame Spectroscopy (Wiley, New York, 1965)
101. S. S. Penner, Quantitative Molecular Spectroscopy (Addison-Wesley, Reading, 1959)
102. D. H. Cotton and D. R. Jenkins, *Trans. Faraday Soc.*, **65**, 1537 (1969)
103. P. J. Th. Zeegers and C. Th. J. Alkemade, *Combustion and Flame*, **15**, 193 (1970)
104. M. J. McEwan and L. F. Phillips, *Combustion and Flame*, **9**, 420 (1965)
105. M. J. McEwan and L. F. Phillips, *Combustion and Flame*, **11**, 63 (1967)
106. L. F. Phillips, Ph.D. Thesis, University of Cambridge (1961)
107. G. C. Collins and H. Polkinhorne, *Analyst.*, **77**, 430 (1952)
108. J. R. Arthur, *Nature*, **164**, 537 (1949)
109. E. M. Bulewicz and T. M. Sugden, *Trans. Faraday Soc.*, **52**, 1475 (1956)
110. C. G. James and T. M. Sugden, *Proc. Roy. Soc.*, **248A**, 238 (1958)
111. C. G. James and T. M. Sugden, *Nature*, **175**, 252 (1955)
112. P. J. Padley, *Trans. Faraday Soc.*, **56**, 449 (1960)
113. L. F. Phillips and T. M. Sugden, *Canadian J. Chem.*, **38**, 1804 (1960)
114. H. P. Broida and D. F. Heath, *J. Chem. Phys.*, **36**, 385 (1962)
115. A. G. Gaydon, *Trans. Faraday Soc.*, **42**, 292 (1946)
116. E. M. Bulewicz and T. M. Sugden, *Proc. Roy. Soc.*, **277A**, 143 (1964)
117. L. F. Phillips and T. M. Sugden, *Trans. Faraday Soc.*, **57**, 914 (1961)
118. P. J. Padley and T. M. Sugden, *Proc. Roy. Soc.*, **248A**, 248 (1958)
119. P. J. Padley and T. M. Sugden, 7th Symp. on Combustion (Butterworths, London, 1959) p. 235
120. T. M. Sugden, *Ann. Rep. Phys. Chem.*, **13**, 369 (1962)
121. A. G. Gaydon and H. G. Wolfhard, *Proc. Roy. Soc.*, **201A**, 570 (1950)
122. P. T. Gilbert, *Proc. Xth Colloq. Spect. Int.*, 171 (1962)
123. F. M. Page, *Disc. Faraday Soc.*, **19**, 87 (1955)
124. P. G. Ashmore, 10th Symp. on Combustion (Combustion Institute, Pittsburg, 1965) p. 377
125. E. M. Bulewicz and T. M. Sugden, *Trans. Faraday Soc.*, **54**, 1855 (1958)
126. G. Dixon-Lewis, M. M. Sutton and A. Williams, *Disc. Faraday Soc.*, **33**, 205 (1962)
127. G. Dixon-Lewis, M. M. Sutton and A. Williams, 10th Symp. on Combustion (Combustion Institute, Pittsburg, 1965) p. 495
128. J. L. T. Rosenfeld and T. M. Sugden, *Combustion and Flame*, **8**, 44 (1964)
129. T. M. Sugden, private communication in ref 130
130 P. J. T Zeegers, Ph.D. Thesis, Utrecht (1965)
131. P. J. T. Zeegers and C. T. J. Alkemade, 10th Symp. on Combustion (Combustion Institute, Pittsburg, 1965) p. 33
132. R. McAndrew and R. C. Wheeler, *J. Phys. Chem.*, **66**, 229 (1959)

133. T. Hollander, Ph.D. Thesis, Utrecht (1963)
134. C. J. Halstead and D. R. Jenkins, 12th Symp. on Combustion (Combustion Institute, Pittsburg, 1969) p. 979
135. C. J. Halstead and D. R. Jenkins, *Combustion and Flame*, **14**, 321 (1970)
136. W. E. Kaskan, *J. Chem. Phys.*, **31** 944 (1959)
137. R. W. Reid and T. M. Sugden, *Disc. Faraday Soc.*, **33**, 213 (1962)
138. J. A. Green and T. M. Sugden, 9th Symp. on Combustion (Academic Press, New York, 1963) p. 607
139. P. F. Knewstubb and T. M. Sugden, *Nature*, **181**, 474 (1958)
140. P. F. Knewstubb and T. M. Sugden, 7th Symp. on Combustion (Butterworths, London, 159) p. 247
141. P. F. Knewstubb and T. M. Sugden, *Proc. Roy. Soc.*, **255A**, 1920 (1960)
142. K. N. Bascombe, J. A. Green and T. M. Sugden, Joint Symp. on Mass Spectrometry (Pergamon, 1962) p. 66
143. H. F. Calcote, *Combustion and Flame*, **1**, 385 (1957)
144. K. Schofield, 10th Symp. on Combustion (Combustion Institute, Pittsburg, 1965) p. 604
145. J. Peeters and A. van Tiggelen, 12th Symp. on Combustion (Combustion Institute, Pittsburg, 1969) p. 437
146. H. F. Calcote, *AGARD Conf. Proc. No. 8.* **1**, Pisa, 1965, p. 142
147. H. A. Wilson, *Rev. Mod. Phys.*, **3**, 156 (1931)
148. R. M. Newman, unpublished work (1970)
149. I. R. King, *Prog. Astronautics Aeronautics*, **12**, 197 (1963)
150. H. F. Calcote, The Dynamics of Conducting Gases (N.W.U.P., Evanston, 1960)
151. I. R. Hurle, T. M. Sugden and G. B. Nutt, 12th Symp. on Combustion (Combustion Institute, Pittsburg, 1969) p. 387
152. E. M. Bulewicz and P. J. Padley, 9th Symp. on Combustion (Academic Press, New York, 1963) p. 647
153. J. A. Green, I. R. Hurle, G. B. Nutt and T. M. Sugden, mentioned in ref. 151
154. H. F. Calcote and S. C. Kurzius, Aerochem. Report T.P. 92, Princeton (1964)
155. K. N. Bascombe, H. Schiff and T. M. Sugden, mentioned in ref. 142
156. P. F. Knewstubb, Ph.D. Thesis, University of Cambridge (1956)
157. R. Kelly and P. J. Padley, *Trans. Faraday Soc.*, **66**, 1127 (1970)
158. A. N. Hayhurst and N. R. Telford, *Nature*, **212**, 813 (1966)
159. A. N. Hayhurst and T. M. Sugden, Proc. 20th IUPAC Congress, Moscow (1965)
160. A. N. Hayhurst and T. M. Sugden, *Proc. Roy. Soc.*, **293A**, 36 (1966)
161 P J. Padley and T. M. Sugden, 8th Symp. on Combustion (Williams and Wilkins, Baltimore, 1962) p. 164
162. A. N. Hayhurst and N. R. Telford, *Trans. Faraday Soc.*, **66**, 2784 (1970)
163. R. G. Soundy and H. Williams, *AGARD Conf. Proc. No. 8*, **1**, (1963) p. 165
164. E. M. Bulewicz, *AGARD Conf. Proc. No. 8*, **1**, (1965) p. 185
165. D. E. Jensen and P. J. Padley, 11th Symp. on Combustion (Combustion Institute, Pittsburg, 1967) p. 357
166. K. Schofield and T. M. Sugden, 10th Symp. on Combustion (Combustion Institute, Pittsburg, 1965) p. 589
167. R. Kelly and P. J. Padley, *Trans. Faraday Soc.*, **65**, 355 (1969)
168. S. C. Brown, Basic Data of Plasma Physics (Wiley, New York, 1961)
169. D. E. Jensen and P. J. Padley, *Trans. Faraday Soc.*, **62**, 2140 (1966)
170. E. M. Bulewicz and P. J. Padley, 9th Symp. on Combustion (Academic Press, New York, 1963) p. 638
171. P. F. Knewstubb and T. M. Sugden, *Trans. Faraday Soc.*, **54**, 372 (1958)
172. F. T. Smith, 8th Symp. on Combustion (Williams and Wilkins, Baltimore, 1962) p. 178

173. A. N. Hayhurst, 10th Symp. on Combustion (Combustion Institute, Pittsburg, 1965) p. 602
174. R. Fowler, Statistical Mechanics (C.U.P., 1936) p. 338
175. C. T. J. Alkemade, Thesis, Utrecht (1954)
176. A. L. Boers, C. T. J. Alkemade and J. A. Smit, *Physica*, **22**, 358 (1956)
177. M. N. Saha, *Phil. Mag.*, **40**, 472 (1920)
178. A. J. Borgers, 10th Symp. on Combustion (Combustion Institute, Pittsburg, 1965) p. 627
179. A. J Borgers, Thesis, Utrecht (1966)
180 T. Hollander, Thesis, Utrecht (1964)
181. T. Hollander, 11th Symp. on Combustion (Combustion Institute, Pittsburg, 1967) p. 356
182. T. Hollander, P. J. Kalff and C. T. J. Alkemade, *J. Chem. Phys.*, **39**, 2558 (1963)
183. S. W. Benson, 9th Symp. on Combustion (Academic Press, New York, 1963) p. 760
184. H. Smith, private communication (1952)
185. A. Karman and L. Guiffrida, *Nature*, **201**, 1204 (1966)
186. F. M. Page and D. E. Woolley, *Analyt. Chem.*, **40**, 210 (1968)
187. P. J. Padley, F. M. Page and T. M. Sugden, *Trans. Faraday Soc.*, **57**, 1552 (1961)
188. A. N. Hayhurst and T. M. Sugden, *Trans. Faraday Soc.*, **63**, 1375 (1967)
189. M. Woodfield, private communication (1970)
190. L. F. Bates, *Proc. Roy. Soc.*, **257A**, 22 (1960)
191. C. A. Coulson and K. Zalewski, *Proc. Roy. Soc.*, **268A**, 437 (1962)
192. T. W. Preist, *J. C. S. Faraday*, 1, **68**, 661 (1972)
193. G. Fowler and T. W. Preist, *J. Chem. Phys.*, **56**, 1601 (1972)
194. A. N. Hayhurst and N. R. Telford, *J. C. S. Faraday*, 1, **68**, 237 (1972)

Chapter 4

Reactions of Atoms and Free Radicals Studied in Discharge-Flow Systems

M. A. A. Clyne

Department of Chemistry, Queen Mary College,
Mile End Road, London, E.1, United Kingdom

4.1. INTRODUCTION

It has been known for many years that certain ground state atoms may be produced by the dissociation of the corresponding molecular gas in a high-voltage electric discharge; such atoms may persist for times up to several seconds at total pressures near 1 torr.[1-3] The main emphasis of studies using the modern discharge-flow method is on quantitative measurements of rate constants of elementary reactions of atoms and small free radicals at temperatures from 200 to 800 K and total pressures between 0·1 and 10 torr. In this method, the time dependences of concentrations are determined by measurement at different distances along the axis of a tube, usually cylindrical; atoms are rapidly pumped along the length of this flow tube under steady state conditions. In the case of a velocity of gas flow which is constant with respect to both axial and radial displacement (plug flow), distance along the tube axis and time are in direct proportion, allowing a simple calculation of reaction rates.

The discharge-flow method as a reliable and general quantitative kinetic technique dates from around 1958. This date marked the establishment of the first simple, specific and reliable method for the determination of atom concentrations—the measurement of oxygen atom concentrations by titration with NO_2.[4] The important methods for the measurement of atom concentrations now available include not only a number of titration and chemiluminescence methods, but also electron paramagnetic resonance (e.p.r.) spectrometry, optical spectrophotometry, particularly atomic resonance spectrometry and mass spectrometry. Major impetus was given to the method around 1958 by the use of high-frequency microwave and r.f. "electrodeless" discharges for the production of atoms. These obviate metal electrodes being in direct contact with the flowing gas, as was necessary in the case of the low-frequency, high-potential discharges used formerly. The use of high-frequency discharges thus eliminates a major source of contamination in the discharge products, and gives a more stable and purer source of atomic species given in refs. 7a, 185, 186.

In addition to the use of spatially-resolved concentration measurements for the determination of rate constants for reactions of ground state atoms, the discharge-flow method has been extensively applied to kinetic and spectroscopic studies of chemiluminescent phenomena. In these cases, the flow parameters in the flow tube are of no great importance, as time resolution is not obtained from axial displacements; consequently, the total pressures and flow rates, and tube diameters may be varied over wide limits, since it is unnecessary to ensure adherence to the conditions for plug flow.

A sophistication of the simple method for the measurement of rates of reaction of ground state atoms is to kinetic studies of simple ground state radicals. This promising field has opened out following several kinetic studies of hydroxyl radical reactions.[187,188] The source of a free radical is generally a rapid reaction between an atomic species and an added reactant; the kinetics of reaction of the radical may then be studied from the variation of concentration with distance along the tube axis. The method depends not only on the use of suitable rapid reactions for the production of radicals, but also on the development of sensitive methods for the measurement of their concentrations. During the last few years, kinetic studies in discharge-flow systems of reactions of a number of short-lived radicals have been reported: OH, CN, ClO, BrO, FO, SO, NCl_2 and the HNO molecule. The mechanisms and rates of radical reactions obtained in some of these studies have already been very useful in the interpretation of complex reactions in which these radicals participate.[6]

Much of the existing work on reactions of molecular radicals, and on atom reactions using e.p.r. spectrometry, optical spectrophotometry, and mass spectrometric detection, has been reported since the appearance of earlier review articles.[7-10] An attempt to emphasize the most recent developments is therefore made in the present article. With the exception of a recent review,[11] relatively little attention has been given to the survey of the growing field of chemiluminescent reactions. This aspect of atom and radical reactions is therefore dealt with rather more fully than is usual in the present article.

In the first section of this chapter the methods of production of atoms and the determination of their concentrations in discharge-flow systems are discussed, with particular reference to two important problems. Firstly, the identification of secondary active species which may accompany the primary active species (a ground state atom) in the products of an electric discharge; and secondly, a critical discussion of various methods for the measurement of atom concentrations.

4.2. THE PRODUCTION OF ATOMS AND THE DETERMINATION OF ATOM AND RADICAL CONCENTRATIONS

4.2.1. Discharge-flow Apparatus

Figure 4.1 shows the essentials of a typical discharge-flow apparatus; this consists of four types of interconnected systems; (a) the discharge tube and the flow tube; (b) apparatus for the measurement of atom and radical concentrations; (c) a system of flowmeters and vacuum valves; and (d) an independent conventional high vacuum line for gas handling.

The mode of operation is as follows. Partially-dissociated gas from the discharge is pumped rapidly into the flow tube, where reactants may be added and measurements may be made, and thence via high conductance cold traps to a relatively fast rotary pump (200 to 1000

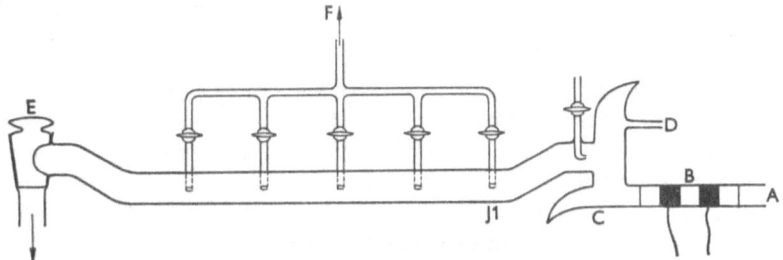

FIG. 4.1. Discharge flow apparatus. *A*, flow of purified nitrogen into discharge tube; *B*, quartz rf discharge tube showing external electrodes; *C*, Wood's horn light trap; *D*, silicone oil manometer inlet; *E*, regulating stopcock and to pump unit; *F*, nitric oxide supply. (After ref. 27b.)

1 min^{-1}). For measurements of the rates of reaction of atoms and radicals, the flow tube is conveniently a cylindrical or rectangular section glass or silica tube of some 20–50 mm i.d., and 10–200 cm in length. With a typical velocity of flow of 200 cm s^{-1} the typical time of reaction in such a flow tube is 0·2 s. Slower reactions may be followed conveniently using a stirred-flow reactor in place of the flow tube. If a relatively large volume of such a reactor is used the residence time can be increased over the maximum time possible with a flow tube. A stirred flow reactor in a discharge-flow system has been used recently[12] for measurements of the rates of the third-order recombination reactions

$$O + O_2 + M \longrightarrow O_3 + M$$
$$O + SO_2 + M \longrightarrow SO_3 + M$$

The dimensions (particularly the diameter) of the flow tube are restricted to those given above by the requirement for plug flow, when kinetic measurements are required. The range of total pressures over which reactions may be studied is then similarly restricted (about 0·5 to 10 torr). For investigations of the kinetics of formation of excited states in chemiluminescent reactions, where time discrimination is not needed, the range of experimental parameters may be considerably extended. For example, the formation of electronically excited NO_2^* in the reaction $O + NO + M \rightarrow NO_2^* + M$ has been studied over the pressure range 10 μmHg to 0·25 torr total pressure.[13,204]

The main variables other than time of reaction in a kinetic study using the discharge-flow method are the partial pressures of reagents, p_i, including that of any third body M. These partial pressures are simply related to the corresponding flow rates of reagents, F_i, to the total pressure p, and to the total flow rate ΣF, if all components approximate to ideal gases: $p_i = pF_i/\Sigma F$. (Note the distinction between *flow rates* (mol s^{-1}) and *flow velocity* (cm s^{-1}).) p is measured directly with pressure gauges attached to the flow tube; the usual types of instruments are the McLeod gauge, transducers, silicone oil manometers and the conductivity gauge. A system of micrometer needle valves (brass, stainless steel or teflon) for the control of flow rates, and of capillary or rotameter flow meters for the measurement of flow rates, is necessary. The accurate calibration of these flow meters is essential to the measurement of rate constants free from systematic errors, and a number of discrepancies reported in the literature are probably due to errors in flow meter calibrations. The most usual methods of calibration are based on measurements of rates of pressure fall in a calibrated volume, or more directly, from rate of loss of mass of reagent.

Reagent flows must be introduced into the flow tube in such a way that the time for complete mixing with the main flow is minimized; the mixing time must be short compared with the half-time for the reaction under study.

Mixing will be least satisfactory when the incoming flow moves with a low velocity under streamline conditions. There is then a negligible radial component of velocity of the incoming reactant molecules, as diffusion at total pressures near 1 torr is relatively slow. Radial concentration gradients may then persist downstream from the reactant inlet. If reagent enters the flow tube via a relatively small diameter orifice, the incoming velocity is sufficiently high to create a small region of turbulence outside the orifice, and the reactant molecules acquire sufficient radial velocity for mixing to be rapid. The simplest arrangement of this type uses an inlet jet (*c.* 1 mm hole diameter) along the axis of the flow tube. The velocity of incoming reactant molecules relative to the main flow is greatest when the flow from the inlet jet opposes the direction of main flow. Various types of multiperforated inlet, on the cylindrical axis of the flow tube, have also been employed; these should give rapid attainment of homogeneous flows near the inlet orifices. Rapid mixing is more difficult to achieve in the rectangular section flow tubes used for absorption spectrophotometry[20] than in cylindrical tubes. The times for complete mixing, as measured by the attainment of a homogeneous optical density in the main flow when a light absorbing reactant was added through a single jet inlet and a multiperforated inlet, have been determined. For a rectangular section tube (45 × 15 mm), mixing was essentially complete in 10 ms and 2 ms for the jet and the multiperforated inlet respectively, with a constant flow velocity of 10 m s^{-1}.[20]

4.2.2. The Electrodeless Discharge

Virtually all recent quantitative investigations of atom reactions in discharge-flow systems have employed either a microwave or radiofrequency discharge in the parent molecules as the source of atoms. The use of a 50 Hz discharge at several kV potential between a pair of metal electrodes in contact with the gas flow has been superseded by these convenient so-called "electrodeless" discharges. The absence of trace impurities originating from the metal surfaces is thereby ensured.

The usual arrangement employs an air-cooled silica discharge tube of about 10 mm internal diameter. Various types of microwave cavities suitable for coupling power ($\sim$25 to 200 W) to the discharge tube are used,[81] whilst radiofrequency power is coupled either by means of a pair of metal sleeves around the outside of the discharge tube, or inductively. It appears that the yields of atoms under the same conditions from microwave (2450 MHz) and r.f. (20 MHz) discharges are similar.

The degree of dissociation of molecular gas by the discharge depends on at least the following factors: power coupled to the discharge, total pressure and flow rate of gas, purity of gas, temperature and nature of the walls of discharge tube; the effects of these parameters have been discussed elsewhere [8,14]. The influence of trace impurities in increasing the yields of atoms is particularly marked.

4.2.3. Production of Atoms

Atoms may be produced by passage of either the parent molecule gas or of dilute mixtures of the parent gas (1 to 10%) with argon (Table 4.1). The resulting partial pressures of atoms are lower in the latter than in the former case, whilst the concentrations of metastable secondary active species accompanying the atoms are also reduced.

The generation of appreciable concentrations of N, O and H atoms by the use of high frequency (or high potential) discharges is now a well-known technique. The concentrations of $N(^4S)$ atoms obtainable with N_2 are generally low (0·1 to 2% of N_2 concentration), unless large amounts of oxygen-containing impurities are present; up to ten times as great a degree of dissociation can be produced by discharges in O_2 and H_2. The lower concentrations of N usually produced do not normally lead to serious limitations in the measurement of rate constants. In fact, it is often desirable to minimize atom concentrations in a discharge-flow system in order to avoid significant temperature rises (due to exothermic reactions of atoms) and non-ideality of the plug-flow assumption. The lower limit in atom concentration imposed by instabilities in the discharge can be extended downwards by use of a discharge by-pass arrangement, such as that shown in Fig. 4.2 for use with chlorine atoms. In this arrangement, a small variable fraction of argon + molecular gas mixture passes through the discharge, whilst the greater part of this mixture by-passes the discharge, rejoining the remaining flow downstream from the discharge. In this manner, low atom concentrations are ensured, (a) by dissociation of a small fraction of total flow; (b) by causing extensive atom recombination in the slow flow which occurs between the discharge and the point at which the by-passed gases rejoin the flow; and (c) by insertion of a metal wire

TABLE 4.1
Production of atoms in discharge-flow systems

Atom	State	Method of production	Range of concentrations (% of total flow)
(a)			
O	$^3P_2$	Discharge in O_2 or O_2 + Ar, He	< 15%
		Reaction N + NO → N_2 + O	
N	^4S	Discharge in N_2 or N_2 + Ar, He	< 2%
H	^2S	Discharge in H_2 or H_2 + Ar, He	< 10%
(b)			
F	$^2P_{3/2}$	Reaction N + NF_2 → 2F + N_2	< 1%
		Discharge in F_2 + Ar, He	< 5%
(c)			
Cl	$^2P_{3/2}$	Discharge in Cl_2 or Cl_2 + Ar, He	< 50%
		Reaction O + Cl_2 → 2Cl + O_2	
(d)			
Br	$^2P_{3/2}$	Discharge in Br_2 or Br_2 + Ar, He	< 50%
		Reactions O + Br_2 → 2Br + O_2	
		Cl + Br_2 → BrCl + Br	
(e)			
I	$^2P_{3/2}$	Reactions Cl + ICl → Cl_2 + I	
		Br + IBr → Br_2 + I	< 10%
		Discharge in I_2	< 1%

In some cases appreciable equilibrium concentrations of low-lying excited states of these atoms are present at equilibrium at 300 K. The relevant excitation energies above ground state and equilibrium populations of these states are given:
(a) $O(^3P_2)(0$ cm^{-1}, 74%), $O(^3P_1)(158$ cm^{-1}, 21%), $O(^3P_0)(226$ cm^{-1}, 5%)
(b) $F(^2P_{3/2})(0$ cm^{-1}, 93%), $F(^2P_{1/2})(404$ cm^{-1}, 7%)
(c) $Cl(^2P_{3/2})(0$ cm^{-1}, 99·2%), $Cl(^2P_{1/2})(881$ cm^{-1}, 0·8%)
(d) $Br(^2P_{3/2})(0$ cm^{-1}, 100%), $Br(^2P_{1/2})(3670$ cm^{-1}, 0%)
(e) $I(^2P_{3/2})(0$ cm^{-1}, 100%), $I(^2P_{1/2})(7590$ cm^{-1}, 0%)

catalyst in the atom flow. An improved version of the discharge-bypass has been specifically designed for production of the very low Cl, Br, I atom concentrations ($\leq 10^{-7}$ mol fraction), containing the minimum possible atom impurities, which are required for kinetics studies of rapid bimolecular atom reactions using atomic resonance fluorescence spectrometry.[189]

As already mentioned, a recent development is the production of ground state $^2P_{3/2}$ halogen atoms in discharge-flow systems. Early reports of insignificant yields of chlorine and bromine atoms by discharges[15] are unjustified. However, the generally greater heterogeneous recombination efficiencies of halogen atoms on glass and silica surfaces[16] ($\gamma_{Cl} \approx 10^{-4}$ on clean pyrex, $\gamma_{Cl} < \gamma_{Br} < \gamma_I$) lead to the necessity of coating the walls of the discharge tube with a surface recombination

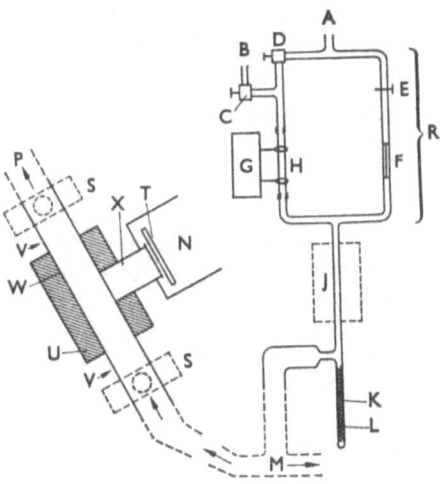

FIG. 4.2. Discharge by pass arrangement. A, carrier gas inlet; B, Cl_2 inlet; C, D, needle valves; E, stopcock; F, capillary; G, rf. generator; H, discharge tube; J, furnace; K, Ni wire; L, Fe slug; M, connections to flow tubes; N, 9558-C photomultiplier cell; O, spectrograph; P, to pump; Q, concave mirror; R, discharge-bypass section; S, 931-A photomultiplier cells; T, filter; U, W, furnace; V, cold section of flow tube; X, sealed-on side arm. (After ref. 38).

inhibiting agent—a relatively involatile oxyacid such as H_3PO_4, H_2SO_4 or 72% aqueous $HClO_4$.[16,17] The use of a similar oxyacid poisoning agent, or of gaseous ClO_2 or O_3, on the walls of the flow tube is also useful in reducing the surface recombination coefficient ($\gamma_{Cl} = 2 \times 10^{-5}$ on H_3PO_4-coated pyrex at 298 K).[17] Whilst $Cl(^2P_{3/2})$ and $Br(^2P_{3/2})$ atoms are readily produced by dissociation in a discharge of pure halogen or of mixtures of halogen with argon or helium, the direct dissociation of iodine is relatively inefficient.[189] However, ground state $I(^2P_{3/2})$ atoms have been formed in discharge-flow systems by the rapid reactions[16,20,190]

$$Cl + ICl \longrightarrow I + Cl_2; \quad \Delta H^\circ_{298} = -31 \text{ kJ mole}^{-1}$$

$$Br + IBr \longrightarrow I + Br_2; \quad \Delta H^\circ_{298} = -15 \text{ kJ mol}^{-1}$$

Ground state $S(^3P)$ atoms have also been produced by chemical reactions in discharge-flow systems;[18] whilst $F(^2P_{3/2})$ atoms have been produced by a discharge in CF_4,[19] in addition to the recent, clearer methods given in Table 4.1.

4.2.4. Limitations of the Discharge-flow Technique

Rate constants measured in discharge-flow systems depend upon measurements of (a) the velocity of flow $\bar{u}$, and (b) the partial pressures of reactants. For a reaction first-order in atom concentration [A], the rate constant k_1 is given by $k_1 = \bar{u} \, \mathrm{d} \ln [A]/\mathrm{d}x = -(RT\Sigma F/Ap)\mathrm{d} \ln [A]/\mathrm{d}x$, where ΣF is the total flow rate, A is the cross-section area of the flow tube, p is the total pressure, and x is the displacement along the tube axis. For simple reactions of higher overall orders, the dependences of rate constants upon the parameters ΣF, A, p and reagent flow rate F_i are summarized in Table 4.2. The importance of accurate measurements of flow rates and of total pressure, particularly for reactions of overall second and third orders, is clear. For example, realistic random errors of $\pm 1\%$ in p, and of $\pm 3\%$ in ΣF and F_i, lead to an error of $\pm 12\%$ in k_3.

TABLE 4.2
Relation of rate constants to measured parameters in a
flow system

Type of reaction		Rate constant[a]
1st order,	$A \rightarrow \tfrac{1}{2}A_2$	$k_1 = -[RT\Sigma F/Ap] \, \mathrm{d} \ln [A]/\mathrm{d}x$
2nd order,	1st order in [A],	$k_2 = -[RT(\Sigma F)^2/F_i Ap^2] \, \mathrm{d} \ln [A]/\mathrm{d}x$
	$A + R \rightarrow$ products	
3rd order,	1st order in [A],	
	$A + R + M \rightarrow$ products	$k_3 = -[R^2 T^2 (\Sigma F)^2/F_i Ap^3] \, \mathrm{d} \ln [A]/\mathrm{d}x$

[a] plug-flow assumption

In fact, considerable care is required to obtain flow rates with a standard error of $\pm 3\%$. Reagent flow rates F_i may be determined by following the rate of fall of pressure $(-\mathrm{d}p/\mathrm{d}t)$ of gas in a volume V: $F_i = -(V/RT)\mathrm{d}p/\mathrm{d}t$. Small flows through the regulating constriction in a micrometer needle valve occur under streamline conditions, and it may be shown (and experimentally demonstrated) that $1/p$ is linear with t; consequently, p is a strongly nonlinear function of t, and it is essential for accurate measurements that $-\mathrm{d}p/\mathrm{d}t$ be measured from the almost linear plot of $1/p$ versus t: $-\mathrm{d}p/\mathrm{d}t = p^2 \, \mathrm{d}(1/p)/\mathrm{d}t$.

It has been pointed out that appreciable systematic errors might occur in some cases when the pressure-drop method of flow calibration is used, on account of adiabatic expansion which leads to a slight decrease in temperature in the gas storage bulbs.

4.2.5. Deviations from Plug Flow

The importance of accurate measurements of flow rates and total pressure in minimizing random errors in rate constants has been demonstrated. The most satisfactory method of reducing systematic errors arising from the physical behaviour of the flow system is to ensure conditions under which deviations from simple flow occur to a negligible extent.

Fehsenfeld and Ferguson[191] have shown that in the case of rapid removal of atoms at the walls of a cylindrical reactor in the streamline flow regime, the expected parabolic radial velocity profile is approximated quite well. Various effects such as radial diffusion, and turbulence introduced by obstructions such as inlet jets in the flow, cause the parabolic profile to be degraded to plug flow in which there is no radial dependence of axial velocity of gas molecules. Poirier and Carr[186] have discussed criteria for the validity of the plug-flow approximation. Additional deviations from plug flow may arise from Poiseuille pressure gradients and axial concentration gradients along the tube, which may lead to appreciable back diffusion of atoms. The first effect is minimized at low linear velocities $\bar{u}$, and the second is minimized at high $\bar{u}$. An optimum value of $\bar{u}$ therefore exists, but it is not always possible to use this value on account of the kinetic constraints of the chemical system under study. For a discussion, see refs. 7a and 185.

The flow pattern along a straight cylindrical or rectangular tube of conventional dimensions will lie well inside the streamline flow region under the usual conditions of flow velocity and pressure p_1. On the other hand, the velocities of incoming reactant flows at inlet jets of 1 mm diameter may often exceed the critical velocity, and the flow patterns in such streams, as well as at bends and obstructions in the main flow tube is likely to be turbulent.

For streamline flow of a gas of viscosity η, such as is expected in straight flow tubes, the pressure gradient between the ends of the tube of length l, is given by $(p_1 - p_2) = 16\eta l RT\Sigma F/2p_1\pi a^4$, provided $(p_1 - p_2) \ll p_1$. For argon at 293 K, this expression reduces to $(p_1 - p_2) = 1{\cdot}3 \times 10^{-6}\ \bar{u}\ l/a^2$; a is the radius of the tube, p_1, p_2 are in torr, $\bar{u}$ in cm s^{-1}, and a and l are in cm. Table 4.3 shows typical values of $(p_1 - p_2)/p_1$ for flows of various gases. The dependence on both a and p_1 is noted.

An exact treatment of radial and axial diffusion effects is difficult. However, approximate discussions[7a,185] are valuable in indicating the limits beyond which these effects become significant. Radial diffusion is unlikely to be important except possibly at high pressures and with large diameter flow tubes when rapid reactions are being studied. The conditions for negligible back diffusion are more stringent, and are given approximately by $Dk/\bar{u}^2 \gg 1$, where D is the coefficient

TABLE 4.3

Pressure gradients in 100 cm long cylindrical flow tubes
(The tabulated values are $(p_1 - p_2)/p_1$ for gas flows of 2×10^{-4} mol s^{-1})

a (cm)	p_1 (torr)	Ar	He	O_2	N_2	H_2	Cl_2	Br_2
0·5	1	0·5	0·4	0·5	0·4	0·2	0·3	0·4
1	1	0·13	0·12	0·12	0·11	0·05	0·08	0·10
2	1	0·03	0·03	0·03	0·03	0·01	0·02	0·02
3	1	0·01	0·01	0·01	0·01	<	0·01	0·01
0·5	2	0·13	0·12	0·12	0·11	0·05	0·08	0·10
1	2	0·03	0·03	0·03	0·03	0·01	0·02	0·02
2	2	0·01	0·01	0·01	0·01	<	0·01	0·01
3	2	<	<	<	<	<	<	<
0·5	5	0·02	0·02	0·02	0·02	0·01	0·02	0·01
1	5	all <0·01						
2	5							
3	5							

< indicates a value of less than 0·01

of diffusion of the atoms, and k is the apparent first-order rate constant for atom decay.

4.2.6. Measurement of Atom and Radical Concentrations

For many years, the measurement of concentrations of atoms and radicals has been a major preoccupation of gas kinetists. A variety of simple methods for the measurement of both relative and absolute concentrations now exists, and has led to continual extensions of the types and complexity of reactions which can be studied. Some methods have been specially developed for the measurement of atom and radical concentrations in discharge-flow systems. These methods are broadly classed as *titration* and *chemiluminescence* (section 3.2.8). Other methods are special adaptations of established techniques for the study of fast gaseous reactions in static systems. These adaptations have often proved difficult to develop, on account of the low concentrations of all reagents and the dynamic nature of a discharge-flow system. Nevertheless, several carefully adapted methods have proved most successful for kinetic studies in discharge-flow systems. These include: *electron paramagnetic resonance* (section 4.2.9), *optical spectrophotometry, particularly atomic resonance* (4.2.10), *mass spectrometry* (4.2.11), *thermal methods* (4.2.12). A number of other methods have also been used, e.g. the Wrede–Harteck gauge[21] which depends upon the differential pressure established across a capillary leak between atoms and their recombination products. However, it is believed that the methods listed above were used in most of the recent studies published.

4.2.7. Absolute and Relative Concentrations

The methods listed above all enable *relative* concentrations of atoms or radicals to be measured. It is a much more difficult problem to measure absolute magnitudes of atoms and radicals in discharge-flow systems, or indeed in any other systems such as flash photolysis experiments. Two principal methods are used for the derivation of absolute concentrations: (a) the combination of spectrometric measurements with calculated transition probabilities; or (b) the use of the stoichiometry of rapid titration reactions. Of these methods, (b) is probably the most frequently used at the present time. Emphasis will be given to the possibilities of absolute concentration measurements in the discussion of the methods which follows.

Fortunately, the requirement of absolute concentration measurements for the measurement of rate constants of elementary reactions is not all-embracing. In the case of a reaction which is first-order with respect to the atom A, conditions can often be chosen such that the overall rate constant k may be derived from relative atom concentrations. For example, consider an overall second-order reaction such as[22]

$$O(^3P) + H_2 \xrightarrow{k} OH + H; \quad \Delta H^\circ_{298} = +8 \text{ kJ mol}^{-1}$$

If the rate of removal of $O(^3P)$ is measured under pseudo first-order conditions such that $[H_2] \gg [O]$, k is given by $k = -[H_2]^{-1} \ln [O]/dt$, and only ratios of oxygen atom concentrations are required. Another case is overall third-order atom + molecule combination reactions; for example, the process

$$O + NO + M \longrightarrow NO_2 + M; \quad \Delta H^\circ_{298} = -306 \text{ kJ mol}^{-1}$$

has been extensively studied[7b] by pseudo first-order analysis. This case is particularly simple: not only is [M] independent of time, but so also is [NO], since nitric oxide is rapidly regenerated in the second step

$$O + NO_2 \longrightarrow NO + O_2; \quad \Delta H^\circ_{298} = -192 \text{ kJ mol}^{-1}$$

Relative concentrations of atoms thus suffice for the determination of rate constants for reactions of established mechanisms which can be studied under pseudo first-order conditions with respect to [A]. There remains a large class of reactions of more complex orders, e.g. atom + atom recombination processes, for which absolute concentrations are necessary. They are also necessary for the analysis of mechanistically unexplored reactions, for the provision of information on reaction stoichiometry.

In the following sections, the principal individual methods for the measurement of concentrations will be briefly discussed, and their applications to selected atoms (H, N, O, Cl, Br) and radicals (OH, CN, ClO, BrO, etc.) will be exemplified.

4.2.8. Titration and Chemiluminescence

Consider an elementary transfer reaction of an atom with a stable molecule, of simple stoichiometry, and sufficiently rapid for at least 99% extent of reaction to occur within the time resolution of a discharge-flow system (1 to 100 ms). Such a reaction constitutes a possible titration reaction for the measurement of atom concentrations if some means of detecting atoms in the system is also available. An atom "indicator" is sometimes provided conveniently by a chemiluminescent emission associated with the titration reaction.

(a) N(^{4}S) *Atoms*

In the type of flow system used for kinetic studies, the principal active species in the products of a discharge in pure N_2 or $N_2 + Ar$ mixtures is the ground state N(^{4}S) atom.[9] There is evidence for only minor concentrations of electronically excited N(^{2}D) atoms in these products.[83] Detectable concentrations of the metastable $A^3\Sigma_u^+$ state of N_2 are also present in these systems.[84] Many possible reactions of this excited state have been postulated in the past, leading at one time to a confused situation. However, the physical and chemical behaviour of $N_2(A^3\Sigma_u^+)$ has now been carefully and systematically studied[192], and it appears that this species normally plays at most a minor role in kinetic investigations of non-radiative reactions of ground state N atoms.

A commonly used titration method for [N] is the measurement of the concentration of NO which must be added to attain equal stoichiometry in the rapid reaction[23]

$$N(^4S) + NO \longrightarrow N_2 + O(^3P); \qquad \Delta H^\circ_{298} = -314 \text{ kJ mol}^{-1}$$

At the stoichiometric "end point", [N] initial = [NO] added. The end point is determined (with a photomultiplier cell, or even visually) as extinction of luminescence between the violet-blue NO glow characteristic of residual N

$$N + O + M \longrightarrow NO^* + M$$

and the green-yellow NO_2 continuum luminescence which occurs in the presence of excess NO

$$NO + O + M \longrightarrow NO_2^* + M$$

The reactions producing NO* and NO_2^* have low rate constants characteristic of third-order recombination reactions, whilst the titration reaction N + NO has a rate constant at 300 K equal to 0·1 of the bimolecular collision frequency ($k = 3 \times 10^{13} \exp(-120/T)$ $\text{cm}^3 \text{ mol}^{-1} \text{ s}^{-1}$).

The N + NO titration reaction is thus a simple method for the measurement of absolute concentrations of N. It is also important because it is frequently used for the production of known concentrations of ground state oxygen $O(^3P)$ atoms in the absence of O_2. This application is exemplified by rate measurements on the reaction $^{25}O + O + N_2 \rightarrow O_2 + N_2$; in this case O was produced by the reaction $N + NO \rightarrow N_2 + O$ in the absence of O_2, thus avoiding the complication of the concurrent reaction $O + O_2 + M \rightarrow O_3 + M$.

Absolute concentrations of $N(^4S)$ atoms have also been determined using measurements of the maximum yield of HCN produced when excess ethylene was allowed to interact with a flow of these atoms. However, it has been found that values of [N] measured in this way were consistently lower (by a factor of 1·4 to 2·5)[26] than those determined by titration with NO. The complexity of the $N(^4S) + C_2H_4$ reaction and other evidence[9] strongly suggests that the NO titration method gives quantitatively correct data, and this view appears to command wide support.[7c,8,9]

Changes in relative concentrations of $N(^4S)$ can be readily determined from measurements of the intensity of various bands of the First Positive emission spectrum of $N_2(B^3\Pi_g - A^3\Sigma_u^+)$, which is proportional to $[N]^2$, but which shows a complex dependence on total pressure.[27]

(b) $O(^3P)$ *Atoms*

Oxygen atoms produced either by a discharge in pure O_2, or in $O_2 + Ar$, or by the reaction $N + NO \rightarrow N_2 + O$ can be titrated with NO_2 to the stoichiometric end point of the reaction

$$O + NO_2 \longrightarrow NO + O_2$$

The indicator in this case is critical extinction of the green NO_2^* chemiluminescence characteristic of oxygen atoms

$$O + NO + M \longrightarrow NO_2^* + M$$

at which point [O] initial $=$ [NO_2] added. A cross check on the $N + NO$ and $O + NO_2$ titration methods has shown that these reactions have the same stoichiometries.[28]

The reaction $O + NO_2 \rightarrow NO + O_2$ has a rate constant[29] of $\simeq 10^{12·5}$ cm^3 mol^{-1} s^{-1} at 298 K; this value is almost an order of magnitude less than that of the rate constant for the $N + NO$ reaction. Whilst the rate of the $N + NO$ reaction is rapid enough for studies in any conventional fast discharge-flow system, this is not always the case for the $O + NO_2$ reaction. NO_2 titration of [O] will overestimate atom concentrations when these are low and where the linear velocity of flow is large. Thus considering initially equal concentrations of O and NO_2 of 1×10^{-10} mol cm^{-3} ($\simeq 0·002$ torr at 300 K), the fractions of reactants remaining unreacted after various times are as follows: 1 ms, 0·87; 10 ms, 0·40; 100 ms, 0·06. The conditions under which a titration is usually performed with less than 100% reaction (i.e. a slight excess of titrant concentration) are somewhat less stringent than these conditions, which thus define a conservative limit. The most favourable methods for carrying out titration reactions have been discussed.[193]

Changes in relative [O] can be determined very simply by measurement of the alteration in the intensity of the air afterglow emission spectrum due to excited NO_2 formed by the radiative combination of O and NO,[7]

$$O + NO + M \longrightarrow NO_2^* + M; \qquad I = I_0[O][NO]$$

Since [NO] remains constant (see above), $I/[NO]$ is directly proportional to [O].[7]

The products of a discharge in oxygen contain not only $O(^3P)$ atoms, but comparable concentrations of the extremely metastable $^1\Delta_g$ state of O_2.[73] Much smaller concentrations of the more energetic and less metastable $^1\Sigma_g^+$ state of O_2 are also present.[85] Although the excitation energies of these singlet states of O_2 (above $v = 0$ of $X^3\Sigma_g^-$) are low (7900 cm^{-1} for $^1\Delta_g$, 13 000 cm^{-1} for $^1\Sigma_g^+$), the presence of relatively large concentrations of $O_2(^1\Delta_g)$ requires considerable caution in interpreting the results of experiments in which $O(^3P)$ atoms are generated by means of a discharge in oxygen. One major source of error is the reaction[86]

$$O_3 + O_2(^1\Delta_g) \longrightarrow O(^3P) + 2O_2(X^3\Sigma_g)$$

which can regenerate O atoms in a stream of recombining atoms

$$O + O_2 + M \longrightarrow O_3 + M$$

Production of $O(^3P)$ atoms by the reaction N + NO is thus a preferred method. The physical and chemical behaviour of singlet oxygen can be readily studied in a discharge-flow system;[85] a considerable amount of information is known about these processes which is summarized later in this article.

(c) $H(^2S)$ *Atoms*

The use of titration methods for [H] is more difficult and requires more experimental care than the methods described above for N and O atoms. However, a number of important investigations have successfully used titration of H for absolute concentration measurements. The most commonly used procedure depends on the extremely rapid reaction

$$H + NO_2 \longrightarrow OH + NO; \qquad \Delta H_{298}^\circ = 122 \text{ kJ mol}^{-1}$$

($k = 2 \times 10^{13}$ cm^3 mol^{-1} s^{-1} at 300 K).[30] The end point indicator can be the infra-red luminescence from HNO* due to the radiative combination process[31]

$$H + NO + M \longrightarrow HNO + M; \qquad I = I_0[H][NO]$$

Most of the intensity of this emission spectrum lies within the (O, O, O)–(O, O, O) band centred at 760 nm[31,32] and thus it is necessary to utilize a photomultiplier with an S20 (or S1) or similar cathode for intensity measurements. The e.p.r. signal[33] or absorption of Lyman-α resonance radiation at 121·6 nm[34] H(^{2}P $\leftarrow$ ^{2}S) are other end-point indicators which have been used. The NO$_2$ titration method is invalid in the presence of large concentrations of H$_2$ unless an extremely rapid flow velocity is used, since the reaction[5,28]

$$OH + H_2 \longrightarrow H_2O + H; \quad \Delta H^\circ_{298} = -63 \text{ kJ mol}^{-1}$$

which can regenerate H in a straight chain reaction has an appreciable rate constant (10^9 cm^3 mol^{-1} s^{-1} at 300 K).[42] However, in any chemically simple system, relative H atom concentrations may be found from measurements of the HNO emission intensity on adding known amounts of NO. The analysis is exactly analogous to that for the NO$_2$ emission, since NO is rapidly regenerated in the step[31] H + HNO $\rightarrow$ H$_2$ + NO: thus, I/[NO] $\propto$ [H].

An alternative titration method for absolute [H] is titration with nitrosyl chloride[38]

$$H + ClNO \longrightarrow HCl + NO; \quad \Delta H^\circ_{298} = -273 \text{ kJ mol}^{-1}$$

The rate constant for this reaction has not yet been reported, but it is known to be comparable in magnitude to those of the other titration reactions mentioned. Although this titration reaction can be carried out in the presence of H$_2$, it is necessary that the flow tube be treated with an oxyacid in order to suppress Cl atom recombination which can interfere with overall stoichiometry through the scheme[35]

$$H + HCl \longrightarrow H_2 + Cl$$
$$Cl + wall \longrightarrow \tfrac{1}{2}Cl_2$$
$$H + Cl_2 \longrightarrow HCl + Cl$$

No low-lying metastable bound states of H$_2$ or H are observed in discharge-flow systems, nor are any expected to exist.

(d) Cl(^{2}P$_{3/2}$) *Atoms*

Absolute concentrations of Cl(^{2}P$_{3/2}$) atoms may also be determined using titration with ClNO[16,36–38]

$$Cl + ClNO \longrightarrow Cl_2 + NO; \quad \Delta H^\circ_{298} = -83 \text{ kJ mol}^{-1}$$

The rate constant for this reaction is high at 300 K and has been reported[189] as $k = (1·8 \pm 0·3) \times 10^{13}$ cm^3 mol^{-1} s^{-1}. The end point indicator is critical extinction of the chlorine atom resonance fluorescence,[189] or of the chlorine afterglow emission spectrum Cl$_2$ B$^3\Pi$(O$_u^+$)—X$^1\Sigma_g^+$, $\lambda > 500$ nm, due to radiative combination of Cl(^{2}P$_{3/2}$) atoms.[37,38]

Alternatively, heat release to a calorimeter wire probe[16] has been used as an indicator of chlorine atoms.

Several workers[36–38] have suggested that intensity measurements of the Cl_2 afterglow emission spectrum can be used to determine relative [Cl], through a relation analogous to that for N atoms, $I = I_0[Cl]^2$. However, it is now clear that the dependence of the intensity of the Cl_2 afterglow on [Cl] is considerably more complex than that given above.[38]

(e) $Br(^2P_{3/2})$ *Atoms*

Titration of $Br(^2P_{3/2})$ atoms with ClNO can be made using Br atom resonance fluorescence,[189] or the Br_2 afterglow spectrum as indicator[16,40]

$$Br + ClNO \longrightarrow BrCl + NO; \quad \Delta H^\circ_{298} = -60 \text{ kJ mol}^{-1}$$

Recently, it has been shown that ClNO titration is also valid for absolute measurements of fluorine atom concentrations.[41]

4.2.9. Electron Paramagnetic Resonance (E.P.R.) Spectrometry

If allowed magnetic or electric dipole transitions between levels can occur in the microwave region in the presence of a field of c. 10 kG, it may be possible to observe electron paramagnetic resonance spectra. Thus, e.p.r. spectra of many atomic and molecular radicals may be observed, as well as those of degenerate singlet states of molecules.

Details of a useful technique, and an outline of the simple theory for the detection of $O(^3P_{2,1,0})$ and $N(^4S)$ atoms, have been described.[43,44] An important advance was the method of calibration of atom signal intensities by the measurement of the signal due to ground state $O_2X^3\Sigma_g^-$ in known concentrations.[43] This has enabled absolute atom concentrations to be derived without the need for difficult measurements of the cavity filling factor. The calibration factors for different oxygen lines have been tabulated.

Most investigations have used a modulated scanning magnetic field and phase sensitive detection of the signal. For relative determinations of atom concentrations, measurements of peak heights suffice under most conditions (see below); but absolute concentrations are derived from the ratio of integrated intensities for the atom and O_2, respectively.

(a) O, N, H, D, Cl *Atoms*

A typical experimental arrangement is shown in ref. 44. In this design, the cavity, which was operated in the TE_{102} mode, accepted a quartz flow tube of 11 mm o.d. The effective sampling length along this tube was about 2 cm. A more recent type[56] (Fig. 4.3) used a cylindrical TE_{011} cavity which could be used for both magnetic and electric dipole transitions, and which could accept a 15 mm i.d. tube. Saturation effects, which lead to nonequilibrium distributions between energy states (and

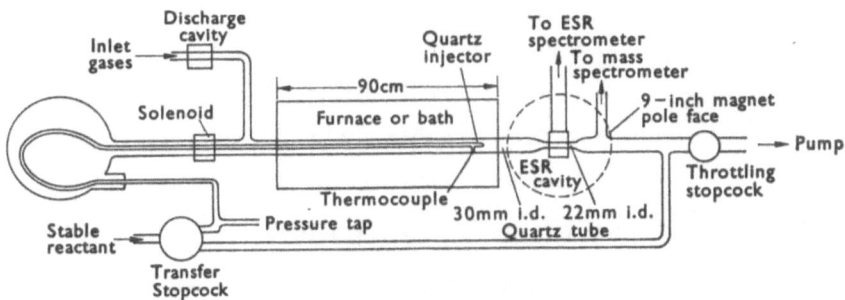

FIG. 4.3. E.P.R. spectrometry and mass spectrometry in a discharge-flow system. (After ref. 33a.)

consequent line broadening) at high microwave powers are unimportant for O atoms or O_2 up to 100 mW in the torr pressure range.[44] However, with N or H atom lines in their parent gases, saturation was easily observed at powers near 0·1 mW. At very low power, the signal increased linearly with (power)$^{1/2}$ as expected, but further increase in power caused the signal to go through a maximum and then decrease in height and broaden as predicted from theory. The procedure with N and H atom lines was thus to maximize the signal with respect to power, and then to reduce the power by at least 10 dB before any measurements were made.[44]

The problem of line widths, as applied to relative atom concentration measurements, has been discussed in a recent paper.[56] For all species (except S-state atoms) in dilute concentration, the true unsaturated line width is determined simply by the total pressure (i.e. by the frequency of collision with any particle), and the signal heights are thus directly proportional to the relative concentration of atoms. This theoretical result has been verified experimentally for $O(^3P_{2,1})$ and $Cl(^2P_{3/2})$ atoms.[47] For S-state atoms ($H(^2S)$ and $N(^4S)$), the true line width depends on the atom concentration via spin-exchange collisions, unless a paramagnetic diluent is present.[46,56] This complication probably does not lead to great difficulty in the majority of commonly encountered chemical systems.

For O and N atoms, a cross-check of absolute atom concentrations measured by e.p.r. with those measured by $O + NO_2$ and $N + NO$ titrations has satisfactorily verified the quantitative validity of both techniques.[56] The e.p.r. method has been fruitful over the last few years in yielding rate constants of a number of simple atom transfer reactions at 300 K; these include the reactions

$$H + HCl \rightleftharpoons H_2 + Cl \text{ (both directions)}^{47}$$

$$O + H_2 \longrightarrow OH + H^{48}$$

$$O + D_2 \longrightarrow OD + D^{48}$$

Atomic reactions involving more complex substrate molecules have also been investigated, such as

$$H + C_2H_4 \rightleftharpoons C_2H_5 \text{ [49,82]}$$

$$O + CH_4 \longrightarrow OH + CH_3 \text{ [48,50]}$$

$$O + C_2H_6 \longrightarrow OH + C_2H_5 \text{ [50]}$$

$$O + NH_3 \longrightarrow OH + NH_2 \text{ [51]}$$

$$D + CH_3Br \longrightarrow CH_3 + DBr \text{ [52]}$$

$$O + C_2H_4 \longrightarrow products \text{ [49,53]}$$

The combination of e.p.r. spectrometry with the fixed observation point method (section 4.4.4) has permitted the extension of kinetic measurements of atom decay rates to elevated temperatures. The results of such studies on the reactions $O + H_2 \rightarrow OH + H$ [48,54] and $Cl + H_2 \rightarrow HCl + H$ [47] are in good agreement with measurements using chemiluminescence and titration. The rate constant for the recombination reaction $H + H + M \rightarrow H_2 + M$ using e.p.r. spectrometry has also been reported. [55,187,188]

(b) OH, OD *Radicals*

For any radical species exhibiting electric dipole transitions, such as $OH(X^2\Pi)$, instead of the magnetic dipole transitions characteristic of simple atoms, it is necessary that the reference gas for calibration has electric dipole transitions, so that the cavity-filling factors are identical. The use of the electric dipole transitions of $NO(X^2\Pi)$ for calibration of absolute OH concentrations has thus been developed. [56] Several kinetic studies of ground state OH radicals have been reported, including the rate constants for the reactions [56,187,188]

$$OH + OH \longrightarrow H_2O + O$$

$$OD + OD \longrightarrow D_2O + O$$

$$HO + wall \longrightarrow$$

The method clearly has much potential for rate investigations of diatomic radicals, a number of whose e.p.r. spectra have been described and interpreted [57] (e.g. NS, ClO, BrO, IO, SO, SF, SeO, SeH).

4.2.10. Optical Absorption Spectrophotometry

The method of kinetic electronic absorption spectrometry is well known as the principal technique for following radicals, reactants and products following flash photolysis. However, this method has been extensively used much in discharge-flow systems only recently. The main problem is the low optical density encountered because of the short path length for absorption and the low concentrations of radicals. (Atomic resonance spectrometry, discussed below, is an important exception.) The low optical densities make it difficult to scan a wavelength range for an unknown band system; the method is thus best suited for the study of radicals whose electronic absorption spectra are well known.

(a) OH($X^2\Pi$) *Radical*

Time resolved absorption spectrophotometry has been used in an important investigation of the reactions of the ground state $X^2\Pi$ OH radical.[5] The source was a microwave discharge in H_2O emitting lines of the (0, 0) vibrational band of the $A^2\Pi$–$X^2\Pi$ system of OH which are strongly absorbed by ground state OH radicals in the flow tube. The optical density was increased with the use of three traversals across the flow tube. Calibration of the optical density of OH at the $Q_2(4)$ line was made from the stoichiometry of the H + NO_2 reaction using known concentrations of added NO_2

$$H + NO_2 \longrightarrow OH + NO$$

(b) CN($X^2\Sigma^+$) *Radical*

Absorption by ground state CN radicals using a CN lamp (actually a discharge in flowing N_2 + Ar mixtures to which methane had been added) has led to kinetic data on the reactions of these radicals.[38] In this case, absolute concentrations of CN were determined from the measured optical densities (integrated over the P branch of the (0, 0) band), using the known oscillator strengths for the (0, 0) to (4, 4) bands of the violet system $B^2\Sigma^+$–$X^2\Sigma^+$. The source of CN radicals was the reaction $O + C_2N_2 \rightarrow CN + NCO$.

(c) ClO, BrO, IO($X^2\Pi$) *Radicals*

A system of concave mirrors first described by White[59] was used[60] to obtain 4, 8, 12 or 16 traversals across a rectangular flow tube, and hence path lengths of up to 70 cm. A stabilized deuterium arc was used as continuum light source. A system of this type (Fig. 3.4) was used for kinetic studies of ClO radicals[61] produced by the rapid reaction

$$Cl + ClO_2 \longrightarrow 2ClO$$

Calibration of the absorption intensity by ClO in the $A^2\Pi$–$X^2\Pi$ band system was by titration with O or NO

$$O + ClO \longrightarrow Cl + O_2$$
$$NO + ClO \longrightarrow Cl + NO_2$$

Reactions of BrO formed by the process

$$Br + O_3 \longrightarrow BrO + O_2$$

and some analogous data on IO have been reported.[20]

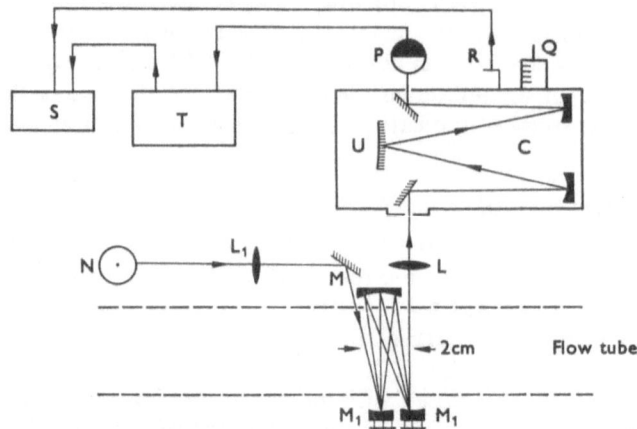

Fig. 4.4. Block diagram of optical system for electronic absorption spectrophotometry. L, L_1, biconvex silica lenses; M, M_1, multiple reflexion cell (eight traversals shown); N, deuterium arc; P, Hakuto R106 photomultiplier cell; Q, wavelength drum and marker (R); S, recorder; T, Vibron electrometer amplifier, U, grating. (After ref. 60.)

(d) NX_2 Radicals

The spectrophotometric method has been extended to certain triatomic radicals: some of the other methods available for the detection of atoms and diatomic radicals are difficult to apply to large radicals.

Kinetic studies on two triatomic radicals (NCl_2[60] and NF_2[62]) have been carried out, and the band spectrum of N_3[60] has been observed in discharge-flow systems. The sources of these radicals were the reactions

$$Cl + NCl_3 \longrightarrow NCl_2 + Cl_2$$
$$N_2F_4 + M \rightleftharpoons 2NF_2 + M$$
$$Cl + N_3Cl \longrightarrow N_3 + Cl_2$$

(e) H, N, O, Cl, Br, I Atoms by Resonance Spectrometry

Basically similar methods may be applied to the kinetic study of atoms, using absorption of resonance radiation by flowing ground state atoms. Table 4.4, which shows the lowest energy allowed transitions of the common atoms, indicates that in all cases the appropriate wavelengths lie in the vacuum ultraviolet region. This fact virtually precludes the use of multiple traversals. However the short path lengths which must be used for absorption are more than counterbalanced by the much greater absorption cross sections of allowed atomic transitions in this wavelength range, as compared with the cross sections for molecular absorption. Consequently, measureable intensities of

TABLE 4.4
Lowest energy allowed transitions of H, O, N, Cl, Br, I

Atom	Lowest energy allowed transition	λ (nm)
H	^2P—^2S	121·6
O	^3S—$^3P_2$	130·2
N	$^4P_{1/2}$—^4S	120·1
Cl	$^4P_{5/2}$—$^2P_{3/2}$	139·0
Br	$^4P_{5/2}$—$^2P_{3/2}$	157·7
I	$^4P_{5/2}$—$^2P_{3/2}$	183·0

absorption ($\sim 1\%$) of incident radiation may be observed from concentrations of atoms below 5×10^{-12} g atom cm^{-3}, provided that the resonance radiation is not too much self-reversed in the source. The sensitivity of the method is similar to that of e.p.r. spectrometry. The main problem of this method appears to be the source. Convenient lamps emitting resonance radiation are discharges through helium or argon with a trace of additive (e.g. O_2 in the case of an O atom lamp). A popular design of lamp is based on that described by Okabe and McNesby.[63] The effects of self-reversal on the absorption intensity, and methods for its reduction in such lamps, have been discussed.[64,65,71]

The method has been applied (both in flow systems and using flash photolysis) to the detection of ground state H[34], O[83a], N[83b], Cl[194], Br[194] and I[194] atoms, as well as to metastable N(^2D)[83b] and Cl($^2P_{1/2}$)[194] excited atoms. Following earlier kinetic investigations of the reactions of H with C_2H_2 and C_2H_4[34], measurements of the rates of several reactions have deen described, including, $O + NO_2 \rightarrow NO + O_2$[194] and $O + Br_2 \rightarrow BrO + Br$.[194]

Calibration procedures for atomic absorption intensities are rather difficult when small atom concentrations are employed, since available titrations may be insufficiently rapid for complete reaction. It is usually inadequate to calibrate only at high atom concentrations when low atom concentrations are to be used in kinetic studies since the Beer–Lambert law of absorption is usually not obeyed over the whole concentration range.[65] This problem also applies to the other main method for the detection of low concentrations—e.p.r. spectrometry.

The related method of atomic resonance fluorescence—the measurement of intensity of fluorescence excited by absorption of resonance radiation—has several advantages over resonance absorption for kinetic studies of reactions of ground state atoms. When the usual strongly self-reversed microwave discharge lamps[194,196] are used as the sources of resonance radiation, resonance fluorescence is much more sensitive than resonance absorption. The following lower limits of concentration detectable by resonance fluorescence have been found in this laboratory[195] for particular instrumental conditions: [Cl] $\geq 5 \times 10^{-15}$ g atom cm^{-3};

$[H] \geq 1 \times 10^{-13}$ g atom cm^{-3}. At atom concentrations less than about 2×10^{-12} g atom cm^{-3}, the dependence of fluorescence intensity upon atom concentration is strictly linear,[189] thus simplifying calibration and kinetic analysis. Calibration of the extremely low atom concentrations required is carried out by production of known atom concentrations by reacting very small metered amounts of a stable reactant with excess of a suitable atomic species; e.g. Cl, Br and I $^2P_{3/2}$ atoms are generated by the following reactions of known rates and stoichiometries:[189,190]

(i) O

$$N + NO \longrightarrow N_2 + O$$

(ii) Cl

$$O + OClO \longrightarrow ClO + O_2$$
$$O + ClO \longrightarrow Cl + O_2$$

$$\text{OVERALL} \quad 2O + OClO \longrightarrow Cl + 2O_2$$

(iii) Br

(a)
$$O + Br_2 \longrightarrow BrO + Br$$
$$O + BrO \longrightarrow Br + O_2$$

$$2O + BrO \longrightarrow 2Br + O_2$$

(b)
$$Cl + Br_2 \longrightarrow BrCl + Br$$
$$Cl + BrCl \longrightarrow Br + Cl_2$$

$$\text{OVERALL} \quad 2Cl + Br_2 \longrightarrow 2Br + Cl_2$$

(iv) I

$$Cl + ICl \longrightarrow Cl_2 + I$$
$$\text{or} \quad Br + IBr \longrightarrow Br_2 + I$$

The high sensitivity of resonance fluorescence detection of atoms allows the time scale of reactions with rates approaching the bimolecular collision frequency to be extended into the millisecond range accessible in the discharge-flow method. In this way, rate constants have been measured at 298 K for the very rapid reaction $Cl + Br_2 \rightarrow BrCl + Br$, $k = 0.5\ Z_{12}$ at 298 K,[190] as well as for the following slower reactions:

$Cl + ClNO \rightarrow Cl_2 + NO$,[189] $Br + ClNO \rightarrow BrCl + NO$,[189]

$O + NO_2 \rightarrow NO + O_2$,[189] $Cl + ICl \rightarrow I + Cl_2$,[190]

$Br + IBr \rightarrow I + Br_2$,[190] $Cl + BrCl \rightarrow Br + Cl_2$,[190]

$Cl + IBr \rightarrow BrCl + I$,[190] $Cl + OClO \rightarrow 2ClO$[195]

and

$H + Cl_2 \rightarrow HCl + Cl$.[195]

(f) Cl, Br *Atoms by Molecular Absorption*

For larger concentrations of atoms (e.g. $\simeq 1 \times 10^{-10}$ g atom cm^{-3}), a simple method is to measure the decrease in absorption intensity by the parent molecule when the discharge which produces atoms is turned on. If absorption in a suitable molecular continuum is employed, the Beer–Lambert law can be expected to hold over a wide range of conditions. The cases of Br and $Cl(^2P_{3/2})$ atoms are particularly suitable, since Br_2 and Cl_2 have relatively large absorption cross sections in conveniently accessible continua ($\sigma_{max} = 2.8 \times 10^{-19}$ cm^2 at 414 nm for Br_2;[67] $\sigma_{max} = 1.0 \times 10^{-19}$ cm^2 at 340 nm for Cl_2[68]). The concentration of Br atoms, for example, is then clearly equal to $2\Delta[Br_2]$, where $\Delta[Br_2]$ is the decrease in $[Br_2]$ when the discharge is energized. A similar method was used for measurement of [I] and [Br] in flash photolysis studies of atom recombination;[69] its extension to O, H and N by measurement of $[O_2]$, $[H_2]$ and $[N_2]$ in the vacuum ultraviolet region appears feasible.

The method of molecular absorption has been used for Br atoms in the author's laboratory to check the stoichiometry of the titration reaction, $Br + ClNO \rightarrow BrCl + NO$.[40] It has also been used to measure the kinetics of radiative[40] and nonradiative[197] third-order recombination of Br atoms, and to follow [Br] in the bimolecular disproportionation of two BrO radicals: these were shown[20] to decay by the reaction $BrO + BrO \rightarrow 2Br + O_2$. It appears that vibrational relaxation of Br_2, which might affect the method, was complete under the conditions used.[20] One advantage of the molecular absorption method is its relative instrumental simplicity. However, the method is clearly inapplicable to any system in which interfering absorption by transients, reactants or products can occur.

4.2.11. Mass Spectrometry

Considerable problems are involved in quantitative measurements of atom and radical concentrations by time-resolved mass spectrometry in discharge-flow systems. The difficulties arise from two causes: (a) the low partial pressures of atoms and radicals which can be sampled, and (b) sampling from flow systems. (a) is manifest as a shortage of signal-to-noise ratio when conventional vacuum techniques and detection methods are used. (b) depends on a representative sample of gas reaching the ion source of the mass analyzer, and may be interfered with by mass separation effects and by recombination of atoms and radicals on the surfaces of the sampling system. Both these problems, and their origins, and an account of the methods used generally, are fully discussed in an excellent review by Foner,[70] who has carried out much pioneering work in this difficult area. In the present brief account, we shall concentrate on the application of the mass spectrometric method to kinetic studies.

The simplest type of study is the mass spectrometric measurement at a fixed station of relatively high concentrations of stable reactants and/or products of atom reactions, when the time of reaction in the flow tube is varied. In this case, available signal-to-noise ratio is adequate for simple techniques to be used, and no sampling problem arises. Much useful work has been carried out in this way, especially when the method is combined with independent measurement of atom and radical concentrations in the same system, e.g. using e.p.r.; the kinetics of reaction of $O(^3P)$ with NH_3 were investigated in this way.[51]

It has been demonstrated that N^{72}, O^{73}, H^{74}, Cl^{198}, Br^{198} and F^{41} ground state atoms sampled through an inlet system from a flow tube may be detected in the ion source of a mass spectrometer. Usually, the atoms are generated by partial dissociation of the parent molecular gas. In these cases, the atoms may be indirectly detected by observation of the decrease in the corresponding molecular mass peak intensity when the discharge is energized.[75] This method can be used with ions produced in the conventional manner by impact with 80 eV electrons. An atom such as $O(^3P)$ may not be easily detected by measurement of the peak intensity at $m/e = 16$ using 80 eV electrons, since large numbers of O^+ ions are then present in the mass spectrum of O_2. If the energy of the electrons (which must be fairly monochromatic) is greater than the ionization potential of O, but less than the appearance potential of O^+ from O_2, the intensity at $m/e = 16$ may be attributed solely to O atoms. The use of monoenergetic low-energy (<15 eV) electrons for the detection of the mass spectra of radicals in general has been fully discussed elsewhere.[70,76] The main problem associated with the use of low energy electrons is their low impact cross sections, which lead to small ion currents and a consequent shortage of signal-to-noise ratio in the detector.

Parent molecular ion intensities provide a useful measurement of relative atom concentrations. If recombination of atoms between the flow tube and the ion source is negligible, absolute atom concentrations may also readily be obtained in this way. Usually, this condition is not realized unless a collision-free sampling path is available. Such collision-free sampling systems have been used; they consist of collinear systems of sampling orifices separated by rapid pumping stations through which a molecular beam may be collimated into the ion source of the mass analyzer. Beam intensity is greatly favoured by formation of a shock front after the gas expands out of the first sampling orifice at supersonic velocities. This leads to a marked degree of collimation in the forward direction. A second sampling orifice (the skimmer nozzle) then selects the central portion of this beam for onward transmission to the ion source.[77] Unfortunately, the design considerations in coupling a supersonic beam system to a low pressure source such as a discharge-flow system are difficult to achieve. Most practical arrangements therefore use a conventional effusive molecular beam sampling system in which the first orifice diameter is not large compared with the mean free path. Foner[70] has shown that a great improvement in available sensitivity (signal-to-noise ratio) may be obtained if such a beam is interrupted with a vibrating reed, and the ion current detected with a phase sensitive amplifier locked in to the mechanical modulator. A system of this type has been used for investigations of the HO_2 radical.[78]

In spite of the instrumental complexity of molecular beam mass spectrometry, an increasing amount of kinetic information on elementary reactions is becoming available from this method. One successful approach is the use of a simple, uncollimated crossed molecular beam system, allowing detection of primary reaction products under virtually collision-free conditions. Following the original description of such a system by Foner,[80] several studies have been described, including an investigation of the reactions of N and O atoms with N_2H_4 (Fig. 4.5).[79] Direct detection of the primary products of these reactions

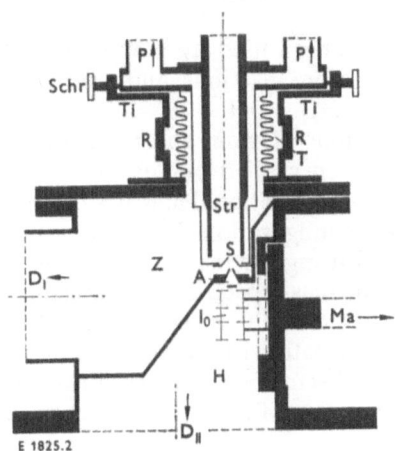

FIG. 4.5. Molecular beam mass spectrometry for kinetic studies in a discharge-flow system. A, skimmer. D_I, 2 × 250 l s^{-1} diffusion pumps. D_{II}, trapped 700 l s^{-1} diffusion pump. H, high vacuum chamber maintained at 10^{-6} torr by D_{II}. I_0, ion source. P, flow tube pump. S, 0·2 mm hole in quartz thimble. Str, flow system. Schr, adjusting screws. Z, intermediate vacuum chamber maintained at 5 × 10^{-4} torr by D_I. (After ref. 41.)

showed that N_2H_2 and H_2O are the primary products of the process,[79]

$$O + N_2H_4 \longrightarrow N_2H_2 + H_2O$$

Similarly, the reaction of O with Br_2 gives the BrO radical[198]:

$$O + Br^2 \longrightarrow BrO + Br$$

A development of the method is the use of photoionization instead of electron impact to discriminate between ionizable species of differing ionization potentials. The use of resonance line radiations in the vacuum ultraviolet for photoionization gives high intensity photon and ion beams with good discrimination against reactants and background,[199,200] which are particularly valuable for identifying hydrocarbon radicals such as the C_2HO radical which is the dominant primary product in the $O + C_2H_2$ reaction.[199]

A two-pinhole, unmodulated beam inlet system has been used by Clyne and Watson[198] to detect the hitherto unknown gaseous FO radical (from $F + O_3 \rightarrow FO + O_2$), and the ClO and BrO radicals, and to measure the rates of a number of elementary reactions involving ClO and BrO radicals, such as,

$$ClO + ClO \longrightarrow Cl + ClOO,$$
$$BrO + NO \longrightarrow Br + NO_2$$
$$Cl + OClO \longrightarrow 2ClO$$

A similar system has been used by Wagner et al.[201] for kinetic studies including FO and F atom reactions, and a variable temperature study[22c] of the reaction $O + H_2 \rightarrow OH + H$.

Kinetic investigations of atom and radical reactions in a discharge-flow system have been carried out using a simple one-stage pinhole sampling hole, and a conventional mass spectrometer equipped with an ion source of open configuration. In these cases, the atoms may either be directly detected using low energy electrons, or may be indirectly detected by using the mass spectrometer to estimate the stoichiometric end point of an atom titration reaction. The latter procedure enables a direct and simple calibration of mass peak intensities. This method has been used with a 60° sector field mass spectrometer in Herron's laboratory to measure the rates of a number of reactions, including

$$O + NO + M \longrightarrow NO_2 + M^{[87]}$$
$$O + NO_2 \longrightarrow NO + O_2^{[87]}$$
$$OH + CO \longrightarrow H + CO_2^{[88]}$$
$$O + HCHO \longrightarrow products^{[89]}$$

Phillips and Schiff have used a 90° mass spectrometer for kinetic measurement of elementary reactions including the reactions

$$N + NO \longrightarrow N_2 + O^{[90]}$$
$$O + NO_2 \longrightarrow NO + O_2^{[90]}$$
$$NO + O_3 \longrightarrow NO_2 + O_2^{[90]}$$
$$N + NO_2 \longrightarrow products^{[91]}$$
$$O + Cl_2O \longrightarrow ClO + ClO^{[92]}$$

Clyne and Walker[183] have used a small quadrupole mass spectrometer to measure the rate constants for reactions including:

$$Cl + CH_4 \longrightarrow CH_3 + HCl,$$

$$Cl + CD_4 \longrightarrow CD_3 + DCl,$$

$$Cl + CH_3Cl \longrightarrow CH_2Cl + HCl, \text{ etc.}$$

Other reactions which have been investigated are

$$O + C_2H_4 \longrightarrow \text{products}^{[89,93]}$$

$$H + \text{butenes} \dashrightarrow \text{products}^{[93]}$$

4.2.12. Thermal Methods

The heat released to a catalytic surface on which atom recombination occurs has been used for many years as the basis of thermocouple and calorimeter probes for the detection and measurement of atom concentrations. The method is nonspecific in that all atoms or radicals in the system can contribute to heat release. Its use has thus been confined to relatively simple systems in which few active species are present. In particular, the rate of heterogeneous recombination of H and O atoms has been studied[94] using a thermocouple gauge and Smith's[95] side-arm method.

A useful development of the heat release principle is the isothermal calorimeter,[96] which has found application for the study of the kinetics of homogeneous atom recombination. The calorimeter forms one arm of a simple d.c. bridge circuit. In the absence of a discharge a known current is passed through the calorimeter wire which then has a resistance R_1, corresponding to temperature T_1. The discharge is now turned on, and the current in the bridge is reduced until R_1 is again reached. The difference in power with the discharge off and on is equal to the rate of transfer of heat to the wire by atom recombination. Absolute flow rates (and hence concentrations), F, of atoms are obtained from the simple relation, $F = W/H$, where W is the power released to the calorimeter wire under isothermal conditions, and H is the atomic enthalpy of recombination of atoms. The use of this relation assumes that (i) all atoms are recombined on the calorimeter surface, (ii) recombination products desorbing from the calorimeter do not possess excess vibrational or electronic energy and (iii) the thermal conductivity of the gaseous environment of the calorimeter is unaffected by turning on the discharge. (iii) is negligible if low atom concentrations are used. There is no evidence that (ii) represents a serious error in the systems so far studied. (i) is difficult to ensure, and the variability of the efficiency of the surface probably accounts for the comparatively few studies carried out with the method. A tandem arrangement of two calorimeters in series has been developed for the study of the reactions[97]

$$H + H + M \longrightarrow H_2 + M$$

$$H + O_2 + M \longrightarrow HO_2 + M$$

in order to allow for the less-than-unity efficiency of the first calorimeter. An independent study of the H + H + M recombination reaction has also been carried out recently using the isothermal calorimeter.[202] This method provides, in principle, a primary standard for atom concentration measurements, by which other methods may be calibrated: e.g. chemical titration of H atoms by the HNO emission method,[31] and by NO_2.[22]

The isothermal wire calorimeter, with a nickel surface rather than the Pt–Rh alloy used with H,[97] has been used for the measurement of halogen atom concentrations.[16] Difficulties are experienced in some cases with the detection of halogen atoms on account of corrosion of the nickel wire.[38] The isothermal calorimeter has also been used, in this case with a metal surface coated with cobalt oxide, for the measurement of relative oxygen atom concentrations,[98] and to detect metastable electronically excited singlet oxygen molecules.[99]

4.3. CHEMILUMINESCENT REACTIONS: ENERGY TRANSFER AND THE KINETICS OF EXCITED STATES

The emission of radiation during exothermic reactions of atoms and simple radicals is favoured under the low-pressure conditions of a discharge-flow system. Two types of excited state are encountered: (a) electronically excited species, and (b) vibrationally excited (i.e. non-equilibrium) electronic ground state species. The former type of reactions has been reviewed in 1968.[11]

4.3.1. Electronic Excitation in Third-Order Atom Recombination Reactions

A well-defined class of reactions which produce electronically excited species is known: the overall third-order atom recombination processes

$$X + Y + M \longrightarrow XY^* + M$$

In these reactions, X and Y may both be atoms (either similar or dissimilar), or Y may be a molecule. Radiative decay and quenching of electronically excited XY^* are normally rapid on the time scale of a discharge-flow system; hence, a stationary state is rapidly established by these (and other) steps involving the excited state. An idealised scheme is

$$X + Y + M \underset{k_{-1}}{\overset{k_1}{\rightleftharpoons}} XY^* + M \tag{1}$$

$$XY^* + M \xrightarrow{k_2} XY + M \tag{2}$$

$$XY^* \xrightarrow{k_3} XY + h\nu \tag{3}$$

This scheme provides a basic description of many of the systems so far investigated. The finer details of the mechanisms are usually complex.

Steady-state analysis gives the intensity of chemiluminescence I in terms of the rate constants for combination, k_1, redissociation k_{-1}, electronic quenching k_2 and radiation k_3:

$$I = k_1 k_3 [X][Y][M]/(k_3 + k_2[M] + k_{-1}[M]) \tag{I}$$

If I is the total number of quanta emitted, the redistribution of energy within the excited state by vibrational relaxation can be neglected. The limiting case when $k_3 \ll (k_2 + k_{-1})[M]$, which often holds as an approximation under the conditions of a discharge-flow experiment, leads to the simple expression (II)

$$I = \{k_1 k_3/(k_{-1} + k_2)\}[X][Y] \tag{II}$$

Kinetic analysis of chemiluminescent reactions of this type is usually directed towards the deduction of values for k_1, k_{-1}, k_2, k_3 and the rate constants of any other processes involved in formation or removal of XY*. Since such studies are steady-state analyses, they do not utilize the time resolution associated with axial displacement along the flow system.

(a) *Homo-atom Recombination Reactions*

The simplest third-order chemiluminescent reactions are those involving two identical atoms:

$$N(^4S) + N(^4S) + M \longrightarrow N_2(B^3\Pi_g) + M \tag{i}$$

$$N_2(B^3\Pi_g) \longrightarrow N_2(A^3\Sigma_u^+) + h\nu$$

$$Cl(^2P_{3/2}) + Cl(^2P_{3/2}) + M \longrightarrow Cl_2(B^3\Pi(O_u^+)) + M \tag{ii}$$

$$Cl_2(B^3\Pi(O_u^+)) \longrightarrow Cl_2(X^1\Sigma_g^+) + h\nu$$

$$Br(^2P_{3/2}) + Br(^2P_{3/2}) + M \longrightarrow Br_2(A^3\Pi(1_u)) + M \tag{iii}$$

$$Br_2(A^3\Pi(1_u)) \longrightarrow Br_2(X^1\Sigma_g^+) + h\nu$$

$$S(^3P) + S(^3P) + M \longrightarrow S_2(B^3\Sigma_u^-) + M \tag{iv}$$

$$S_2(B^3\Sigma_u^-) \longrightarrow S_2(X^3\Sigma_g^-) + h\nu$$

Of these reactions, (i) has been extensively investigated,[27] and quantitative data are also available[35-37] or (ii). Kinetic study of (iii), the bromine afterglow emission spectrum, first studied by Gibbs and Ogryzlo,[100] is complicated by appreciable radiation from the $B^3\Pi(O_u^+)$ state, as well as the $A^3\Pi(1_u)$ state[101] of Br_2. Data on the chemiluminescent combination reaction (iv) of two $S(^3P)$ atoms have also been published.[18] The reactions $O + O + M$ and $H + H + M$ do not combine into electronically excited states with allowed transitions to the ground states of O_2 and H_2.

(b) *Hetero-atom Recombination Reactions*

The large class of hetero-atom combination reactions has not been at all fully investigated. Examples of such reactions which lead to the formation of electronically excited states as detected in discharge-flow systems are:

$$N(^4S) + O(^3P)(+M) \longrightarrow NO^* \ (+M)^{102,203}$$

NO* can be any of a number of excited states.

$$O(^3P) + H(^2S)(+M) \longrightarrow OH(A^2\Sigma^+)(+M)^{103}$$
$$Br(^2P_{3/2}) + Cl(^2P_{3/2})(+M) \longrightarrow BrCl(B^3\Pi(0^+))(+M)^{104}$$
$$I(^2P_{3/2}) + Cl(^2P_{3/2})(+M) \longrightarrow ICl(A^3\Pi(1))(+M)^{104}$$
$$I(^2P_{3/2}) + Br(^2P_{3/2})(+M) \longrightarrow IBr(A^3\Pi(1))(+M)^{101}$$

Of these reactions, quantitative kinetic data are available only for the combination of $N(^4S)$ with $O(^3P)$ atoms, which unfortunately leads to numerous excited states of NO.[102]

(c) *Atom-plus-Molecule Recombination Reactions*

The experimental problems associated with the study of reactions first-order in atom concentrations are generally simpler than the cases so far discussed. Hence, often more complete data for such processes are available at the present time:

$$O(^3P) + NO + M \longrightarrow NO_2^* + M^{13,28,110,204}$$
$$O(^3P) + CO + M \longrightarrow CO_2^* + M^{28}$$
$$O(^3P) + SO(X^3\Sigma^-) + M \longrightarrow SO_2^* + M^{114}$$
$$H(^2S) + NO + M \longrightarrow HNO^* + M^{31}$$

In particular, the formation of NO_2^* by combination of O with NO has been the subject of several full kinetic studies over a wide range of conditions.

The type of kinetic information available from the study of chemi-luminescent combination reactions will be exemplified with reference to three systems which have received attention recently: $N + N + M$, $Cl + Cl + M$ and $O + NO + M$.

(d) *Example:* $N(^4S) + N(^4S) + M$

The yellow emission spectrum emitted during combination of two ground state N atoms is well known to consist of bands from selectively populated vibrational levels ($0 \leq v' \leqslant 12$) of the allowed First Positive $B^3\Pi_g - A^3\Sigma_u^+$ system of N_2. It has been shown that weaker infra-red bands of the $B'\ ^3\Sigma_u^- - B^3\Pi_g$ system,[27a] and far ultraviolet bands of the Lyman–Birge–Hopfield $a\ ^1\Pi_g - X^1\Sigma_u^+$ magnetic dipole transition[105] and the forbidden $a'\ ^1\Sigma_u^- - X^1\Sigma_g^+$ system[106] are also present in the afterglow spectrum. All the excited states of these transitions correlate with non-ground state ($^4S + {}^4S$) atomic products.

Thrush[11,27b] has reviewed the kinetics of the $B^3\Pi_g–A^3\Sigma_u^+$ emission bands of the nitrogen afterglow spectrum. The intensity of emission, I, is proportional to $[N]^2$ and, in pure N_2, it is independent of $[M]$ in the 1 to 10 torr pressure range. Between 0·25 and 1 torr, there was found to be a reduction in the number of quanta emitted per recombination as the pressure was increased; this is consistent with the reported half quenching pressure for $B^3\Pi_g$ of 0·25 torr in pure N_2.[126] Above 1 torr the observed independence of I upon $[M]$ is due to cancellation of the $[M]$-dependences of third-order recombination and of electronic quenching by ground state $N_2 = M$; it cannot be satisfactorily assigned

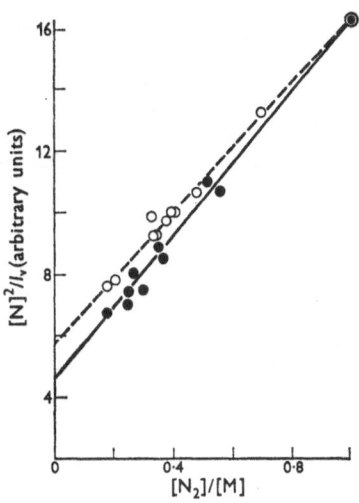

FIG. 4.6. Variation of $[N]^2/I_y$ with mole fraction of N_2 in argon ($\circ$) and helium ($\bullet$) at 298 K. (After ref. 27b.)

to a simple two-body recombination of two $N(^4S)$ atoms. The intensity of emission was enhanced by addition of Ar or He inert gas, and linear Stern–Volmer quenching plots of $[N]^2/I$ with $[N_2]/[M]$ were obtained with these gases (Fig. 4.6).

On the basis of all the then available data, Campbell and Thrush[27b] proposed the following mechanism for the nitrogen afterglow. The $A^3\Sigma_u^+$ state is populated by three body combination. Measurements of absolute afterglow intensities were combined with quenching data, as described for $O + NO + M$, to show that formation of $A^3\Sigma_u^+$ by three body recombination accounted for about 50% of the total rate of recombination of N atoms. The vibrational energy distribution in $A^3\Sigma_u^+$ is governed by the population process, by collisional redissocia-tion, and by vibrational relaxation. Several types of collision-induced crossings from the $A^3\Sigma_u^+$ to the $B^3\Pi_g$ state were thought to be responsible for populating the $B^3\Pi_g$ state. With N_2 as carrier gas, electronic quenching ensures that the observed vibrational distribution in the $B^3\Pi_g$ state is close to its initial distribution, whereas for He and Ar as carrier gases, vibrational relaxation in the $B^3\Pi_g$ state modifies this distribution. Figure 4.7 shows potential energy curves for N_2.

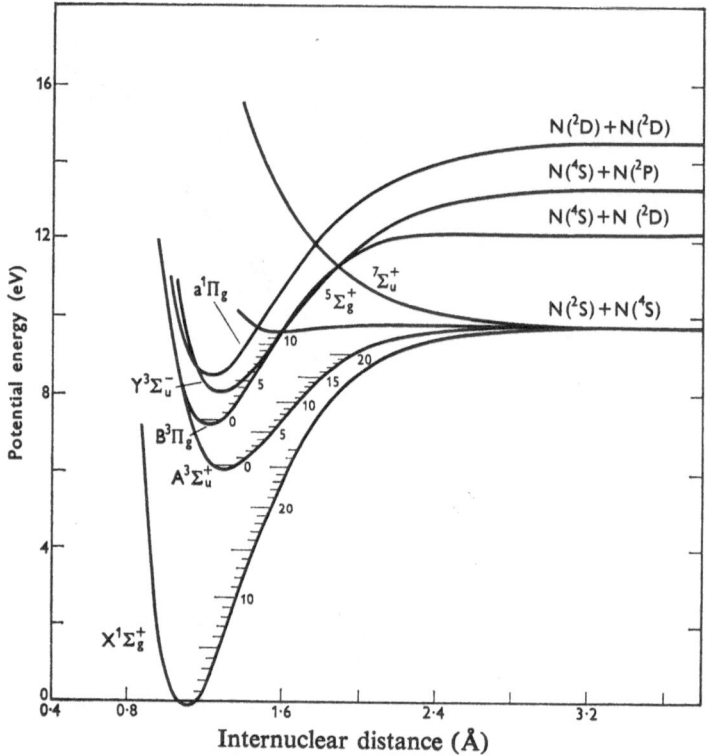

Fig. 4.7. Potential curves of nitrogen (Gilmore 1965). (After ref. 27b.)

However, Benson[107] has presented arguments in favour of an earlier hypothesis in which $B^3\Pi_g$ is populated by a crossing from a shallow $^5\Sigma_g^+$ state which can be formed directly from $N(^4S) + N(^4S)$. It was argued that the large transition probability (10^{-1} to 10^{-2}) for non-radiative transitions of the type that occurs in N_2 predissociation, $N_2(B^3\Pi_g) \rightleftharpoons N_2(^5\Sigma_g^+)$, ensures that the hot rotational levels of the $v' = 12$ state of $B^3\Pi_g$ will be in equilibrium with free $N(^4S)$ atoms without the intervention of a third body. However, the kinetic implications of Benson's mechanism are similar to those of Campbell and Thrush.

Nitrogen atom radiative recombination has been reinvestigated recently, in one case over the pressure range 0·5 to 500 mTorr in a

220 m^3 vessel,[192] and in the other case at higher pressures in the flow tube.[229] In the low pressure regions, relaxation of primary vibrational distributions within the excited B$^3\Pi_g$ ($v' = 13$) and $a^1\Pi_g$ ($v' = 6$) states formed is slight or negligible.[192] It is thought that in this region the excited B and a states are formed by two body association [of N(^{4}S) + N(^{4}S)] involving inverse predissociation via a $^5\Sigma_g^+$ state.[192,229]

Recent work[27b] has reconsidered the possibility that chemical reactions of the metastable A$^3\Sigma_u^+$ state of N$_2$, which is present in the products of a discharge in N$_2$, might account for anomalous behaviour observed in the reactions of active nitrogen. Generally, there is little evidence that this state is important in determining the reactions of active nitrogen, although the simple step

$$N_2(A^3\Sigma_u^+) + N \longrightarrow N_2(X^1\Sigma_g^+) + N$$

is thought to be rate-determining in nitrogen atom recombination at high pressures (20 to 760 torr).[127]

(e) *Example:* Cl($^2P_{3/2}$) + Cl($^2P_{3/2}$) + M

The kinetics of radiative recombination of halogen atoms are more complex than those of the N$_2$ afterglow emission spectrum. The extra complexity partly arises from the fact that appreciable populations of two or more excited states of the molecular halogen may be produced.[40,101,104] For example, the low-lying $^3\Pi$ state of Br$_2$, whose energy levels are intermediate between coupling cases (i) and (iii), separates into a number of sub-states O_u^+, O_u^-, 1_u, 2_u, having similar energies. In principle, several of these sub-states may be populated by atom recombination subject to correlation rules. However, only the $^3\Pi(O_u^+)$ state, which correlates with Br($^2P_{3/2}$) + Br($^2P_{1/2}$) and the $^3\Pi(1_u)$ state, which correlates with Br($^2P_{3/2}$) + Br($^2P_{3/2}$), are known.

The Cl$_2$ afterglow emission spectrum is arguably the simplest of the halogen atom recombination reactions, since only one state—the B$^3\Pi(O_u^+)$—contributes substantially to the emission.[104] This state is probably populated from two ground state (Cl($^2P_{3/2}$) atoms by crossing from another excited state, possibly $^1\Pi_u$. The kinetics of the Cl$_2$ afterglow emission in Ar or Cl$_2$ carrier gas were fitted by the empirical relation,[38] $I_\lambda = I_{0\lambda}[Cl]^{n_\lambda}$, where I_λ is the intensity at wavelength λ(nm), and n_λ is a parameter which decreases from a value of 2 for emission from levels $v' > 10$ (520 nm) to a value of 1 for the lowest vibrational levels of the excited state studied, $v' \sim 0$ (960 nm). (Previous reports[36,37] had erroneously concluded that the simple rate law $I_\lambda = I_{0\lambda}[Cl]^2$ held.) Qualitatively, the dependence of n_λ upon λ was manifested[38] by a shift of spectral distribution to the red (lower v') as [Cl] → 0 (Fig. 4.8). A similar red shift occurred with increase in total pressure [M], using both Ar and Cl$_2$ as carrier gases. Under the conditions studied, predominant quantum output was from the lower vibrational levels of the excited state. Thus, the total intensity from all levels of the excited state ΣI_λ

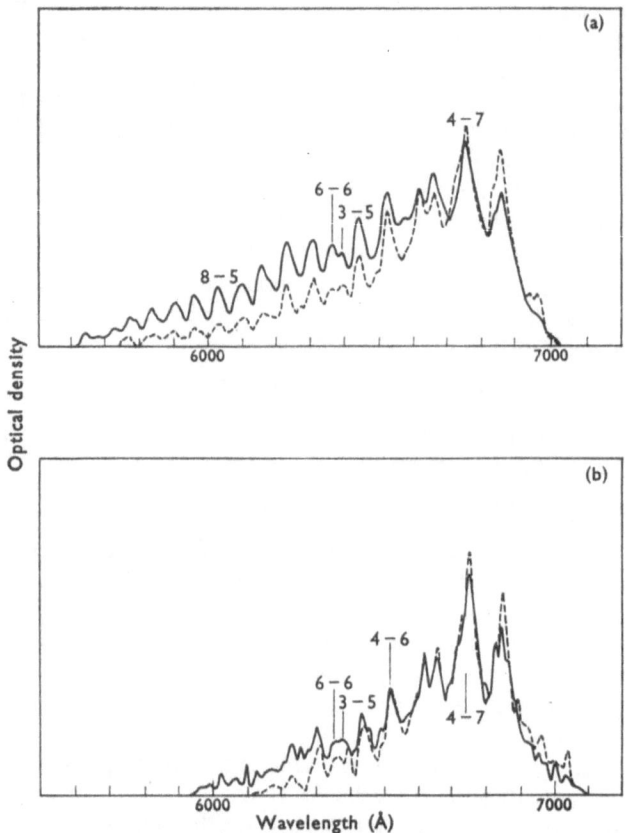

FIG. 4.8. Effect on the intensity distribution of the Cl_2 afterglow spectra of variable [Cl] and [M]. Spectra were obtained using argon carrier gas at 298 K; microdensitometer traces are shown in the Figure. The long wavelength cut-offs are due to the negligible sensitivities beyond these wavelengths of the photographic emulsions used.

		[atom](10^{-10} mol cm^{-3})	[M](10^{10} mol cm^{-3})
(a) $Cl_2\,^3\Pi_{ou^+} \rightarrow\,^1\Sigma_g^+$	full line	12·0	1·40
	dotted line	2·4	
(b) $Cl_2\,^3\Pi_{ou^+} \rightarrow\,^1\Sigma_g^+$	full line	8·6	0·48
	dotted line		4·70

was calculated using emission from the high levels as a correction to the main term, I_{9600}. The kinetics of ΣI_λ were described in terms of the expression, $\Sigma I_\lambda = I_0[Cl]^n$, with $2 > n > 1$. The intermediate value of n between 1 and 2 suggested that significant removal of $Cl_2(^3\Pi(O_u^+))$ occurred through a process first-order in [Cl] specific quenching by chlorine atoms (3)

$$Cl + Cl + M \rightleftharpoons Cl_2^*(^3\Pi(O_u^+)) + M \qquad \text{(i)}$$

$$Cl_2^* \longrightarrow Cl_2 + h\nu \qquad \text{(ii)}$$

$$Cl_2^* + Cl \longrightarrow Cl_2 + Cl \qquad \text{(iii)}$$

$$Cl_2^* + M \longrightarrow Cl_2 + M \qquad \text{(iv)}$$

On the other hand, the results showed that electronic quenching rates by Ar or Cl_2 (step iv) are insignificant under the conditions used.[38] The [M]-dependence of ΣI_λ shows that reaction (i) is overall third-order,[38,154] as would be expected on other grounds. Work by Browne and Ogryzlo[154] is basically in agreement with the results of Clyne and Stedman,[38] although they conclude[154] that a more complex mechanism is required to fully explain their results.

Studies of the kinetics of both the $B^3\Pi(O_u^+)$[62] and the predominant $A^3\Pi(1_u^+)$[40] states of Br_2 formed in the radiative recombination of $Br\,^2P_{3/2}$ atoms have recently been reported. For $Br_2\ A^3\Pi(1_u)$, the results are similar to those for $Cl_2\ ^3\Pi(O_u^+)$.

(f) *Example:* $O(^3P) + NO(X^2\Pi) + M$

The spectrum of the yellow-green luminescence due to combination of O with NO consists of an apparent continuum superimposed on which are reported to be a number of diffuse bands.[108] The intensity maxima of these emission bands correspond well with those of the absorption bands of NO_2 at 300 K.[108] The high energy cut-off at 397·5 nm of the luminescence corresponds closely with the energy of the reaction.[108]

$$O + NO + M \longrightarrow NO_2 + M$$

There is consequently no doubt that the excited species formed is NO_2, and strong evidence can be adduced that this species is, in fact, an electronically excited state of NO_2, probably a $^2B_1$ state correlating, together with the ground state, with the same $^2\Pi$ state of linear NO_2.[11]

Although it has been suggested that, in the 1 torr pressure region the O + NO chemiluminescent reaction, among others, is a simple bimolecular combination process,[109] the weight of evidence is in favour of third-order kinetics.[11,110] In confirmation of third-order kinetics, study[13] at this pressure of the kinetics of emission in the $10–250 \times 10^{-3}$ torr pressure region has shown that the intensity becomes pressure-dependent below $60 \times 10^-$ torr. Good agreement was obtained with this value independently.[110] Above 60×10^{-3} torr, cancellation of M-effects causes the intensity to become independent of [M]. It was suggested[13] that losses of atomic oxygen as a result of rapid radial diffusion and wall recombination at low pressures, which were neglected in earlier work,[109] led to the finding[109] of apparent second-order kinetics for the O + NO + M reaction. However, at sufficiently low pressures, Becker, Groth and Thran,[204] using a 220 m^3 spherical vessel (at pressures down to 0·1 mtorr, conclusively showed that both third order and second order mechanisms apply, with the second order term predominant below 1 mtorr.

The simple mechanisms consisting of reactions (1), (2) and (3), (plus the second order recombination term at low pressures), together with vibrational quenching steps, give a satisfactory description of the O + NO + M reaction.[28] Since the excited state of NO_2 in the O + NO + M emission is believed to be the same as that excited in fluorescence by irradiation of NO_2,[113] the rate constant ratio for fluorescence and electronic quenching (k_3/k_2) obtained from fluorescence quenching experiments may be applied to the O + NO + M system.[28] In that system, equation (II) shows that, $I = k_1k_3/(k_{-1} + k_2)[O][NO]$. Combination of the measured absolute value of the composite constant $I_0 = k_1k_3/(k_{-1} + k_2)$ with the value of k_3/k_2 from fluorescence studies gives k_1^*, the third-order rate constant for forming NO_2 via NO_2^*, $k_1^*[M] = I_0(k_3 + k_2[M])/k_3$. Such a procedure showed that formation of excited NO_2 contributes a fraction approaching unity of the total rate of formation of all states of NO_2 by third-order recombination.[111] Related arguments[204] indicated a similar, but higher, fraction of reaction via NO_2^*.

It is most important that the kinetics of the O + NO + M radiative reaction have been thoroughly studied. This is because the intensity of emission from this reaction has been measured absolutely,[112] and it thus serves as a useful secondary actinometric standard for the determination of absolute intensities of other chemiluminescent processes.

4.3.2. Electronic Excitation in Bimolecular Transfer Reactions

Extremely few well-established cases are known in which emission is observed from an electronically excited state formed in a bimolecular transfer reaction of ground state reactants. On the other hand, many bimolecular processes transfer electronic excitation energy to a ground state molecule (see section 4.4 for examples).

For electronic excitation to occur during chemical reactions of ground state species, the release of energy must be adequate to populate states which are accessible via the potential energy surfaces of the reaction. Symmetry restrictions on the correlations between reactants and products may be operative.[11] These considerations appear to place a severe limitation on the number of simple reactions which can lead to electronic excitation. Two cases which do involve the formation of electronically excited products have been discussed by Thrush,[11] and will be described briefly below.

The exothermic reactions of ozone with NO and with the ground state $SO(X^3\Sigma^-)$ radical lead to electronically excited products

$$NO + O_3 \longrightarrow NO_2 + O_2; \quad \Delta H^\circ_{298} = -200 \text{ kJ mol}^{-1}$$

$$SO + O_3 \longrightarrow SO_2 + O_2; \quad \Delta H^\circ_{298} = -443 \text{ kJ mol}^{-1}$$

In the case of the NO + O_3 reaction, the degeneracy associated with the nonzero orbital angular momentum of NO allows two doublet potential surfaces directly correlating NO + O_3 with a bent ground state and a new linear electronically excited state of NO_2.[11] As might be expected, the frequency factors for reaction over both surfaces are similar, although formation of excited NO_2 involves a higher activation energy (18 kJ mol^{-1}, as compared with 10 kJ mol^{-1} for ground state NO_2).[113]

If no potential surface leads to the radiating state, then such a state may be formed by a radiationless transition to the surface which yields excited products, occurring in the region of the transition state. This is thought to be the case for the SO + O_3 reaction in which the $^3B_1$ and ^1B excited states as well as the ground state of SO_2 are observed as products[11,114]

$$SO + O_3 \longrightarrow SO_2 + O_2; \quad \Delta H^\circ_{298} = -443 \text{ kJ mol}^{-1}$$

As would be expected on this basis, the formation of electronically excited SO_2 involves a higher activation energy *and* a lower frequency factor than formation of ground state SO_2.[114]

4.3.3. Singlet Oxygen

Low-lying electronically-excited $^1\Delta_g$ and $^1\Sigma_g^+$ states of O_2, which are metastable with respect to transitions to the $X^3\Sigma_g^-$ ground state, are well known. These species not only have very long radiative life times ($\sim$10 s for $^1\Sigma_g^+$, 4000 s for $^1\Delta_g$),[85] but also electronic quenching is inefficient. It is consequently possible to detect these species and study the kinetics of physical and chemical processes in which they are involved using a discharge-flow system. The topic has been fully reviewed elsewhere.[85]

Detection of $O_2(^1\Delta_g)$ has been made using a thermal probe,[123] photometry on the (0, 0) band of the $^1\Delta_g$–$X^3\Sigma_g^-$ transition at 1·27 μm, e.p.r. spectrometry,[125] by photoionization,[128] and by photoelectron spectroscopy.[205] $O_2(^1\Sigma_g^+)$ is perhaps most easily detected by the fairly intense emission (0, 0) band of the $^1\Sigma_g^+$–$X^3\Sigma_g^-$ system at 7600 Å. Substantial concentrations of $O_2(^1\Delta_g)$ ($\sim$10% of total flow) are present in the products of a microwave discharge in O_2, whilst $\sim$0·1% of $O_2(^1\Sigma_g^+)$ are also present in these products.

There is a significant, although low, probability that two excited singlet oxygen molecules can collide. Such collisions appear to lead to emission ('dimol' emission) resulting from the simultaneous loss of energy from both excited molecules; diffuse bands of this type have been reported to have intensity maxima near 6340 Å and 7030 Å. Although the intensity of the $\lambda 6340$ Å dimol emission was originally reported[124a] to be first-order in $[O_2 \, {}^1\Delta_g]$, it now seems clear[124b,129] that it is second-order in $[O_2 \, {}^1\Delta_g]$, as expected:

$$O_2({}^1\Delta_g) + O_2({}^1\Delta_g) \longrightarrow 2O_2(X^3\Sigma_g^-) + h\nu$$

Singlet oxygen $({}^1\Delta_g)$ shows a number of novel chemical reactions, which have been identified and studied in discharge-flow systems. One such reaction is the energy-pooling process, which leads to $O_2({}^1\Sigma_g^+)$:

$$O_2({}^1\Delta_g) + O_2({}^1\Delta_g) \longrightarrow O_2({}^1\Sigma_g^+) + O_2(X^3\Sigma_g^-)$$

The rate constant for this reaction has been measured; although there are divergences between the values obtained, it seems likely that the lower value of 1×10^6 cm³ mol⁻¹ s⁻¹ is correct.[85] Another reaction is the decomposition of ozone, forming ground state $O({}^3P)$ atoms, which accounts for some anomalous features of the kinetics of oxygen atom recombination in the products of a discharge in O_2

$$O_2({}^1\Delta_g, {}^1\Sigma_g^+) + O_3 \longrightarrow 2O_2 + O({}^3P)$$

Energy transfer from singlet oxygen, by electronic quenching as well as by transfer of electronic excitation energy to a molecule which can emit, has also been studied.[85] One example is the excitation by $O_2({}^1\Delta_g, {}^1\Sigma_g^+)$ of the $B^3\Pi(O_u^+) \leftarrow X^1\Sigma_g^+$ systems of Br_2 and I_2, and of the $B^3\Pi(O^+)$ systems of BrF and IF,[206] in systems containing recombining atoms.[130]

4.3.4. Vibrational and Rotational Excitation in Bimolecular Transfer Reactions

It is well known that rapid exothermic bimolecular elementary reactions can lead to product molecules in which the initial rotational and vibrational energy distributions are dissimilar to those of equilibrium at laboratory temperatures, T_0. A nonrigorous but useful description of such distributions is that the rotational and vibrational "temperatures", T_R and T_V (obtained by fitting the observed distributions to Boltzmann distributions for T_R and T_V) are such that $T_R > T_0 < T_V$. It is possible for complete vibrational population inversion to occur in the initial products of reactions. This corresponds to the case when the population of an excited vibrational level $\theta(v)$ exceeds that of the vibrational ground state, $\theta(v) > \theta(0)$, and leads to the unrealistic description $T_V < 0$. An intermediate case—partial inversion—is also observed when $T_V > 0$; as rotational relaxation is more rapid than vibrational relaxation, $T_R \sim T_0$, and thus inversion exists over a limited range of rotational quantum numbers in respect of $P(J)$ or $R(J)$ transitions to the vibrational ground state. Laser action has been observed, as predicted, in cases of either complete or partial inversion.[115] Vibration–rotation emission from HF produced in the $H_2 + UF_6$ reaction shows strong laser action.[116]

Polanyi has described criteria for different types of elementary reaction in respect of the state of vibrational excitation of the products.[117] It appears that many more reactions lead to vibrational excitation than to electronic excitation of molecular products. A range of such reactions has been studied in a series of fundamental investigations by Polanyi and his co-workers using an adaptation of the discharge-flow method. These include the reactions

$$H + Cl_2 \longrightarrow HCl + Cl$$
$$Cl + HBr \longrightarrow HCl + Br$$
$$Cl + HI \longrightarrow HCl + I$$

In each case, the product molecule HCl is vibrationally excited. The fraction of the energy liberated in these reactions which appears as vibration-rotational excitation of HCl is greater for attack by Cl than for attack by the much lighter H atom. At least part of the decreased vibration in HCl from $H + Cl_2$ is explicable by the fact that the light attacking atom approaches the centre of mass of the system too rapidly to allow energy release by repulsion between the two Cl atoms whilst H and Cl are still at an extended separation.[117]

Systematic experimental studies of the initial vibrational and rotational energies in HCl produced by simple atom transfer reactions have been carried out using a type of low pressure discharge-flow system. Populations of energy levels are inferred directly from measurements of the intensities of infra-red emission spectra of HCl. In the first method (I) used,[118] the rate of vibrational relaxation was measured directly from the decay of population with distance along the axis of a large diameter flow system, through which the products of reaction were rapidly passed at low pressures. Initial distributions were then inferred by extrapolation of the decays to zero time. In a second method (II), relaxation was arrested. In this case, the walls of the flow chamber were cooled to 77 K, ensuring an efficient sink for any HCl molecules which collide. The pressure was kept sufficiently low to ensure that the mean free path (10–100 cm) was substantially greater than the diameter of the chamber (15 cm). Initial distributions could thus be directly measured (see Fig. 4.9).

Both experimental methods[118] lead to rate constants $k(v)$ for the population of vibrational levels v, e.g.

$$H + Cl_2 \xrightarrow{k(v)} HCl(v) + Cl$$

The vibrational populations of HCl formed in reactions of the type

$$H + X_2 \longrightarrow HX + X \ (X = Cl, Br)$$

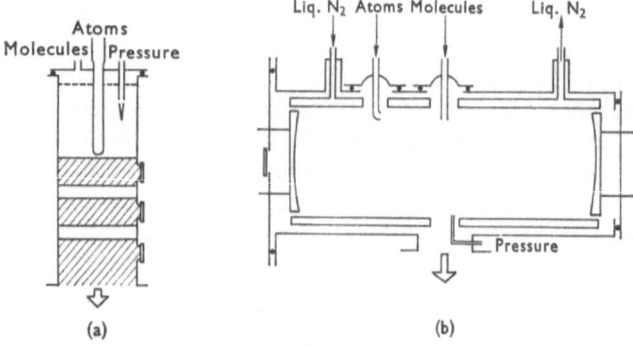

FIG. 4.9. (*a*) Reaction vessel for method I; shaded regions indicate internal gold coating. (*b*) Reaction vessel for method II (to the same scale as (*a*)). (After ref. 118.)

have been used to identify the correct potential energy surfaces and collision dynamics when the reactions are simulated by trajectory calculations.[118] The results for reactions of the types $H + X_2$ and $X + HY$ (X, Y are halogen atoms) have shown that a large fraction of available energy enters the vibrational degree of freedom in the newly-formed H—Y bond. Recently, populations in HCl and DCl produced by the reactions of H and D with SCl_2 and S_2Cl_2 and of H with $SOCl_2$, SO_2Cl_2 and Cl_2O have also been measured.[141]

Exothermic elementary reactions of F $^2P_{3/2}$ atoms with a range of hydrogen-containing molecules RH have been shown to produce vibrationally excited HF(*v*) molecules,

$$F + RH \longrightarrow HF(v) + R,$$

and detailed energy transfer studies in HF(*v*) similar to those of Polanyi have been described.[208]

(a) *Vibrationally Excited* OH(X²Π)

Infra-red emission from OH(X²Π, $v \le 3$) has also been observed and analysed,[119] although no detailed measurements of relaxation rates have been made; the source of radicals was the reaction sequence

$$H + O_2 + M \longrightarrow HO_2 + M; \qquad \Delta H^{\circ}_{298} = -196 \text{ kJ mol}^{-1}$$

$$H + HO_2 \longrightarrow OH(v = 0) + OH(v \le 3); \quad \Delta H^{\circ}_{298} = -160 \text{ kJ mol}^{-1}$$

Much greater vibrational excitation in OH was observed[120] (up to $v = 10$) in the reaction of atomic hydrogen with ozone in a discharge-flow system at pressures near 1 torr:

$$H + O_3 \longrightarrow OH + O_2; \qquad \Delta H^{\circ}_{298} = -321 \text{ kJ mol}^{-1}$$

Similar emission by HF ($v \leq 10$) has been observed under these conditions,[198] and at lower pressures,[207] from the reaction

$$H + F_2O \longrightarrow HF + OF$$

In the case of OH from the H + O_3 reaction, energy flux calculations were made[121] assuming the collisional deactivation probabilities were proportional to the vibrational radiative probabilities. This model indicated that there was appreciable initial population of all vibrational levels with $v \leq 9$.

(b) Vibrational Excitation in NO_2 and N_2O

The extension of the infra-red emission technique of Polanyi to polyatomic molecules is difficult on account of the rapidly increasing spectral complexity which accompanies an increase in atomicity. However, the case of NO_2, which is observed with both electronic (section 4.3.2) and vibrational excitation, has been given attention:[122]

$$NO + O_3 \longrightarrow NO_2 + O_2$$

It appears that direct population of high vibrational levels of the ground state is not primarily responsible for the initial distribution of vibrational energy in the NO_2 molecule formed. Evidence has been presented[122] that vibrationally excited NO_2 is mainly produced by radiation or collisional quenching of electronically excited NO_2. Data on the effects on vibrational relaxation rates of N_2, NO_2 and CO_2 have also been reported.[122]

Extensive vibrational excitation in N_2O produced directly by the rapid reaction $N + NO_2 \rightarrow N_2O + O$ has been reported; about 37% of the energy released in the reaction enters vibration of N_2O.[131]

4.4. Chemiluminescent Reactions: Spectroscopic Studies

Before concluding this account of the information obtainable from studies of chemiluminescent reactions in discharge-flow systems, it is appropriate to indicate the increasingly wide applications of this area to the study of the energy levels of simple molecules and radicals.

4.4.1. Instrumental Aspects

In principle, specific excitation of a particular molecule state or states, as is observed in chemiluminescent reactions, provides an ideal source for high resolution spectroscopic studies. There is little or no overlapping by unwanted spectra, and transitions usually occur over a wide range of excited state vibrational levels, often including some which are inaccessible by absorption from the ground electronic state. However, the low intensity of chemiluminescent phenomena leads to difficulty in recording spectra at the high resolution which is necessary for accurate measurements of vibrational and rotational energy levels.

However, two promising techniques are now available for the facilitation of such studies. Firstly, the use of laser components, such as multilayer dielectrics as mirror coatings of extremely high reflectivity ($\leq 99.99\%$), in order to increase the apparent luminosity of the source. Secondly, the use of improved detection methods when photoelectric recording is employed. Photon counting and/or phase-sensitive amplification combined with photomultiplier cells with good signal-to-noise ratios, such as E.M.I.'s 9558 and 6256S cells, offer a great improvement in sensitivity. Image intensifier tubes of good gain and satisfactory resolution are also now available.

Nevertheless, at the present time virtually all spectroscopic studies of chemiluminescent reactions in flow systems have been made under conditions which do not satisfy the ideal criterion of near-full rotational resolution. The generally low source luminosities have had the consequence that many spectroscopic studies were carried out with photographic spectrographs of good light-gathering powers. The conventional prism and grating spectrographs of medium or long focal length (≥ 1 metre) are generally unsuitable, since they usually possess low apertures and moderate to high dispersion, which lead to a low output luminosity. Instruments intended for Raman spectrometry and astrophysical spectrometry are usually excellent also for recording chemiluminescent spectra. Bass and Kessler[132] have described a high aperture grating spectrograph for the visible and near infra-red, based on commercial camera components, which has been used successfully in Broida's laboratory,[133] and more recently in the writer's laboratory, for recording chemiluminescent spectra. The writer's current spectrograph uses an f/4·5 collimator lens and an f/1·5 camera lens, to give a substantially diminished image, and it uses a single lens reflex camera. A 300 line mm^{-1} grating blazed in the infra-red (3·5 μm) is found useful, since a range of spectral orders (and hence wavelengths) may be obtained without adjustment of the grating; it also allows optimum dispersion and resolution to be obtained.

4.4.2. The $^3\Pi(O^+)$–$X^1\Sigma^+$ System of BrCl

As has already been pointed out, one important advantage of chemiluminescence as a source of spectra is often the absence of many overlapping band systems which may defy analysis. One case where this advantage has been decisive is the identification for the first time of the visible band system of BrCl.[104] An extensive system of long progressions of red-degraded bands was observed with a Bass–Kessler spectrograph from 675 nm to at least 950 nm when either $Br(^2P_{3/2})$ and $Cl(_2P_{3/2})$ atoms were mixed, or during the chain reaction when molecular bromine and chlorine dioxide were mixed.[104] Many of the bands showed well-separated weaker bands with heads lying at longer wavelengths than the main band heads. Tentatively identifying the stronger and weaker bands, as due to molecules containing ^{35}Cl and ^{37}Cl, respectively, the values of the isotopic splittings between the band heads were found to confirm the assignment of the bands to BrCl. Isotope splitting between ^{79}Br and ^{81}Br species could not be readily observed, on account of the small difference in reduced masses between such species. The vibrational numbering in both ground $X^1\Sigma^+$ and excited states was also fixed (to ± 1 unit for v'', ± 2 units for v') by the magnitudes of the isotopic splittings, and thus the values of the vibrational constants (cm^{-1}), $\omega_e' = 243 \pm 3$, $\omega_e' x_e' = 6\cdot7 \pm 0\cdot2$, were determined. The highest

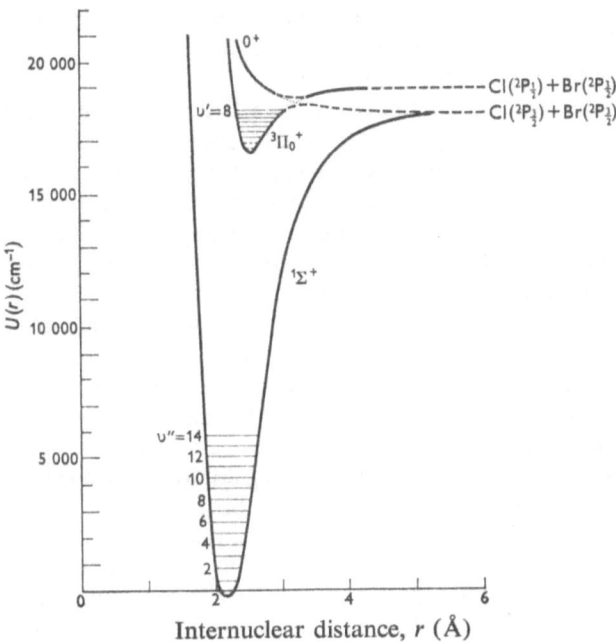

FIG. 4.10. Morse potential energy functions for BrCl (broken lines indicate the estimated positions of unobserved states). The internuclear distance, r_0'' of BrCl in its ground state ($^1\Sigma^+$) is known from microwave spectroscopy (Smith, Tidwell & Williams 1950). (After ref. 104.)

vibrational energy level observed in the excited state was $v' = 8$, lying at 18 153 cm^{-1} above $v'' = 0$ of the $X^1\Sigma^+$ ground state. Since the energy of dissociation of $X^1\Sigma^+$ to $Br(^2P_{3/2}) + Cl(^2P_{3/2})$ is 18 007 $\pm$ 50 cm^{-1}, the excited state cannot correlate with ground state atoms except via a potential maximum (Fig. 4.10). The excited state was thus identified[104] as $^3\Pi(O^+)$, since this state is predicted to dissociate to $Br(^2P_{3/2}) + Cl(^2P_{1/2})$, or conceivably to $Br(^2P_{1/2}) + Cl(^2P_{3/2})$.

Subsequently, the same system of BrCl has been observed in absorption[134] using a long path length (8 m) of chlorine and bromine mixtures heated to 50°C. Extremely careful choice of conditions was necessary in this case to reduce overlapping by Br_2 and Cl_2 bands, and it is thus not surprising that the BrCl bands had not previously been observed in absorption. The high resolution analysis of several absorption bands of BrCl[134] has confirmed that the excited state is $^3\Pi(O^+)$, and also confirmed the vibrational numbering found from the low resolution chemiluminescent emission spectra.

4.4.3. $Br(^2P_{3/2}) + Br(^2P_{3/2}) + M$ Recombination Luminescence

The case of the bromine afterglow emission spectrum is another example of the usefulness of even low resolution spectra in investigating the energy levels of simple molecules.

Gibbs and Ogryzlo[100] first described the appearance of the band spectrum of the red and near infra-red luminescence emitted during the recombination of ground state $Br^2P_{3/2}$ atoms from a microwave discharge. Bands with $v' \leq 3$ of the $B^3\Pi(O_u^+)$-$X^1\Sigma_g^+$ transition of Br_2 were assigned at the short wavelength end of the spectrum, together with a number of unidentified bands at longer wavelengths. Clyne and Coxon[101] observed the same bands, and whilst confirming the presence of weak $B^3\Pi(O_u^+)$ bands, found that the remaining, much more intense, band system was the $A^3\Pi(1_u)$-$X^1\Sigma_g^+$ system of Br_2. This latter system was very extensive, and covered the spectrum from 645 nm to at least 980 nm. Many of the band heads were the same as absorption bands of Br_2 assigned by Darbyshire to the "extreme red system", A–X. To longer wavelengths, gradually increasing differences appeared between the absorption band heads and those observed in emission. These divergences showed the requirement for a reassignment of vibrational numbering of the absorption bands. The emission bands involved transitions from lower vibrational levels of the excited state ($v' \geq 1$) than had been observed in absorption ($v' \geq 6$), and thus allowed a more reliable estimate of the vibrational constants of the $A^3\Pi(1_u)$ state. High resolution analysis of the A–X system from the bromine afterglow spectrum should provide easily analysable simple spectra, which cannot be obtained from the absorption spectrum of Br_2, in which the weak A–X is heavily overlapped by the more intense $B^3\Pi(O_u^+)$-$X(^1\Sigma_g^+)$ system.

Recent work on the weak B–X bands in the bromine afterglow emission spectrum using improved spectral resolution indicates that the prominent transitions arise from vibrational levels, up to $v' = 9$, of the B state of Br_2.[62] The energy of two $^2P_{3/2}$ bromine atoms at 300 K lies just above $v' = 1$ of the $B^3\Pi(O_u^+)$ excited state. The surprising observation of bands arising from $v' \leq 9$ is due to enhancement of the intensity of emission from these sparsely-populated levels at 300 K by the rapid increase in the Franck–Condon factors with increasing v' in this spectral region.[62]

4.4.4. Excitation by Impact with Metastable Noble Gas Atoms

A novel and promising development of the discharge-flow method (largely developed in Setser's laboratory) is the excitation of molecules by impact with a flow of excited $Ar(^3P_{2,0})$ atoms produced by a low power d.c. discharge.[135,136,209] The method is useful both for steady state studies of the kinetics of energy transfer, and for spectroscopic studies. Transitions from the $^3P_2$ and $^3P_0$ to the ground state of Ar are forbidden, and thus these atomic states are metastable. They persist for several milliseconds at concentrations estimated to be 0·01% of total argon flow, when produced in a rapid flow system using a low power hollow cathode discharge. Both the $^3P_2$ and $^3P_0$ states (which have excitation energies of 93 144 cm^{-1} and 94 554 cm^{-1} respectively) have been detected in absorption in a system of this type.[137] (Ar $^3P_2$ has also been detected by e.p.r. spectrometry.[219]) Figure 4.11 shows the apparatus which has been used for studies with argon metastables. The still more energetic He metastable atoms have also been produced in a similar manner. Kr and Xe metastable atoms, which are less energetic than argon atoms, can be generated by energy transfer when krypton or xenon is added to a stream of $Ar(^3P_{2,0})$ atoms.[139]

Metastable $^3P_{2,0}$ argon atoms possess sufficient excitation energy (mean energy = 1121 kJ mol^{-1}) to dissociate all known chemical bonds in collisions. Thus the result of impact of these atoms with many molecules is typically dissociative excitation, in which the products are electronically excited molecular fragments. Such cases are impact with compounds containing CN and OH groups,[136] for example

$$Ar^* + BrCN \longrightarrow CN(B^2\Sigma^+, A^2\Pi) + Br + Ar$$

$$Ar^* + H_2O \longrightarrow OH(A^2\Sigma^+) + H + Ar$$

Emission spectra from the electronically excited radicals produced give rise to the characteristic diffusion flames observed in the regions where Ar* and the reactant molecule mix (Fig. 4.11).

In the cases of N_2 and CO, which possess large dissociation energies, no dissociation occurs on impact with Ar*, but energy transfer from Ar* causes extensive electronic excitation. The steady state distributions of energy in both N_2[137] and CO excited by Ar* have been investigated in detail. In the case of the most fully allowed transitions observed, little or no rotational (or vibrational) relaxation is expected to occur within the radiative life-time at a pressure of 0·5 torr. In such a case, the observed energy distribution is that initially imparted to the substrate molecule by impact with Ar*. In some cases, study of the dependence of rotational energy distribution upon total pressure also enables rates of rotational relaxation to be derived; an example is the $C^3\Pi_u$ state of N_2.[137]

High resolution spectrometry is practicable with some of the emission spectra produced by Ar* impact, although the low intensities observed

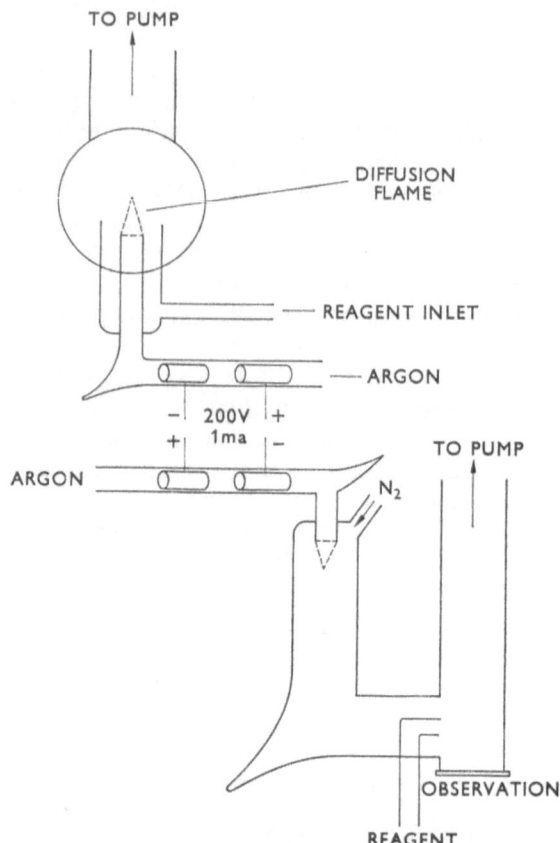

FIG. 4.11. Argon metastable flow apparatus (upper figure). Lower figure shows arrangement for studies with $N_2 A^3\Sigma_u^+$ produced by $Ar^* + N_2$. (After ref. 137.)

make the techniques difficult. Such a study has been carried out for the excitation of N_2 by Ar^* impact,[137] which was shown to give radiative transitions of N_2 from $C^3\Pi_u \to B^3\Pi_g \to A^3\Sigma_u^+ \to X^1\Sigma_g^+$. Independent channels exist for the production of $C^3\Pi_u$ and $B^3\Pi_g$, but $A^3\Sigma_u^+$ is populated by radiative cascade and quenching from the higher energy states. The $C^3\Pi_u$–$B^3\Pi_g$ Second Positive system of N_2 shows unusual intensity variations in the rotational structure (Fig. 4.12), which indicate an anomalous initial distribution of rotational energy within $C^3\Pi_u$. These anomalies are: (i) population of high rotational levels, i.e. $N' \leq 54$ in level $v' = 0$; (ii) alternation of intensities between even and odd rotational (even and odd N') triplet levels; (iii) unequal intensities associated with the spin sub-levels designating in coupling case (a): $^3\Pi_2$, $^3\Pi_1$, $^3\Pi_0$; and (iv) anomalous intensity alternation within the

Λ-doublets of a given rotational line.[137] Some interesting and quite detailed hypotheses for these unusual distributions have been put forward.[137] The explanations for the intensity alternations (ii), (iii) and (iv) seem convincing. The usual simple collision dynamics model can be satisfactorily applied to the anomalous coarse initial rotational energy distributions (i) in $N_2(C^3\Pi_u)$ following impact with Ar.* In this model, energy and angular momentum are conserved during the course of the impact. The impact parameter of the collision,† b, from angular momentum conservation for the observed distribution within $N_2(C^3\Pi_u)$, covers a wide range of values, corresponding to a wide range of rotational states.[137] A similarly wide range of impact parameters apparently also exists for the formation of $OH(A^2\Sigma^+)$ from Ar* + H_2O.

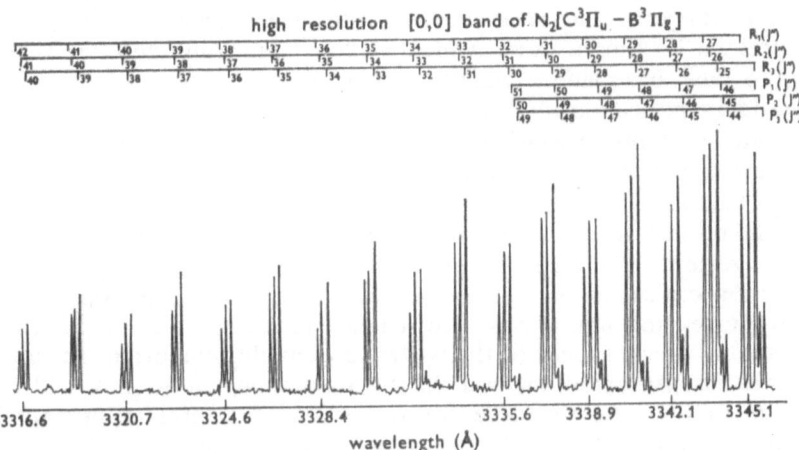

FIG. 4.12. Rotational structure of the (0, 0) band of $N_2 C^3\Pi_u \rightarrow B^3\Pi_g$, excited by Ar* + N_2. Note anomalous intensity alternations in the branches. (After ref. 137).

The energy and mass of the impacting atom can be varied by using Kr* or Xe* in place of Ar*. Correlations of initial energy distributions with these parameters, and with the nature of the impacting particle (e.g. using photons derived from resonance lines of Ar, Kr, Xe, which can have identical energies, or electrons) can thus be made.

A further interesting aspect of the metastable atom technique is the production of $N_2(A^3\Sigma_u^+)$ excited molecules in the absence of the N atoms, which are invariably present in the products of an ordinary microwave of r.f. discharge in nitrogen. Energy transfer processes involving $N_2(A^3\Sigma_u^+)$ molecules have been characterized in this way.[140]

† b is the minimum displacement between the centroid of the collision system and the linear projection of the trajectory of argon atom at infinity.

Excitation of molecular ions, such as O_2^+ A $^2\Pi_u$ and HBr^+ A $^2\Sigma^+$, by Penning ionization with He 2 $^3S_1$ metastables, has been observed,[212] and it is likely that many new states of ions will become accessible through processes of this type.

The applications of the method as a clean source of molecular spectra have only now begun to be investigated, and it seems likely that in the future many new systems of simple molecules and radicals will be identified using impact of metastable argon atoms with suitable stable molecules.

4.5. KINETICS OF REACTIONS OF GROUND STATE ATOMS

Studies of the kinetics of reactions of atoms and simple free radicals in discharge-flow systems have increased so much in number during the last few years that it is impracticable to cover the whole of the recent literature. Kaufman, in an Annual Review[70] covering the three years up to 1968, has made a valiant attempt to deal comprehensively with the data reported during that period. His article contains valuable tables of rate constants to which the reader is referred. Several other useful reviews[142] partly cover the literature up to 1968, on the whole less fully than Kaufman's article.

The present article is intended to complement rather than duplicate these reviews. The emphasis in sections 4.5 and 4.6 of this work will be on a critical discussion of the methods of kinetic analysis, their advantages and limitations, and detailed accounts of a few selected reactions which appear to the writer to exemplify important features of the discharge-flow method. In this discussion, an attempt will be made to illustrate the need for rigorous and careful experimental procedures if the powerful nature of the discharge-flow method for kinetic studies of simple reactions is to be fully utilized. The simple principles underlying the method have sometimes in the past tended to obscure the considerable difficulties in experimental procedure and in interpretation, and some of the published results have not always been of a high standard of reliability.

4.5.1. Recombination of Atoms

We first consider the possible recombination reactions removing atoms A in the absence of added reactants. These are: (i) *heterogeneous recombination*, usually first-order in atom concentration,† $A + \text{wall} \xrightarrow{k_2} \frac{1}{2}A$; and (ii) *homogeneous recombination*, which is in second-order in [A] and overall third order, involving a third body M,

$$A + A + M \xrightarrow{k_3} A_2 + M$$

† In the case of N(^4S) atoms, the occurrence of an additional heterogeneous recombination reaction, second-order in [A], has been demonstrated.[144-146]

The observed rate of decay of [A] then lies between first and second-order in [A]

$$-d[A]/dt = k_1[A] + 2k_3[A]^2[M] \qquad (I)$$

The relative contributions by heterogeneous and homogeneous decays, given by the ratio $k_1/2k_3[A][M]$, is thus dependent upon [A] and [M]; at sufficiently high total pressure and atom concentrations, homogeneous recombination predominates, and vice versa. This ratio is also strongly dependent upon the nature of the surface and upon the radius of the flow tube, r, since $k_1 = \gamma \bar{c}/2r$ for a cylindrical tube, where $\bar{c}$ is the mean velocity of the atom, and γ is the probability of recombination following collision with the surface (recombination coefficient).

Under typical conditions in a 30 mm pyrex or silica tube, at 1 torr total pressure and at 300 K, k_1 is typically of the order of $0 \cdot 1 \text{ s}^{-1}$, corresponding to $\gamma \sim 10^{-5}$. With $k_3 = 10^{15} \text{ cm}^6 \text{ mol}^{-2} \text{ s}^{-1}$ and $[A] = 5 \times 10^{-10}$ g atom cm^{-3}, $k_3[A][M]$ would be about $0 \cdot 025 \text{ s}^{-1}$.

4.5.2. Determination of k_1 and γ

At sufficiently low pressures and atom concentrations, recombination of atoms is entirely heterogeneous, and k_1 then approximates to $-\text{d} \ln [A]/\text{d}t$. The apparently simple method of investigating heterogeneous recombination of atoms by measurements of [A] along a flow tube has not, however, led to any great number of reliable determinations. The difficulty lies in obtaining a reproducible surface under flow conditions, since traces of impurities, such as transition metal oxides, can affect γ very greatly; also, for this reason, γ may vary with distance along the tube. The latter effect is particularly unsatisfactory, since it leads to apparent deviations from first-order kinetics. The side-arm method described by Smith[95] has some advantages. The interpretation of data obtained from this method is, however, unfortunately sometimes ambiguous. The agreement between values of γ obtained by different workers is thus qualitative rather than quantitative in many cases; the reported temperature coefficients in particular are the subject of unexplained divergences. The present article will not deal with heterogeneous reactions; this subject has been reviewed elsewhere.[138]

4.5.3. Determination of k_3

Values for k_3 may be found by data analysis according to equation (I). For the reasons explained above, particular care is needed when the term $k_1[A]$ contributes an appreciable fraction to the overall rate of decay of atoms; it is then necessary to check the constancy of k_1 during the experiments. However, errors arising from variations in γ can normally be minimized by use of high [A] and [M]. Nevertheless, the accurate determination of values of third-order atom recombination rate constants k_3 remains one of the most difficult measurements in a discharge-flow system. The difficulty is not peculiar to the discharge-flow method, but applies to other techniques used, for example, in flash photolysis—namely, the requirement for absolute atom concentration measurements (see section 4.2.7). Yet this field is one in which the discharge-flow method has already shown considerable success and the development of improved methods for the measurement of absolute atom concentrations should lead to investigations of a number of unexplored reactions.

(a) Determination of k_3: Results

The following list includes most of the homo-atom third-order recombination reactions whose rate constants have been measured at room temperatures.

$$O + O + M \longrightarrow O_2 + M$$
$$N + N + M \longrightarrow N_2 + M$$
$$H + H + M \longrightarrow H_2 + M$$
$$Cl + Cl + M \longrightarrow Cl_2 + M$$
$$Br + Br + M \longrightarrow Br_2 + M$$
$$I + I + M \longrightarrow I_2 + M$$

Of these rate constants, all except the last cited have been determined in discharge-flow systems. Bromine and iodine atom recombination rates have been measured using flash photolysis over a wide temperature range, and for a wide variety of third bodies M.[69] Temperature coefficient measurements using discharge-flow methods have been reported for N, H and Cl atom recombinations. Data on Br atom recombination at 298 K have also been obtained using these methods.[197] Limited data on the effect of third body on the rate constants k_3 have been obtained in these cases. However, there is much scope for the acquisition of good data with systematic variation of T and M, as well as extensions to hetero-atom recombination reactions. At present, quantitative kinetic data are available for only one reaction of the latter type

$$N + O + M \longrightarrow NO + M$$

Measurements of dissociation rates in shock-heated gases have enabled the determination of atom recombination rates to be made at elevated temperatures.[210] Overlap of the ranges of temperatures and third bodies in shock tube experiments and in discharge-flow studies would enable dependences of k_3 on these parameters to be well established, thus providing raw material for the testing and further development of theories of third-order atom recombination reactions.[6a] Such data, with overlapping ranges of T and M, would also provide verification further quantitative of the methods of workers with shock tube and with discharge-flow systems.

At present, adequate tests for models of atom recombination are provided by data for the $I + I + M$ and $Br + Br + M$ reactions. Johnston[147] has pointed out that these limited results are just as well explained by ultra-simple energy transfer and bound complex (IM) theories, as by serious detailed theories, such as those of Keck,[148] Light[149] and others. Useful theoretical progress can now be expected as a more substantial body of kinetic results is becoming available.

The cases of recombination of $N(^4S)$ and of $Cl(^2P_{3/2})$ atoms will now be discussed briefly, as examples of what is now known about recombination reactions from discharge-flow studies.

(b) $N(^4S) + N(^4S) + M \xrightarrow{k_4} N_2 + M$ \hfill (4)

The recombination of $N(^4S)$ atoms probably presents fewer experimental difficulties than those of other atoms. Hence, it is not surprising that there has been an unusually large number of investigations of this reaction. In most of these studies, advantage was taken of the convenient and well-established $N + NO$ titration reaction for the measurement of absolute atom concentrations. However, the agreement of the values for k_4 obtained by earlier workers[150–152] was not satisfactory (Table 4.5). Later investigations[144,145] have shown that nitrogen atom recombination in "clean" pyrex or silica tubes involves not only first-order heterogeneous and third-order homogeneous recombination, but

TABLE 4.5
$N(^4S) + N(^4S) + N_2 \rightarrow N_2 + N_2$
Values for $k_4 = -\frac{1}{2}[N_2]^{-1}[N]^{-2}\,d[N]/dt$
$(cm^6\,mol^{-2}\,s^{-1})$ at 300 K and near 5 torr total pressure

Reference	k_4
150†	$2\cdot8 \times 10^{15}$
151†	$4\cdot5 \times 10^{15}$
152†	$2\cdot4 \times 10^{15}$
155†	$3\cdot6 \times 10^{15}$
156†	$2\cdot7 \times 10^{15}$
144	$2\cdot2 \times 10^{15}$
145	$2\cdot9 \times 10^{15}$
146	$2\cdot7 \times 10^{15}$

† No account was taken of second-order heterogeneous recombination in these studies.

also an unexpected second-order heterogeneous recombination process on the walls of the tube; this process is effectively suppressed by coating the tube with phosphoric acid.[146] Values of k_4, for $M = N_2$, at 298 K, obtained by kinetic analysis taking account of second-order heterogeneous recombination, are in excellent agreement (Table 4.5). Data now cover the range 90 to 611 K, although it would be desirable for rate constants to be determined at a larger number of intermediate temperatures. Rate constants at 300 K are also available for a number of third bodies: $M = N_2$, Ar, He, H_2, CO_2, N_2O, H_2O.[145]

(c) $Cl(^2P_{3/2}) + Cl(^2P_{3/2}) + M \xrightarrow{k_5} Cl_2 + M$

The three published investigations of the kinetics of recombination of ground state $Cl(^2P_{3/2})$ atoms[17,37,153] have used absolute measurements of [Cl] from titration with ClNO, or alternatively, an isothermal wire calorimeter. Both these methods require considerable care if systematic errors are to be avoided.[17,193]

In an earlier publication of kinetic data on the recombination reaction, Bader and Ogryzlo[153] postulated the simple reaction (5)

$$Cl + Cl + Cl_2 \xrightarrow{k_5} Cl_2 + Cl_2 \qquad (5)$$

as the major homogeneous recombination step for Cl atoms in a flow system. However, Hutton and Wright[37] suggested that predominant recombination in Cl_2 carrier gas occurs through an intermediate, Cl_3

$$Cl + Cl_2 \rightleftarrows Cl_3$$
$$Cl + Cl_3 \longrightarrow Cl_2 + Cl_2$$

The decay of Cl was followed in their work by measurements of the intensity I of chlorine afterglow emission spectrum, assumed to be given by $I = I_0[Cl]^2$. The evidence for this scheme was that the apparent third-order rate constant k_5 showed an increase with increasing [Cl]; $(1/k_5)$ was found to be a linear function of [Cl]. However, Clyne and Stedman[17] found that the kinetics of homogeneous recombination of $Cl(^2P_{3/2})$ atoms, in argon or chlorine third body, were second-order in [Cl] and first-order in [M]. They argued that Hutton and Wright's conclusion was incorrect since these workers had assumed $I = I_0[Cl]^2$ at all values of [Cl] and [M], whereas the kinetic order of I in both Ar and Cl_2 third bodies actually decreases with an increase in [M].[38] Also, the data for variation of k_5 with Cl were based partly on results in which chlorine atoms represented as much as 50% of the total flow. No account was taken of the nonisothermal nature of the flow, and the effects of alteration of the flow rate by atom recombination. Reinterpretation[17] of Hutton and Wright's data using recent data for the kinetic order of the intensity,[38] and neglecting data at very high [Cl], led to the conclusion that these data were consistent with the simple mechanism favoured by Bader and Ogryzlo and Clyne and Stedman Table 4.6 shows a summary of the data for k_5, including the re-interpreted results of Hutton and Wright. It is noted that although the temperature variation of k_5 is known from 195 to 500 K for M $= \mathrm{Cl_2}$ and Ar, no systematic study of the dependence of k_5 upon the type of third body M has yet been reported. At 300 K, the equilibrium population of the low lying J-excited $^2\mathrm{P}_{1/2}$ state of Cl, which possesses an excitation energy of 882 cm^{-1}, is 0·8% of that of ground state $^2\mathrm{P}_{3/2}$ atoms. Thus, recombination data based on $-\mathrm{d[Cl]}/\mathrm{d}t$ at this temperature refer predominantly to recombination of $^2\mathrm{P}_{3/2}$ atoms, whilst at higher temperatures an increasing contribution to recombination by

TABLE 4.6

$$\mathrm{Cl + Cl + M_2 \rightarrow Cl_2 + M_2}$$

Values for $k_5 = -\tfrac{1}{2}[\mathrm{M}]^{-1}[\mathrm{Cl}]^{-2}\,\mathrm{d[Cl]}/\mathrm{d}t$

(cm^6 mol^{-2} s^{-1}) at 298 K

Reference	M	k_5
153	$\mathrm{Cl_2}$	3·1 × 10^{16} [a]
37[b]	$\mathrm{Cl_2}$	1·5 × 10^{16}
17	$\mathrm{Cl_2}$	2·0 × 10^{16}
153	Ar	4·4 × 10^{15} [a]
37	Ar	4·2 × 10^{15}
17	Ar	4·4 × 10^{15}

[a] Measured at 313 K. Corrected to 298 K using temperature coefficient of ref. 17.

[b] Reinterpreted by ref. 17 (see section 4.5.3(c)).

$^2\mathrm{P}_{1/2}$ atoms can be expected. It appears feasible to determine the rates of recombination of both $^2\mathrm{P}_{3/2}$ and $^2\mathrm{P}_{1/2}$ chlorine atoms separately, using methods capable of state discrimination—e.p.r. spectrometry, or vacuum u.v. resonance line absorption (sections 4.2.9a, 4.2.10e).

(d) *Atom Recombination Reactions—Summary*

Experimental difficulties have largely accounted for the small number of full investigations of oxygen atom and hydrogen atom recombination rates. There is much scope for future work in this area, especially in the case of O(^{3}P) recombination, the rate constant for which, at 298 K, at present, depends largely on a single investigation[25] for M $=$ Ar. Recently developed methods for the measurement of atom concentrations—particularly e.p.r. and resonance absorption—have not yet been fully exploited in the area of atom recombination, and the prospects are encouraging. For the case of halogen atom recombination kinetics, measurement of atom concentrations by molecular halogen absorption in flow systems is now routinely practicable[20] and should be useful.

4.5.4. Reactions of Atoms with Added Reagents

If a bimolecular elementary reaction of an atom with a molecule of added reagent (R) occurs,

$$A + R \xrightarrow{k_2} \text{products} \tag{2}$$

the overall rate of removal of atoms is

$$-d \ln [A]/dt = k_1 + k_2[R] + 2k_3[A][M]$$

If [R] is independent of time, k_2 may be found from the observed rate of decay of atoms, provided that k_1 and k_3 are known. One simple case is when the initial concentration of added reagent [R] is much greater than [A]. Another simple case is when reaction (2) is actually an overall third-order atom recombination reaction, for example, the reaction scheme

$$O + NO + M \xrightarrow{k_6} NO_2 + M$$

$$O + NO_2 \xrightarrow{\text{rapid}} NO + O_2$$

In this case, since both [NO] and [M] are independent of reaction time, the reaction is pseudo-first-order in [A] with $k_2[R] = k_6[NO][M]$.

(a) *The Fixed Observation Point Method*

Whilst the rate constant k_2 may thus be deduced from measurement of [A] at different distance (times) along the length of the flow tube, a more satisfactory technique for the measurement of pseudo-first-order rate constants k_2 is the fixed observation point method. In this method, the atom concentrations [A] remaining at a fixed observation point are measured when the same (excess) concentration of reagent [R] is added in turn at each of several inlets along the flow tube.

The analysis has been described by Thrush,[22a,111] and subsequently using a slightly different formulation by Westenberg.[22b] In the analysis, it is supposed that the three concurrent reactions (1), (2) and (3) remove atoms, so that

$$-d \ln [A]/dt = 2k_3[A][M] + (k_1 + k_2[R])$$

This equation may readily be integrated. Let $[A_1]$ and $[A_2]$ be the concentrations of atoms remaining at the fixed observation point when the same flow of R is added in turn at each of two inlets; let t be the time of flow between the inlets. Then it can be shown that, provided $2k_3[A][M] \ll (k_1 + k_2[R])$, the simple equation (II) is obtained

$$\ln [A_1]/[A_2] = -k_2[R]t \tag{II}$$

Equation (II) has several useful properties. (1) The observed decay of atoms is independent of concurrent first-order atom decay processes, e.g. heterogeneous decay, and of second-order atom recombination (k_3[M]), provided first-order decay dominates this recombination. It may be helpful to increase the first-order decay (e.g. by adding NO to O atoms, $NO + O + M \rightarrow NO_2 + M$) in order to suppress second-order recombination. (2) Any variation of k_1 with distance along the flow tube does not affect the results, provided k_1 is not a function of [R]. (3) The fixed observation point may be at room temperatures, whilst the reaction zone containing the reactant inlets is at an elevated temperature. The results then give a value for k_2 at the elevated temperature. These useful properties have been exploited in series of studies over wide temperature ranges of rate constants for atom + molecule recombination reactions,[28,31,159,161] and for atom + molecule transfer reactions.[22a,45,168,171,195]

4.5.5. Atom + Molecule Recombination Reactions

Table 4.7 shows a summary of atom-plus-molecule recombination reactions whose kinetics have been studied in discharge-flow systems; the list is not exhaustive; only selected data are tabulated. Since these recombination reactions are first-order in [A], only relative concentrations need be measured in order to obtain values for the rate constants. This fact, and the elimination of concurrent atom recombination processes by use of the fixed observation point method, account for the relatively large number of detailed rate studies in the literature. The superficially simple atom + O_2 reactions are, for various reasons, more complex than the atom + NO reactions. The three reactions $H + NO + M$, $O + NO + M$ and $Cl + NO + M$ possess the following simple mechanism in the 1 torr total pressure region

$$A + NO + M \longrightarrow ANO + M \text{ (slow)}$$
$$A + ANO \longrightarrow A_2 + NO \text{ (fast)}$$

hence

$$-d[A]/dt = 2k[A][NO][M]$$

This series of reactions has been studied in some detail: some data on M-effects, and temperature coefficients, are available for the reactions mentioned. Where a rate constant has been measured by different workers, e.g. for the $O + NO + M$ reaction, the agreement is excellent (see Table 4.7). Good agreement is also observed with results from flash photolysis,[214] and pulse radiolysis,[215] for the $H + O_2 + M$, $H + NO + M$, $O + O_2 + M$, $O + NO + M$ reactions.

Returning to the three atom + O_2 reactions listed in Table 4.7, the kinetics of the $H + O_2 + M$ reaction have been studied[22a,97a,211] in some detail for $M = H_2$, Ar and at 224, 244 and 293 K. The mechanism of this reaction is complex, since rapid reaction of H with the HO_2

TABLE 4.7

Atom + molecule recombination reactions at 298 K

Reaction	$-\Delta H^\circ_{298}$ (kJ mol^{-1})	M	k (cm^6 mol^{-2} s^{-1})	References
H + O$_2$ + M → HO$_2$ + M	197 ± 10	Ar	8 × 10^{15}	22a, 97a[211][a][d]
O + O$_2$ + M → O$_3$ + M	106	Ar	2·1 × 10^{14}	157–159[a][b]
Cl + O$_2$ + M → ClOO + M	29 ± 5	Ar	2 × 10^{14}	164
H + NO + M → HNO + M	208	Ar	1·0 × 10^{16}	31, 111
O + NO + M → NO$_2$ + M	306	Ar	2·2 × 10^{16}	28, 87, 160[a][c]
Cl + NO + M → ClNO + M	159	Ar	3·4 × 10^{16}	161, 17[e]
Cl + NO$_2$ + M → ClNO$_2$ + M	142 ± 5	Ar	1 × 10^{17}	162, 62
O + SO + M → SO$_2$ + M	552	Ar	3 × 10^{17}	163
O + SO$_2$ + M → SO$_3$ + M	348	Ar	1·1 × 10^{15}	12
Cl + NF$_2$ + M → ClNF$_2$ + M	—[f]	Ar	9·8 × 10^{16}	213
Br + NF$_2$ + M → BrNF$_2$ + M	—[f]	Ar	3·6 × 10^{16}	213

(a) Mean of references cited.

(b) Values were as follows: (cm^6 mol^{-2} s^{-1}): 1·45 × 10^{14} (ref. 157); 2·9 × 10^{14} (ref. 158); 1·89 × 10^{14} (ref. 159).

(c) Values were as follows: (cm^6 mol^{-2} s^{-1}): 1·9 × 10^{16} (ref. 28); 3·0 × 10^{16} (for M = N$_2$, ref. 87); 2·2 × 10^{16} (ref. 160).

(d) Reference 97a gives k (for M = H$_2$) = 2·1 × 10^{16} cm^6 mol^{-2} s^{-1} at 293 K.

(e) Value of ref. 161 amended in ref. 17.

(f) Not known; see ref. 213.

radicals formed leads to significant quantities of both OH radicals and O atoms:[22a, 211]

$$H + O_2 + M \xrightarrow{k_6} HO_2 + M$$
$$2OH$$

$$H + HO_2 \longrightarrow H_2 + O_2 \qquad (6)$$

$$H_2O + O$$
$$OH + OH \longrightarrow H_2O + O$$

Secondary reactions of these species leads to a complex reaction stoichiometry whose uncertainty limits the precision of measurement of k_6. The $O + O_2 + M$ reaction also involves mechanistic problems, which were not fully appreciated in earlier work, and thus some such work is not quantitatively reliable. One difficulty stems from the low rate constant for the secondary step (7a)

$$O + O_2 + M \xrightarrow{k_7} O_3 + M \qquad (7)$$

$$O + O_3 \xrightarrow{k_{7a}} O_2 + O_2 \qquad \cdot(7a)$$

In many systems at 300 K, k_{7a} will not be rapid enough for $[O_3]$ to be in a stationary state; ozone will appear as product largely unreacted with O atoms. Hence, $-d \ln [O]/dt = k_7[O][O_2][M]$. The presence of a trace of $H(^2S)$ atoms in the oxygen atom stream (e.g. from dissociation of H_2O) causes a rapid chain removal of ozone:

$$O + O_2 + M \xrightarrow{k_7} O_3 + M$$

$$H + O_3 \xrightarrow{k_8} OH + O_2$$

$$OH + O \longrightarrow H + O_2$$

The rate of removal of ozone will then be $k_8[H][O_3]$, and since $k_8 \gg k_{7a}$, $-d \ln [O]/dt = 2k_7[O][O_2][M]$. Thus, the apparent third-order rate constant for $O + O_2 + M$ would increase by a factor of two when a trace of moisture is added to dry O_2. A further problem arises from the presence of significant concentrations of $O_2(^1\Delta_g)$ when $O(^3P)$ atoms are produced by a discharge in oxygen. As $O_2(^1\Delta_g)$ is known to react with O_3, forming $O(^3P)$, the presence of singlet oxygen can thus lead to suppression of the decay of oxygen atoms:

$$O + O_2 + M \xrightarrow{k_9} O_3 + M$$

$$O_3 + O_2(^1\Delta_g) \longrightarrow 2O_2 + O$$

These problems have been overcome either by generating $O(^3P)$ by pyrolysis of ozone at 1000 K or above,[157,158] or by a discharge through

11

very dilute, carefully purified mixtures of oxygen with argon.[159] The agreement between the values thus obtained for k_9 is satisfactory (Table 4.7); the agreement with values obtained by different methods in the 100 torr pressure range is also satisfactory.

4.5.6. Atom–Molecule Transfer Reactions

Of all the categories of reactions considered in this article, the large group of atom–molecule reactions has received the greatest amount of attention from workers using the discharge-flow method. Reactions of $O(^3P)$ oxygen atoms have been particularly often studied. Studies of atom–molecule reactions exemplify many of the points discussed in this article. Several useful tables of rate constants for such reactions have been published recently,[7a,143,165,166] and it is therefore felt desirable in this article to select only a few reactions for discussion. The first category chosen is the reaction of atoms with H_2; these studies are representative of some of the most extensive and complete investigations so far made, and many were carried out using the fixed observation point method. The second category—the reactions of atoms with ozone—are much less fully studied, but interesting mechanistic features emerge from these studies, which are of general applicability to the discharge-flow method. The third category concerns a group of related halogen atom plus halogen molecule reactions, which involve low or negligible energy barriers, and which are of interest in molecular beam kinetics. The section is concluded with a brief account of the reaction of H atoms with C_2H_4.

4.5.7. Reactions of Atoms with H_2

(a) *Reactions* $^{1,2}H + {}^{1,2}H_2 \longrightarrow {}^{1,2}H_2 + {}^{1,2}H$

The hydrogen atom–hydrogen molecule exchange reaction is the classic case for which extensive theoretical calculations have been made. However, only recently have new extensive experimental studies of the kinetics of the exchange reactions been carried out:

$$H + H_2 \xrightarrow{k_a} H_2 + H$$
$$D + D_2 \xrightarrow{k_b} D_2 + D$$
$$D + H_2 \xrightarrow{k_c} DH + H$$
$$H + D_2 \xrightarrow{k_d} HD + D$$

A type of flow system using atoms produced by dissociation on a hot tungsten filament, and analysis using thermal conductivity probes, has been used to determine[167] the rate constants of all four of these reactions in the range 350 to 450 K (a wider range for k_a and k_c). A more direct experimental method, using the fixed observation point method and e.p.r. detection of H and D, has been described[168] for the measurement of k_c (250 to 750 K) and k_d (300 to 750 K). Figure 4.13 shows the close agreement of the two sets of results, except at the lower end of the temperature range, where Westenberg and de Haas[168] find less curvature in their plots than in those of LeRoy and co-workers.[167] Interpretation

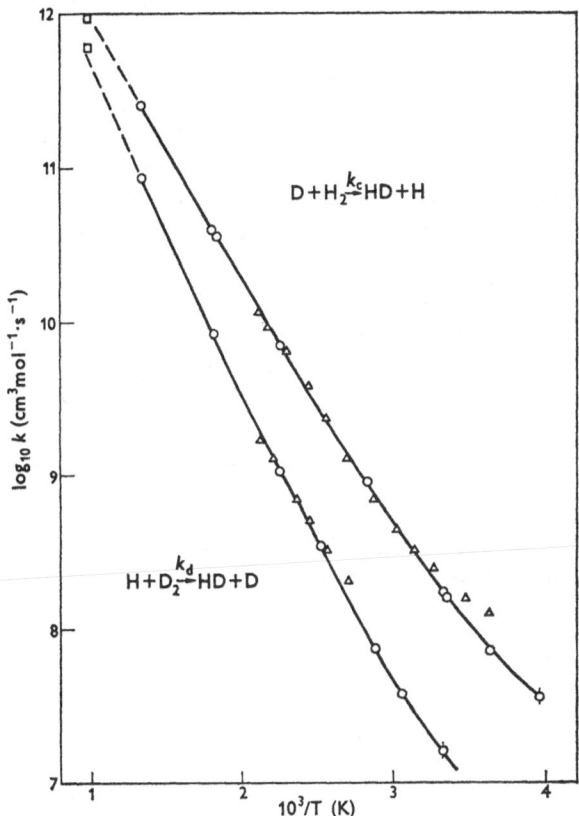

FIG. 4.13. Arrhenius plot of k_c and k_d experimental data. ○, ref. 168; △, □, data of Le Roy and coworkers.[167] (After ref. 168.)

of the data in terms of Weston's transition state model calculations[169] for the complexes DH_2 and HD_2 was made.[168] The results of this approach led to rate constants already greater than the observed values with neglect of tunnelling; inclusion of tunnelling corrections would cause a still greater divergence between the calculated and observed rate constants. The interpretation is therefore unprofitable, and more refined calculations taking account of improved potential energy surfaces seem to be in order.[218]

(b) *Reaction* $O + H_2 \longrightarrow OH + H$

The reaction $O + H_2 \xrightarrow{k_6} OH + H$ is one of the most thoroughly investigated elementary reactions; three independent groups have reported systematic studies of k_6 over a range of temperatures using the discharge-flow method. Two of these studies employed the fixed observation point method (one with chemiluminescence detectors,[22a] the other with e.p.r. spectrometry).[171] The third systematic study used a combination of mass spectrometry and e.p.r. spectrometry.[22c] A fourth study[170] using mass spectrometric analysis and a stirred reactor probably gave less reliable data, although the results are in good agreement with the other three investigations. The agreement between the first three studies is extremely good (Table 4.8). These studies, and other investigations using different techniques (shock tube, flames, etc.) at higher

TABLE 4.8

Rate constants for the $O + H_2$ reaction

$$k_6 = A \exp(-E_a/RT) \text{ cm}^3 \text{ mol}^{-1} \text{ s}^{-1}$$

T (K)	A	E_a (kJ mol^{-1})	References
409–733	$1\cdot2 \times 10^{13}$	39	22a
373–478	$1\cdot3 \times 10^{13}$	39	22c
500–900	$3\cdot2 \times 10^{13}$	43	171[(a)]
Mean 370–900	$1\cdot9 \times 10^{13}$	40	[(b)]

[(a)] This reference supersedes earlier work by the same workers quoted in ref. 22b.

[(b)] Mean of the three values cited in Table 4.8. Cf. recommended value of ref. 42, $k_6 = 1\cdot74 \times 10^{13} \exp[-39/RT] \text{ cm}^3 \text{ mol}^{-1} \text{ s}^{-1}$ based on all data including ref. 22b (ref. 171 was published subsequently).

temperature and/or pressures, have been reviewed.[42] The simplest procedure in discharge-flow studies of the $O + H_2$ reaction is to ensure pseudo-first-order conditions ($[H_2] \gg [O]$), and then to measure $-d[O]/dt$. In these circumstances, OH radicals are rapidly scavenged by excess oxygen atoms, $O + OH \to H + O_2$, leading to the overall stoichiometry, $2O + H_2 \to 2H + O_2$, and the simple rate law, $-d \ln[O]/dt = 2k_6[H_2]$. If $O(^3P)$ atoms are generated by means of a discharge in O_2, side reactions can occur between H atoms (produced in the $O + H_2$ reaction) and the large excess of molecular oxygen present

$$H + O_2 + M \longrightarrow HO_2 + M$$

$$O + HO_2 \longrightarrow O_2 + OH \text{ etc.}$$

These side reactions lead to accelerated removal of $O(^3P)$ atoms, and hence to systematic errors in k_e. For this reason, all recent studies have used O atoms free from O_2, e.g. from the reaction $N + NO \rightarrow N_2 + O$.

The ratio of the rate constants k_e/k_f has also been determined in a full study,[171] following earlier preliminary work.[22a]

$$O + D_2 \xrightarrow{k_f} OD + D$$

Although the rate constants and their temperature coefficients for the reactions of O with H_2 and D_2 are well known, no attempt has yet been made to construct a detailed model for the transition state complex. The potential energy surfaces for these reactions are also largely unexplored.

(c) *Reaction* $Cl + H_2 \underset{k_g}{\overset{k_f}{\rightleftarrows}} H + HCl$ *in both directions*

Following earlier studies on the kinetics of the $Cl + H_2$ reaction,[172] e.p.r. spectrometry and the fixed observation point method have been used with the discharge-flow technique to find k_f and k_g separately.[45] This is one of the few cases where the rates of both forward and backward reactions may be determined by the same method in the same temperature range; it is facilitated by the near zero value for the reaction enthalpy change $(\Delta H^\circ_{298} = +3 \text{ kJ mol}^{-1})$. Curiously, k_f/k_g determined in this manner was found to be a factor of 2 to 3 less than the calculated equilibrium constant for the $Cl + H_2$ reaction.[45] This discrepancy is definitely outside the limits of error of the calculated equilibrium constant, and appears to be outside the limits of error of the rate determinations. It has been suggested that the discrepancy arises from the inapplicability of microscopic reversibility to this reaction, since it was argued that there may be a greater probability of obtaining rotationally hot HCl in the forward reaction than rotationally hot H_2 in the backward reaction.[45] The discrepancy between k_f/k_g and K_{eq} for the reaction

$$Cl(^2P_{3/2}) + H_2 \underset{k_g}{\overset{k_f}{\rightleftarrows}} HCl + H$$

can be explained in an alternative manner. In the cases of the exothermic reactions

$$H + HBr \rightleftarrows H_2 + Br$$
$$H + HI \rightleftarrows H_2 + I$$

the potential energy surfaces for $H + HX$ lead not only to $H_2 + X(^2P_{3/2})$, but also to detectable extents of $H_2 + X(^2P_{1/2})$. Let us suppose that the reaction $H + HCl$ can similarly lead to $H_2 + Cl(^2P_{1/2})$ (slightly endothermic, $\Delta H^\circ_{298} = +6 \text{ kJ mol}^{-1}$) as well as to ground state products

$$H + HCl \underset{k_f^*}{\overset{k_g^*}{\rightleftarrows}} H_2 + Cl(^2P_{1/2}); \qquad \Delta H^\circ_{298} = +4 \text{ kJ mol}^{-1}$$

Let K_{eq}^* be the equilibrium constant for this reaction ($K_{eq}^* = k_f^*/k_g^* = 35\,K_{eq}$ at 350 K). In terms of this scheme, the observed rate constants in the e.p.r. experiments are k_f for the forward reaction, and $k_{obs} = (k + k_g^*)$ for the backward (H + HCl) reaction. If k_g^* is not negligible compared with k_g, $k_f/(k_g + k_g^*)$ will be appreciably less than K_{eq}, as is observed. Taking a mean value for $k_f/|K_{(eq}k_g + k_g^*)| = 2.5$ from the experimental data, it is deduced that $k_g = 0.4k_{obs}$, $k_g^* = 0.6k_{obs}$. This interpretation implies that at the mean temperature of 350 K, 60% of the reaction H + HCl leads to $H_2 + Cl(^2P_{1/2})$ and 40% to $H_2 + Cl(^2P_{3/2})$. This conclusion further leads to the ratio $k_f^*/k_f = 20$ at the same temperature, suggesting that the reaction $Cl(^2P_{1/2}) + H_2$ has an activation energy about 6 kJ mol^{-1} less than that of $Cl(^2P_{3/2}) + H_2$.[217]

It appears difficult to test this explanation by direct experimentation. It is unlikely that appreciable concentrations of $Cl(^2P_{1/2})$ above the equilibrium value (0.8% of $^2P_{3/2}$) could be observed in the reaction of H with HCl under normal conditions, since this reaction is probably slower than the rate of spin-orbit relaxation of $^2P_{1/2}$ atoms at 1 torr total pressure. However, the explanation put forward here appears to the writer less arbitrary than that suggested by Westenberg and de Haas.

The value reported for $k_f = 1.2 \times 10^{13} \exp(-2200/T)$ cm^3 mol^{-1} s^{-1} in the range 251 to 456 K is in fair agreement with work[173] using a static method with ICl as the source of Cl, $k_f = (4.7 \pm 2.0) \times 10^{13} \exp[(-2600 \pm 200)/T]$ cm^3 mol^{-1} s^{-1} over a narrower range of T^{-1} (610 > T > 478 K). The value reported for k_f over the range 497 > T > 195, $k_g = 2.3 \times 10^{13} \exp(-1800/T)$ cm^3 mol^{-1} s^{-1}, is also in fair agreement with a study[35] using determinations of [H] from HNO chemiluminescence and the fixed observation point method at four temperatures between 195 and 373 K, $k_g = 1.0 \times 10^{13} \exp(-1600/T)$ cm^3 mol^{-1} s^{-1}. Hence, *direct* measurements of k_f and of k_g by different groups of workers are in agreement, whilst values calculated from the back reaction and the equilibrium constant disagree with directly determined values by a significant factor. This was not realized by Benson et al.[173] in their review of the kinetics of the Cl + H_2 reaction, which appeared at the same time as the paper of Westenberg and de Haas.[45] It appears to the present writer that the most meaningful expressions for k_f and k_g are obtained by considering the data for k_f[218] and k_g separately, and if the ranges of temperature and number of data are taken into account, the following "best" expressions are suggested in the range 250 to 610 K:

$$k_f = 3.7 \times 10^{13} \exp(-2400/RT) \text{ cm}^3 \text{ mol}^{-1} \text{ s}^{-1}$$
$$k_g = 6.2 \times 10^{13} \exp(-2050/RT) \text{ cm}^3 \text{ mol}^{-1} \text{ s}^{-1}$$

(d) *Reactions of Other Atoms with H_2*

The reactions of ground state $N(^4S)$, $Br(^2P_{3/2})$ and $I(^2P_{3/2})$ atoms with hydrogen are endothermic and hence possess large activation energies

(≈ 50 kJ mol^{-1}), and thus are too slow for kinetic study in conventional discharge-flow systems. The rate constants of the reverse reactions

$$H + HBr \rightleftharpoons H_2 + Br$$

$$H + HI \rightleftharpoons H_2 + I$$

are known to approach the hard-sphere bimolecular collision frequency at 300 K; however, an appreciable fraction of the Br and I atoms is formed in the spin-orbit $^2P_{1/2}$ excited state.[118] Reactions of this type, which form the metastable $^2P_{1/2}$ halogen atoms, could be used as sources of excited atoms for kinetic studies in discharge flow systems. Another source of Br($^2P_{1/2}$) atoms is the rapid reaction sequence

$$O + Br_2 \longrightarrow Br + BrO$$

$$O + BrO \longrightarrow Br + O_2$$

which we have studied recently in this laboratory using resonance line absorption.[194]

Kinetic studies of reactions of ground state fluorine F($^2P_{3/2}$) atoms in discharge flow systems have been reported, including the reaction:[41]

$$F + H_2 \xrightarrow{k_f} HF + H$$

Detection of fluorine atoms was by means of mass spectrometry with molecular beam sampling. The value of k_f was reported to be $10^{14\cdot2} \exp(-700/T)$ cm^3 mol^{-1} s^{-1} in the range 300 to 400 K.

4.5.8. Reactions of Atoms with Ozone

The reactions of endothermic compounds with atoms are useful sources of energy rich species and free radicals. Ozone and chlorine dioxide are the compounds which have been most fully studied, but atom reactions with the difficult compounds nitrogen trichloride and the azides of chlorine and bromine are also interesting.[60] The reactions of H(^2S), O(^3P), N(^4S), F, Cl, Br and I($^2P_{3/2}$) have all received study by the discharge-flow method, and these cases are chosen to exemplify the features of this class of reactions.

(a) *Reaction* H(^2S) + O$_3$

The reaction

$$H(^2S) + O_3 \xrightarrow{k_f} OH(X^2\Pi, v = 0) + O_2$$

has a standard enthalpy change $\Delta H^\circ_{298} = -321$ kJ mol^{-1}. Much of the energy of this reaction is transferred to vibrational excitation ($v \leq 10$) of OH(X$^2\Pi$); the reaction zone appears a dull red colour due to emission of overtone vibration–rotation bands of OH ($\Delta v \leq -5$), which occur in the visible spectral region (see section 4.3.4(a)).

The kinetics of this reaction have been investigated using simple mass spectrometric analysis.[30] Low concentrations of reagents ($\sim 5 \times 10^{-12}$ mol cm^{-3}) were used with second-order analysis for measurements of the rate constant for the extremely rapid primary reaction

$$H + O_3 \xrightarrow{k_l} OH + O_2$$

$k_l = (1 \cdot 6 \pm 0 \cdot 4) \times 10^{13}$ cm^3 mol^{-1} s^{-1} at 300 K. Under these conditions, it was reported that the $H:O_3$ stoichiometry was unity.[30] In this study, the kinetic analysis depended on the 1:1 reaction stoichiometry which does not hold at higher concentrations. The error limits of $\pm 25\%$ for k_l therefore seem rather optimistic; a factor of two is probably more realistic. The observed O_3/H stoichiometry of $c.$ 3 at higher concentrations of reagents ($\sim 5 \times 10^{-10}$ mol cm^{-3}) remains difficult to explain quantitatively. Phillips and Schiff[30] described experiments which they interpreted in terms of a rapid reaction of vibrationally excited OH with O_3

$$OH + O_3 \longrightarrow OH + O_2 + O$$

Additional reactions are needed to explain the O_3/H stoichiometry. As the $H + O_3$ reaction is extremely rapid and exothermic, the reaction zone will attain a much higher temperature than ambient. Hence, it appears likely that the reaction

$$OH + H_2 \longrightarrow H_2O + H$$

which has an activation energy[5] of $c.$ 25 kJ mol^{-1} can rapidly propagate a chain capable of removing extra ozone.

(b) *Reaction* $O(^3P) + O_3$

The reaction of ground state oxygen atoms with ozone

$$O(^3P) + O_3 \xrightarrow{k_m} 2O_2; \quad \Delta H^\circ_{298} = -391 \text{ kJ mol}^{-1}$$

is an important bimolecular process in the upper atmosphere. However, the Arrhenius parameters for this reaction remain uncertain. Schiff,[17] has reviewed data on k_m from ozone photolysis and thermal decomposition, and data from discharge flow experiments up to 1963. As has been pointed out in the discussion of the $O + O_2 + M$ reaction (section 4.4.5), much difficulty arises from the extreme sensitivity of $-d[O_3]/dt$ to traces of H, which catalyse ozone removal through the chain

$$H + O_3 \longrightarrow OH + O_2$$

$$O + OH \longrightarrow O_2 + H$$

Ozone is also decomposed by any singlet oxygen $O_2(^1\Delta_g, {}^1\Sigma_g^+)$ present. In an attempt to overcome these problems, $O(^3P)$ atoms for reaction with ozone have been produced by decomposition of nitrous oxide on a heated Nernst filament[175] (cf. Kaufman and Kelso[157]). A value of $(8\cdot8 \pm 0\cdot3) \times 10^9$ cm^3 mol^{-1} s^{-1} for k_m at 296 K has been reported using this method.[175] It is significantly lower than the value of Phillips and Schiff,[90] $k_m = (1\cdot5 \pm 0\cdot4) \times 10^{10}$ cm^3 mol^{-1} s^{-1} at 298 K. This was obtained using an ordinary discharge source of O(^{3}P), and is thus likely to be erroneously high.

(c) *Reaction* N(^{4}S) + O$_3$

Similar considerations apply to the exothermic reaction

$$N(^4S) + O_3 \xrightarrow{k_n} NO + O_2; \qquad \Delta H^\circ_{298} = -525 \text{ kJ mol}^{-1}$$

Broida and Garvin[176] have studied the interesting chemiluminescent properties of this reaction when traces of H atoms were also present with N(^{4}S). NO(A$^2\Sigma^+$–X$^2\Pi$) and (B$^2\Pi$–X$^2\Pi$) bands, as well as various systems of OH and NH were observed. Measurements of ozone concentration by absorption spectrometry showed rapid catalysis of ozone removal by a trace of hydrogen atoms: the following simple cycle for chain removal of ozone was suggested[176]

$$H + O_3 \longrightarrow OH + O_2$$
$$N + OH \longrightarrow NO + H$$

It is uncertain to what extent such a catalytic scheme may have affected the only reported value[90] for k_n, $3\cdot4 \pm 0\cdot8 \times 10^{11}$ cm^3 mol^{-1} s^{-1} at 298 K, which was published at the same time as Broida and Garvin's work.

(d) *Reactions* X(^{2}P$_{3/2}$) + O$_3$; X = F, Cl, Br, I

Measurements of the rate constants for the reactions of ground state halogen atoms with ozone are not subject to catalysis by traces of H atoms, since regeneration of H is thermochemically improbable. Therefore, such measurements would be expected to be straightforward. The rate constant k_o has been determined, using absorption spectrophotometry, as $k_o = (4 \pm 2) \times 10^{11}$ cm^3 mol^{-1} s^{-1} at 294 K,[220] following a preliminary analysis[20] which gave $k_o \simeq 2 \times 10^{11}$ cm^3 mol^{-1} s^{-1}:

$$Br + O_3 \xrightarrow{k_o} BrO + O_2.$$

A more extensive investigation, using mass spectrometry with molecular beam sampling, and first order kinetic analysis for the measurement of $-\text{d} \ln [O_3]/\text{d}t$ in the presence of excess [Br], gave $k_o = (7 \pm 1) \times 10^{11}$ cm^3 mol^{-1} s^{-1} at 298 K.[198] The same method has been used for the rate constant of the more rapid Cl + O$_3$ reaction,

$$Cl + O_3 \longrightarrow ClO + O_2;$$
$$k^{298} = (1\cdot1 \pm 0\cdot2) \times 10^{13} \text{ cm}^3 \text{ mol}^{-1} \text{ s}^{-1}.^{(198)}$$

Preliminary data on $I + O_3$ by absorption spectrophotometry gave a rate constant of $10^{(11\cdot7\pm0\cdot3)}$ cm^3 mol^{-1} s^{-1} at 294 K for this reaction.[20]

The reaction of F atoms with O_3 is of interest, in that it has recently been shown to produce the hitherto unknown gaseous FO radical (see section 4.6.5). Mass spectrometric analysis has been used to measure the rate constant for this reaction,[221]

$$F + O_3 \longrightarrow FO + O_2;$$
$$k = 1\cdot7 \times 10^{13} \exp(-230/T) \text{ cm}^3 \text{ mol}^{-1} \text{ s}^{-1}.$$

4.5.9. Reactions X + YZ → XY + Z; X, Y = halogen; Z = halogen or NO.

The high sensitivity of atomic resonance fluorescence spectrometry has been exploited in a study of the rates of a series of halogen atom ($^2P_{3/2}$) plus halogen molecule reactions,[189,190] most of which are extremely rapid at 298 K:

$Cl + Br_2 \longrightarrow BrCl + Br$;	k(cm^3 mol^{-1} s^{-1}) =	$(7\cdot2 \pm 0\cdot9) \times 10^{13}$
$Cl + BrCl \longrightarrow Cl_2 + Br$;		$(8\cdot7 + 1\cdot2) \times 10^{12}$
$Cl + ClNO \longrightarrow Cl_2 + NO$;		$(1\cdot8 \pm 0\cdot3) \times 10^{13}$
$Cl + ICl \longrightarrow Cl_2 + I$;		$(4\cdot8 \pm 0\cdot6) \times 10^{12}$
$Br + ClNO \longrightarrow BrCl + NO$;		$(6\cdot0 \pm 1\cdot2) \times 10^{12}$
$Br + ICl \longrightarrow BrCl + I$;		$(1\cdot8 \pm 0\cdot5) \times 10^{10}$
$Br + IBr \longrightarrow Br_2 + I$;		$(2\cdot1 \pm 0\cdot4) \times 10^{13}$

The use of low atom and molecule concentrations enabled the time scale for reactions with rate constants approaching the bimolecular collision frequency to be extended into the millisecond range accessible by the discharge-flow method.

A number of these halogen reactions have been studied in crossed molecular beam experiments,[222] and the availability of direct measurements of overall rate constants from the discharge-flow method should be valuable in understanding the dynamics of the collision processes involved. Beam studies of $Cl + Br_2$,[222] the most fully investigated of the above reactions, confirm a high integral cross section for this reaction approaching the bimolecular collision frequency. It does not appear possible to reconcile earlier, indirect, much lower values for the rate constants of $Cl + Br_2$, $Cl + BrCl$, $Cl + ICl$, $Cl + ClNO$, based on analysis of photochemical reactions,[223] with the direct measurements from resonance fluorescence.[189,190]

4.5.10. Addition Reaction of H Atoms with C_2H_4

The kinetics of addition reactions of H(^2S) atoms with olefins have been extensively investigated both experimentally and theoretically, notably in a series of studies by Rabinovitch and coworkers.[177] Absolute rate constants for the addition steps are difficult to determine by the methods of bulk kineticist, and it is in this area that techniques such as the discharge-flow method can make a useful contribution. The types of elementary reactions involved in the H + propene system have been summarized:[10]

$$H + C_3H_6 \rightleftharpoons C_3H_7^* \longrightarrow CH_3 + C_2H_4$$
$$\downarrow M$$
$$C_3H_7$$

The vibrationally excited adduct $C_3H_7^*$ may thus either (i) redissociate to $H + C_3H_6$, or (ii) dissociate to products $CH_3 + C_2H_4$, or (iii) become stabilized to C_3H_7 by collision with M. The rate constants for (i) and (ii) are expected[177] to depend on the magnitude of the energy of $C_3H_7^*$, and thus a simple kinetic scheme for $-d[H]/dt$ and the rates of formation of products cannot be rigorously justified. However, Westenberg and de Haas, in a survey of their own and other data on the kinetics of the $H + C_2H_4$ reaction,[33a] considered that the data and their analyses at the present time warranted only a simple steady state formulation with neglect of the energy dependence of the rate constants. An apparent second-order rate constant was defined by $k_p = -[C_2H_4]^{-1}$ $d \ln [H]/dt$; the simple treatment of the reaction scheme for stabilization and dissociation of $C_2H_7^*$ predicted a linear correlation of $(k_p)^{-1}$ with $[M]^{-1}$, with an intercept at $[M] \to 0$ such that $k_p = k_f$, is

$$H + C_2H_4 \xrightarrow{k_a} C_2H_5^*$$

Pseudo-first-order analysis of $-d[H]/dt$ in the $C_2H_4 + H$ system, using e.p.r. detection and the fixed observation point method, showed that whilst k_p was independent of $[C_2H_4]$ and reaction time as expected, it increased with a decrease in [M] over the range 0·5 to 2·5 torr total pressure of argon or helium, at 297 and 525 K. A satisfactory fit of the data to linear correlations of k_p^{-1} with $[M]^{-1}$ was obtained. Earlier data for helium third body, using Lyman-α 121·6 nm line absorption in a discharge-flow system[34a] to measure [H], and at higher pressures in flash photolysis using Lyman-α fluorescence,[178] gave similar increases in k_p with decreasing total pressure. However, the slopes of the k_p^{-1} versus $[M]^{-1}$ plots were appreciably greater than those of Westenberg and de Haas. There is consequently a considerable scatter in the results for k_a, with a mean value of $2 \pm 1 \times 10^{11}$ cm³ mol⁻¹ s⁻¹ at 297 K from these studies. The agreement of the values of k_p for M = Ar of Westenberg and de Haas[33a] with a study by Brown and Thrush[33b] is satisfactory within a factor of 1·5 in the range of overlap of argon pressures. The latter study was conducted in a manner similar to that of Westenberg and de Haas, except that second-order analysis and a narrower range of total pressures (1·6 to 3·1 torr) were used.

M. A. A. CLYNE

More recent flash photolysis results[224] using resonance fluorescence to follow [H] gave $k_q = 8.2 \times 10^{11}$ cm^3 mol^{-1} s^{-1}, by extrapolation to infinite He pressure, in fair agreement with earlier work,[178] and independent of stoichiometric corrections. It was suggested[224] that extrapolations (from low pressure measurements) based on plots of k_p^{-1} against $[M]^{-1}$ may be in error due to significant curvature. Also, stoichiometric factors obtained under low pressure conditions might not[224] be applicable to higher pressures.

4.6. STUDIES WITH MOLECULAR RADICALS

There is a considerable potential for the determination of rate constants for free radical reactions in discharge-flow systems. Such measurements have so far been carried out in any detail only for the OH, ClO, BrO, NF$_2$ and, to a more limited extent, the CN radial. However, these few studies have made a significant contribution to the theory of reactivity of these radicals.

As for the study of atom reactions, investigations of radicals require a source of radicals in an inert carrier gas. The detection and determination of radicals usually requires spectrometric methods; the simpler titration and chemiluminescence methods are useful in conjunction with spectrometry. For radicals, it is usually not possible to use the fixed observation point method (section 4.4.4(a)) with the consequent simple pseudo-first-order analysis which is so useful in the study of the rates of atom reactions. Nevertheless, rate measurements have been made in discharge-flow systems with the following radicals OH, OD, ClO, BrO, IO, FO, CN, SO, HS, NCl$_2$ and NF$_2$.

4.6.1. Production of Radicals

The simplest technique for the production of molecular radicals is to pass a suitable stable gas in an inert carrier through a discharge. This method is exemplified by the production of SO($X^3\Sigma^-$) radicals by a discharge in SO$_2$ + Ar mixtures,[179] but it is not always satisfactory. Firstly, the yields of radicals may be undetectably low, and secondly, any radicals produced may be accompanied by large amounts of undesirable secondary active species. For example, although discharge in H$_2$O vapour gives small amounts of OH[5], these are produced not by the discharge, but by reactions outside the discharge, probably through the scheme H + O$_2$ + M $\rightarrow$ HO$_2$ + M, followed by H + HO$_2$ $\rightarrow$ 2OH. The presence of H and O$_2$ in this system thus leads to complications in the reaction schemes for OH, and kinetic studies based on this method for the formation of OH must consequently be regarded with scepticism.

The preferred method for the production of OH is the extremely rapid reaction between NO$_2$ and H atoms

$$H + NO_2 \longrightarrow NO + OH; \qquad \Delta H^\circ_{298} = -122 \text{ kJ mol}^{-1}$$

The other product, NO, is inert towards OH under the conditions used, and can only slowly react with H in a third-order recombination reaction, $H + NO + M \xrightarrow{\text{slow}} HNO + M$; $H + HNO \xrightarrow{\text{fast}} H_2 + NO$. Also, OH is inert towards NO_2. The $H + NO_2$ reaction is, in principle, an almost ideal method of forming OH radicals for kinetic studies. (a) The reactants are stable compounds and ground state radicals. (b) Neither the radical formed, nor any other product of primary reaction, can react in rapid secondary steps with the original reactants. (c) The reaction forming radicals is rapid and of simple mechanism.

For the production of diatomic radicals, the usual source is an exothermic reaction of an atom with a triatomic molecule. Endothermic compounds are often found to be suitable sources of radicals, since their reactions with atoms are likely to satisfy the criteria (a) to (c) mentioned above. The following processes have been used to produce diatomic radicals for kinetic studies in discharge-flow systems:

OH and OD($X^2\Pi$): (a) $H + NO_2 \longrightarrow OH + NO$
                      (b) $H + O_3 \longrightarrow OH + O_2$

CN($X^2\Sigma^+$):   (a) $O + C_2N_2 \longrightarrow CN + NCO$
                      (b) reactions of N with C_2H_2, CH_2Cl_2, CCl_4, C_2N_2, CH_3CN

SO($X^3\Sigma^-$):   (a) $O + OCS \longrightarrow CO + SO$
                      (b) discharge in $Ar + SO_2$ mixtures
                      (c) $S + O_2 \longrightarrow SO + O$

FO($X^2\Pi$):   (a) $F + O_3 \longrightarrow FO + O_2$

ClO($X^2\Pi$):   (a) $Cl + OClO \longrightarrow 2ClO$
                      (b) $Cl + O_3 \longrightarrow ClO + O_2$

BrO($X^2\Pi$):   (a) $Br + O_3 \longrightarrow BrO + O_2$
                      (b) $O + Br_2 \longrightarrow BrO + Br$

IO($X^2\Pi$):   (a) $I + O_3 \longrightarrow IO + O_2$

HS($X^2\Pi$):   (a) $H + H_2S \longrightarrow HS + H_2$

Some of these methods involve complications of secondary reactions of primary products with atoms. The preferred methods of production are probably: OH, (a); CN, (a); SO, (a); ClO, (a) or (b); BrO, (a); IO, (a).

The difficulty of avoiding complex secondary reactions is naturally greater for triatomic and larger species than for diatomic radicals. However, the following sources of polyatomic radicals are among those which have been used for kinetic work in discharge-flow systems:

HO_2:   discharge in H_2O_2

$$H + O_2 + M \longrightarrow HO_2 + M$$

NCl_2: $Cl + NCl_3 \longrightarrow NCl_2 + Cl_2$

NF_2: $N_2F_4 + M \longrightarrow NF_2 + NF_2 + M$

N_3: $Cl + N_3Cl \longrightarrow N_3 + Cl_2$

In addition, a considerable group of other diatomic (and a few triatomic) radicals have been detected, mostly by e.p.r. spectrometry,[57] although no kinetic studies on these radicals have yet been carried out.

4.6.2. Detection and Estimation of Radical Concentrations

One example in which the titration and chemiluminescence methods so useful in atom kinetics have proved suitable for the estimation of radical concentrations is the SO radical.[179] The sum of the concentrations of O and SO in the products of a discharge in SO_2 may be estimated by means of the extinction of the green NO + O and violet SO + O emissions by titration with NO_2:

$$O + NO_2 \longrightarrow NO + O_2$$
$$SO + NO_2 \longrightarrow NO + SO_2$$

Since the emission intensities I_a of the NO + O glow and I_s of the SO + O glow are given by $I_a = I_{0a}[O][NO]$ and $I_s = I_{0s}[O][SO]$, measurements of intensity, combined with the titre $[O] + [SO] = [NO_2]$, gave [SO] absolutely. An alternative method[114] took advantage of the chemiluminescence of $SO_2(^1B, \,^3B_1)$ excited states produced in the reaction, $SO + O_3 \rightarrow SO_2 + O_2$. Since it has been shown that the intensity of SO_2 emission, I_3, was given by $I_3 = I_{03}[O_3][SO]/[M]$, $I_3[M]/[O_3]$ is directly proportional to [SO].

Usually, titration methods are used to calibrate spectrometrically determined radical concentrations in absolute terms. Most kinetic studies of radicals have used either e.p.r. spectrometry or electronic absorption spectrophotometry. A survey of these methods has been given in section 4.2 of this article; selected examples will be discussed later in this section.

4.6.3. Kinetic Analysis of Radical Reactions

The simplest reaction undergone by most small radicals is second-order decay, leading to disproportionation or recombination; for example:

$$OH + OH \longrightarrow H_2O + O$$
$$BrO + BrO \longrightarrow 2Br + O_2$$
$$NF_2 + NF_2 + M \longrightarrow N_2F_4 + M$$

The rate constants for reactions of this type are usually large at 300 K, and these rates (especially of disproportionation reactions) may dominate other processes in discharge-flow systems, when radical concentrations [R] are high ($\approx 5 \times 10^{-10}$ mol cm^{-3}). The kinetics of removal of R are thus often intermediate between first and second order in [R] when a reactive substrate S is present,

$$-d \ln [R]/dt = k_a[S] + k_b[R]$$

First-order analysis by use of the fixed observation point method is precluded in such a case, and $-d[R]/dt$ must be found by measurement of [R] along the axis of the flow tube at a series of points from the origin of the reaction zone. This may be accomplished for absorption spectrophotometry by moving the detection system on a railway parallel to the tube axis (Fig. 4.4); or, also, in some cases, by using a moveable inlet (or a series of fixed inlets) to enable the reaction origin to be shifted relative to a fixed detector, as in e.p.r. spectrometry (Fig. 4.3).

4.6.4. Reactions of OH($X^2\Pi$) Radicals

Both methods of kinetic analysis referred to in section 4.6.3, a moveable absorption spectrophotometric detector,[180] and fixed e.p.r. detector with moveable inlets,[56] have been used to study the kinetics of reactions of OH radicals. Mass spectrometric analysis has also been used to measure the rate[88] of the reaction $OH + CO \rightarrow H + CO_2$. In all these studies, OH radicals were produced by the reaction of H with NO_2.

Del Greco and Kaufman established the mechanism of the decay reaction of OH, using time-resolved electronic absorption spectrophotometry with an OH lamp.[5] Rapid bimolecular disproportionation of 2OH occurs, leading to oxygen atoms,

$$OH + OH \xrightarrow{k_2} H_2O + O; \qquad \Delta H^\circ_{298} = -71 \text{ kJ mol}^{-1} \qquad (2)$$

However, the oxygen atoms detected in the decay reaction of OH were in a small steady state concentration, in agreement with predominant removal of most of the O atoms produced in reaction (2) by the still more rapid reaction (3),

$$O + OH \xrightarrow{k_3} H + O_2; \qquad \Delta H^\circ_{298} = -70 \text{ kJ mol}^{-1} \qquad (3)$$

(The reverse reaction (-3) is the main branching step in the $H_2 + O_2$ thermal chain reaction.) Assumption of a pseudo-stationary-state value of O, i.e. established rapidly by reactions (2) and (3), gave the result, $-d[OH]/dt = -3k_2[OH]^2$. At long reaction times, t, such that $t \gg 1/k_2[OH]_0$, the result $[O] = \frac{1}{3}k_2t$ was obtained. Values for the second-order rate constant k_2 were obtained near 300 K using e.p.r. and absorption spectrophotometry. Kaufman and Del Greco's revised value[7b] for k_2 was $8\cdot5 \times 10^{11}$ cm^3 mol^{-1} s^{-1} which may be compared with the earlier e.p.r. value[56b] of $(1.2 \pm 0.3) \times 10^{12}$ cm^3 mol^{-1} s^{-1}, subsequently redetermined as $(1.6 \pm 0.2) \times 10^{12}$ cm^3 mol^{-1} s^{-1}. The arguments advanced in favour of the latter, higher value[56b] obtained by e.p.r. spectrometry are rather unconvincing, and it appears to the writer that the best value for k_2 is a mean of the values of refs. 7b and 56c, i.e. $k_2 = (1.2 \pm 0.3) \times 10^{12}$ cm^3 mol^{-1} s^{-1}. In any event, the agreement between the various values found for k_2 is rather satisfactory.

Values for k_3 have also been reported at 300 K, using direct measurements of [O] ($k_3 = 2.5 \times 10^{13}$ cm^3 mol^{-1} s^{-1}),[22a] and from the acceleration in $-d[OH]/dt$ caused by the addition of O atoms ($k_3 = 2.0 \times 10^{13}$ cm^3 mol^{-1} s^{-1}).[5] The agreement between these values is again satisfactory, and allows a reliable calculation of the rate constant for the reverse reaction, the chain branching reaction, $H + O_2 \rightarrow OH + O$, to be made.

Several workers have considered the implications of the decay scheme for OH discussed above for the stoichiometry of the $H + NO_2$ reaction.[5,22a,56a,174] The latter reaction constitutes one of the main methods for the measurement of [H] in discharge-flow systems. However, a limitation is that in the presence of appreciable amounts of H_2, a secondary reaction occurs leading to regeneration of H and consequent overestimation of initial [H] from the NO_2 titre:

$$H + NO_2 \longrightarrow OH + NO$$

$$OH + H_2 \xrightarrow{k_4} H_2O + H$$

In the absence of appreciable amounts of H_2, the overall stoichiometry of the $H + NO_2$ reaction measured at long reaction times ($t > 10^{-12}$ $[NO_2]_{0s}$), as in a titration experiment, will be given by the sum of the reactions

$$H + NO_2 \longrightarrow OH + NO$$
$$OH + OH \longrightarrow H_2O + O$$
$$O + OH \longrightarrow H + O_2$$

$$\overline{\tfrac{2}{3}H + NO_2 \longrightarrow NO + \tfrac{1}{3}O_2 + \tfrac{1}{3}H_2O}$$

E.p.r. spectrometry has been used, as for OH, to measure k_{2d} for the isotopically related reaction[56d]

$$OD + OD \xrightarrow{k_{2d}} D_2O + O$$

The ratio k_2/k_{2d} at 300 K was 1.6.

More recently, it has been shown that appreciable first order heterogeneous recombination of OH occurs on the silica or oxy-acid coated walls of a flow tube, with $k_{wall} \sim 20$–80 s^{-1} in most cases.[187,188,225] This reaction would be important at low [OH], and appreciable at the higher values of [OH] where purely second-order decay was previously reported.[76,56c] Consequently, both Kaufman and Del Greco's[76] and Westenberg and de Haas'[56c] values are probably too high, and are currently being redetermined.[225]

The kinetics of reaction of OH with the added substrates H_2, CH_4, CD_4 and of OD with CO, have also been investigated using e.p.r. spectrometry.[56] Less precise rate constants were obtained in these cases than when k_2 was measured. For example, in the case of reaction (4)

$$OH + H_2 \longrightarrow H_2O + H; \quad \Delta H^\circ_{298} = -63 \text{ kJ mol}^{-1} \quad (4)$$

the rate of removal of OH is $-d[OH]/dt = k_2[OH]^2 + k_4[OH][H_2]$; the kinetics lie between first and second order in [OH], and hence k_4 must be extracted from the rate equation making due allowance for the term $k_2[OH]^2$. As Table 8 of ref. 7c shows, the results for the reactions $OH + H_2$ and $OH + CO$ are nevertheless in good agreement with those obtained using different methods.

Most of the kinetic studies on OH using discharge-flow systems have been carried out at 300 K. Very few reliable values for activation energies have been obtained by other methods. Hence, although the rate constants for some simple reactions of OH at 300 K are well established, there is an urgent need for systematic studies of temperature dependences.

4.6.5. Reactions of Halogen Oxide XO Radicals

The ClO and BrO($X^2\Pi$) radicals are other examples of diatomic radicals whose kinetic behaviour has been studied in some detail using the discharge-flow method. The best sources of these radicals for kinetic work were found to be the reactions

$$Cl(^2P_{3/2}) + OClO \longrightarrow 2ClO;^{164,181} \quad \Delta H^\circ_{298} = -23 \text{ kJ mol}^{-1}$$

$$Cl(^2P_{3/2}) + O_3 \longrightarrow ClO + O_2;^{164} \quad \Delta H^\circ_{298} = -162 \text{ kJ mol}^{-1}$$

$$Br(^2P_{3/2}) + O_3 \longrightarrow BrO + O_2;^{164,20} \quad \Delta H^\circ_{298} = -129 \text{ kJ mol}^{-1}$$

$$O + Br_2 \longrightarrow BrO + Br;^{20} \quad \Delta H^\circ_{298} = -23 \text{ kJ mol}^{-1}$$

However, the ClO radical was also formed in the products of the reactions $O + OClO$ (excess),[181] and $Br + OClO$,[164] which latter reaction also gave BrO.[164] The IO($X^2\Pi$) radical in both vibrational levels $v' = 0$ and $v'' = 1$ was detected as a product of the reaction $I(^2P_{3/2}) + O_3 \rightarrow IO + O_2$, and limited kinetic data on this species were obtained.[20] Detection of all three radicals, ClO, BrO and IO was by means of multiple traversal electronic absorption spectrophotometry of the intense ultraviolet $A^2\Pi \leftarrow X^2\Pi$ band systems.[20]

Absolute concentrations of ClO were obtained from measured

absorption intensities using titration with O atoms or NO.[181] Both these reactions

$$O + ClO \longrightarrow Cl + O_2; \qquad \Delta H^{\circ}_{298} = -229 \text{ kJ mol}^{-1}$$

$$NO + ClO \longrightarrow Cl + NO_2; \qquad \Delta H^{\circ}_{298} = -196 \text{ kJ mol}^{-1}$$

are extremely rapid ($k > 10^{12}$ cm^3 mol^{-1} s^{-1} at 300 K), and the end points of titration were observed as critical removal of the ClO absorption by added O or NO. The absorption cross section of ClO in the A ← X continuum at 257·7 nm (which adjoins the band system) was thus determined as $(2·26 \pm 0·05) \times 10^{-18}$ cm^2. As for OH radicals, the decay of ClO was kinetically second-order, with a rate constant k_5 given by $k_5 = (7·2 \pm 0·4) \times 10^{11} \exp [(-1250 \pm 150)/T]$ cm^3 mol^{-1} s^{-1}. In the cases of ClO and BrO, no first order heterogeneous recombination has been detected. Recently, a more accurate determination of the temperature coefficient of k_5, $k_5 = 10^{11·86} \exp \times [(-1160 \pm 50)/T]$ cm^3 mol^{-1} s^{-1}, has been carried out from 259 to 710 K using a moveable OClO inlet systems and a fixed absorption detector.[62]

The products of decay of ClO radicals reacted with added H$_2$ at 300 K, but on addition of bromine, which acts as an effective scavenger through the rapid transfer reaction,[190] Cl + Br$_2 \to$ BrCl + Br, the ClO + H$_2$ reaction was completely suppressed. Subsequently, Cl atoms were detected directly by atomic absorption spectrometry at λ 138·0 nm, as a product of the decay reaction of ClO radicals.[66] The following scheme was proposed[164]

$$ClO + ClO \overset{k_5}{\rightleftharpoons} Cl + ClOO$$
$$ClOO + M \rightleftharpoons Cl + O_2 + M$$
$$ClOO + Cl \longrightarrow Cl_2 + O_2$$

Subsequent work on the photolysis of Cl$_2$ + O$_2$ mixtures at higher pressures using modulated molecular spectrometry[182] led to the first identification of the gaseous unstable ClOO radical, and confirmed the validity of this mechanism for ClO decay. The same work[182] also indicated the importance of an overall third-order decay reaction at higher pressures (10 torr) than those used in discharge-flow work:

$$ClO + ClO + M \rightleftharpoons Cl_2O_2 + M$$
$$Cl_2O_2 + M \longrightarrow Cl_2 + O_2$$

Such a "high pressure" mechanism had been proposed earlier[164] to account for discrepancies between the results of discharge-flow measurements of k_2 and those at higher pressures using flash photolysis.[226]

The decay reaction of BrO radicals was also found to be second-order, but was much more rapid than that of ClO radicals at 300 K.[20]

This fact precluded the use of a simple titration reaction for the calibration of BrO absorption intensities in terms of absolute concentrations, and led to small concentrations of radicals in the flow system. An improved version of the absorption system proved satisfactory, however, for measurements of absorption intensities down to 0·02%. Absolute BrO concentrations were obtained[20] by measurement of the amounts of O_3 consumed in the reaction, $Br + O_3 \rightarrow BrO + O_2$. Br atoms were detected in the products of decay of BrO, and measurement of Br by absorption of Br_2 showed that a simple scheme was followed

$$BrO + BrO \xrightarrow{k_6} 2Br + O_2; \qquad \Delta H^{\circ}_{298} = -28 \text{ kJ mol}^{-1}$$

The value of k_6 was $10^{(13.5 \pm 0.2)}$ exp $[(-450 \pm 300)/T]$ cm^3 mol^{-1} s^{-1} between 293 and 573 K; the frequency factor is surprisingly large, and is closely similar in magnitude to the hard-sphere bimolecular collision frequency for 2BrO. The rate constant for the reaction

$$BrO + NO \longrightarrow Br + NO_2; \qquad \Delta H^{\circ}_{298} = -71 \text{ kJ mol}^{-1}$$

is also large and was found to have a value of $10^{(13.1 \pm 0.1)}$ cm^3 mol^{-1} s^{-1} at 293 K.[198]

Mass spectrometric studies in flow systems of ClO, BrO and FO radicals reactions[198] have led to results basically in agreement with the work of Clyne and Cruse[20] using absorption spectrophotometry. Absolute concentrations of ClO and BrO were measured by conversion of the radical into stoichiometric quantities of NO_2 through addition of an excess of NO,[198] e.g.,

$$BrO + NO \longrightarrow Br + NO_2.$$

The result for k_5 (ClO decay) by mass spectrometry was within the limits of error of the spectrophotometric data, i.e. $k_5 = (1·4 \pm 0·2) \times 10^{10}$ cm^3 mol^{-1} s^{-1} at 298 K,[198] but k_6 (BrO decay) was found to be appreciably less than that reported by Clyne and Cruse,[20] $k_6 = (3·9 \pm 0·6) \times 10^{12}$ cm^3 mol^{-1} s^{-1} at 298 K,[198] indicating that the O_3 method used previously[20] to obtain absolute [BrO] may be inaccurate. The temperature coefficient of $k_6^{(20)}$ is unaffected by this change, and the revised Arrhenius expression $k_6 = 10^{(13.2 \pm 0.2)}$ exp $[(-450 \pm 200(/T]$ is not suggested.

The reaction of F atoms with O_3 has been shown to produce the FO radical (section 4.6.5). Identification of this radical was shown mass spectrometrically by Clyne and Watson[227] from evidence based on the different appearance potentials of FO$^+$ from FO and F_2O, the absence of chlorine ions in the mass spectrometer which occur at the same mass peak as FO$^+$ (m/e 35, ^{35}Cl, ^{19}F, ^{16}O), and the observed second-order decay of the FO radical in the flow system. Wagner, Zetzsch and Warnatz[228] have also reported that FO is a product of the $F + O_3$ reaction, but no details of the identification were given.

From a kinetic analysis of F, FO and O_3 concentrations in the F + O_3 systems, these workers[221] have obtained an estimate of the rate constant for second order decay of FO radicals, $k \sim 1 \times 10^{13}$ cm^3 mol^{-1} s^{-1} at 298 K. The suggested mechanism of this reaction, FO + FO → 2F + O_2,[221] is analogous to that found for the BrO + BrO and ClO + ClO[164] reactions. Although Wagner et al.[221] reported detectable first order heterogeneous recombination for FO radicals, our own studies[198,227] do not confirm this finding.

The decay of IO was similarly bimolecular and rapid, and the second-order rate constant was estimated to be $10^{(12.5\pm0.3)}$ cm^3 mol^{-1}s^{-1} at 293 K.[20]

Studies at 300 K of the possible reactions of ClO with H_2 and CH_4, and of BrO with H_2, CH_4, C_2H_6, C_2H_4 and O_3 showed no detectable rates of these reactions.[20,164] It appears that, contrary to earlier views based partly on flash photolytic investigations of the ClO radical, ClO and BrO are very inefficient chain carriers and are largely inert towards stable singlet molecules at laboratory temperatures. The reactivity previously attributed to ClO is probably at least partly due to the high reactivity of the Cl atoms formed during bimolecular decay of the ClO radical. For example, a rapid chain reaction between ClO and H_2 propagated by Cl atoms has been characterized[164] at 293 K:

$$Cl + H_2 \longrightarrow HCl + H$$

$$H + ClO \longrightarrow HCl + O$$

$$O + ClO \longrightarrow Cl + O_2$$

overall
stoichiometry $\quad \overline{2ClO + H_2 \longrightarrow 2HCl + O_2}$

4.6.6. Reactions of NX$_2$ Radicals

A system of diffuse absorption bands (λ = 280 to 320 nm) has been used to detect NCl$_2$ radicals formed in the rapid reaction[60]

$$Cl + NCl_3 \longrightarrow NCl_2 + Cl_2$$

Calibration of absolute concentrations of NCl$_2$ was either by measurement of the concentration of NCl$_3$ reacted, or by photometric measurement of the yield of NO on adding to a stream of NCl$_2$ radicals an excess of oxygen atoms from a second discharge, [NO] produced = [NCl$_2$]; satisfactory correlation of the two methods was obtained.[60]

Decay of NCl$_2$ radicals was closely similar to that of BrO and ClO radicals, in that the kinetics were second-order in radical concentrations, and gave atoms as reaction products. The rate constant k_7 for decay of NCl$_2$ was given by $k_7 = 10^{(11.7\pm0.1)}$ exp $(0 \pm 200/T)$ cm^{3}mol^{-1} s^{-1} from 259 to 373 K. The following scheme for this reaction was suggested

$$NCl_2 + NCl_2 \longrightarrow N_2Cl_3 + Cl$$

$$N_2Cl_3 + M \longrightarrow N_2Cl_2 + Cl + M$$

Disproportionation of N_2Cl_3 in this way would be analogous to the easy disproportionation of N_2H_3 radicals which has been reported.[79] No evidence in support of bimolecular or termolecular combination reactions of $2NCl_2 \rightarrow N_2Cl_4$ could be obtained.[60]

The NF_2 radical is much simpler to generate than the NCl_2 radical, and it is surprising that so little attention has been paid to the simple reactions of this species. As expected from the thermochemistry of N_2F_4, virtually complete dissociation of tetrafluorohydrazine has been found when N_2F_4 is passed through a heated tube at 500 K and 5 torr pressure.[183] In recent studies,[183] we have used this source of NF_2 in a flow system. Measurements of $[NF_2]$ were made by absorption spectrometry in the continuum at 260 nm, and by mass spectrometry. No problem of calibration of absolute radical concentrations arises in view of the simple method of generating radicals. (The same method of forming NF_2 has been described recently in concurrent independent work by Wagner and co-workers.[41]) Decay of NF_2 in argon has been found[183] to be second-order in $[NF_2]$ and first-order in $[M]$, $NF_2 + NF_2 + M \underset{k_{-8}}{\overset{k_8}{\rightleftharpoons}} N_2F_4 + M$. Under the conditions used (293 K and 0·5–3 torr total pressure), equilibrium lies to the right hand side. Values for k_8 for a range of third bodies M have been found.[213] The agreement with extrapolations from higher temperature studies of k_{-8} from shock tube work[184] was satisfactory. In the case of NF_2 radicals, decay thus occurs predominantly by recombination, and no evidence for disproportionation leading to F atoms analogous to the behaviour of NCl_2 radicals could be obtained. The reactions of Cl and Br with NF_2 radicals have also been characterized:[213] $Cl + NF_2 + M \rightarrow ClNF_2 + M$, $Br + NF_2 + M \rightarrow BrNF_2 + M$, and their rate constants have been measured.[213] The $H + NF_2$ reaction leads to emission of $B^3\Pi_g \rightarrow A^3\Sigma_u^+$ bands possessing a closely similar vibrational distribution to that of the N_2 afterglow spectrum formed by recombination of ground state $N(^4S)$ atoms

$$H + NF_2 \longrightarrow HF + NF$$

$$H + NF \longrightarrow HF + N$$

This is the only ground state reaction so far reported which apparently gives $N(^4S)$ atoms. The reactions of NF_2 with $N(^4S)$ atoms is a good source of F atoms,[41,227] and provides the basis for further kinetic studies.

I thank A. A. Westenberg for helpful correspondence, and A. A. Westenberg, F. Kaufman, K. H. Becker, R. E. Huie, D. Gutman and K. D. Bayes, for communication of results prior to publication.

REFERENCES

1. Rayleigh (R. J. Strutt), *Proc. Roy. Soc.*, **A151**, 567 (1935); ibid., *Proc. Roy. Soc.*, **A176**, 1 (1940)
2. P. Harteck and U. Kopsch, *Z. physik. Chem.*, **B12**, 327 (1931)
3. M. L. Spealman and W. H. Rodebush, *J. Am. Chem. Soc.*, **57**, 1474 (1935)
4. F. Kaufman, *Proc. Roy. Soc.*, **A247**, 123 (1958)
5. F. P. Del Greco and F. K. Kaufman, *Discussions Faraday Soc.*, **33**, 128 (1962)
6. N. Cohen and J. Heicklen, Comprehensive Chemical Kinetics, Vol. 2., (Elsevier, 1969)
7. (a) F. Kaufman, *Progr. Reaction Kinetics*, **1**, 1 (1961); (b) *Ann. Geophysique*, **20**, 106 (1964); (c) *Ann. Rev. Phys. Chem.*, **20**, 45 (1969)
8. I. M. Campbell and B. A. Thrush, *Ann. Rept. Chem. Soc.*, **67**, 17 (1965)
9. B. Brocklehurst and K. R. Jennings, *Progr. Reaction Kinetics*, **4**, 1 (1967)
10. B. A. Thrush, *Progr. Reaction Kinetics*, **3**, 65 (1965)
11. B. A. Thrush, *Ann. Rev. Phys. Chem.*, **19**, 371 (1968); ibid., *Chem. in Britain*, 287 (1966)
12. M. F. R. Mulcahy, J. B. Steven, J. C. Ward and D. J. Williams, 12th Symp. Combustion, 323 (1968); *J. Phys. Chem.*, **71**, 2124 (1967)
13. N. Jonathan and R. Petty, *Trans. Faraday Soc.*, **64**, 1240 (1968); F. Kaufman and J. R. Kelso, Symp. Chemiluminescence, Duke University (1965)
14. A. M. Bass and H. P. Broida (Editors), Formation and Trapping of Free Radicals (Academic Press, New York, 1960). T. M. Shaw, *J. Chem. Phys.*, **30**, 593, 1366 (1959)
15. G. M. Schwab, *Z. Physik. Chem.*, **B27**, 452 (1934)
16. E. A. Ogryzlo, *Can. J. Chem.*, **39**, 2556 (1961)
17. M. A. A. Clyne and D. H. Stedman, *Trans. Faraday Soc.*, **64**, 2698 (1968)
18. R. W. Fair and B. A. Thrush, *Trans. Faraday Soc.*, **65**, 1208 (1969)
19. A. Carrington and D. H. Levy, *J. Chem. Phys.*, **44**, 1298 (1966); ibid., *J. Chem. Phys.*, **47**, 3801 (1967); ibid., *Chem. Commun.* 641 (1967); ibid., *Mol. Phys.*, **15**, 187 (1968); ibid., *J. Chem. Phys.*, **52**, 309 (1970)
20. M. A. A. Clyne and H. W. Cruse, *Trans. Faraday Soc.*, **66**, 2214 (1970)
21. E. Wrede, *Z. Instrumentenk.*, **48**, 201 (1928); P. Harteck, *Z. Physik. Chem.*, **A139**, 98 (1928); J. C. Greaves and J. W. Linnett, *Trans. Faraday Soc.*, **55**, 1338 (1959)
22. (a) M. A. A. Clyne and B. A. Thrush, *Proc. Roy. Soc.*, **A275**, 559 (1963);
    (b) A. A. Westenberg and N. de Haas, *J. Chem. Phys.*, **46**, 490 (1967);
    (c) K. Hoyermann, H. G. Wagner and J. Wolfrum, *Berichte physik. Chem.*, **71**, 599 (1967)
23. F. Kaufman and J. R. Kelso, *J. Chem. Phys.*, **28**, 510 (1958)
24. Recommended value from D. L. Baulch, D. D. Drysdale and D. G. Horne, High Temperature Reaction Rate Data, No. 4, Dec. 1969, University of Leeds
25. J. E. Morgan and H. I. Schiff, *J. Chem. Phys.*, **38**, 1495 (1963)
26. G. J. Verbeke and C. A. Winkler, *J. Phys. Chem.*, **64**, 319 (1960)
27. (a) K. D. Bayes and G. B. Kistiakowsky, *J. Chem. Phys.*, **32**, 992 (1960);
    (b) I. M. Campbell and B. A. Thrush, *Proc. Roy. Soc.*, **A296**, 201 (1967)
28. M. A. A. Clyne and B. A. Thrush, *Proc. Roy. Soc.*, **A269**, 404 (1962)
29. I. W. M. Smith, *Discussions Faraday Soc.*, **44**, 194 (1967); A. A. Westenberg and N. De Haas, *J. Chem. Phys.*, **50**, 707 (1969)
30. L. F. Phillips and H. I. Schiff, *J. Chem. Phys.*, **37**, 1233 (1962)
31. M. A. A. Clyne and B. A. Thrush, *Trans. Faraday Soc.*, **57**, 1305 (1961); ibid., *Discussions Faraday Soc.*, **33**, 139 (1962)

32. M. J. Y. Clement and D. A. Ramsay, *Can. J. Phys.*, **39**, 205 (1961)
33. (a) A. A. Westenberg and N. de Haas, *J. Chem. Phys.*, **50**, 707 (1969);
    (b) J. M. Brown and B. A. Thrush, *Trans. Faraday Soc.*, **63**, 630 (1967)
34. (a) J. V. Michael and R. E. Weston, *J. Chem. Phys.*, **45**, 3632 (1966);
    (b) J. R. Barker, D. G. Keil, J. V. Michael and D. T. Osborne, *J. Chem. Phys.*, **52**, 2079 (1970)
35. M. A. A. Clyne and D. H. Stedman, *Trans. Faraday Soc.*, **62**, 2164 (1966)
36. L. W. Bader and E. A. Ogryzlo, *J. Chem. Phys.*, **41**, 2926 (1964)
37. E. Hutton and M. Wright, *Trans. Faraday Soc.*, **61**, 78 (1965)
38. M. A. A. Clyne and D. H. Stedman, *Trans. Faraday Soc.*, **64**, 1816 (1968)
39. W. G. Burns and F. S. Dainton, *Trans. Faraday Soc.*, **48**, 52 (1952)
40. M. A. A. Clyne, J. A. Coxon and A. R. Woon-Fat, *Disc. Faraday Div.*, **53**, 82 (1972)
41. K. H. Homann, W. C. Solomon, J. Warnatz, H. G. Wagner and C. Zetzsch, *Berichte physik. Chem.*, **74**, 585 (1970)
42. Recommended value from D. L. Baulch, D. D. Drysdale and D. G. Horne, High Temperature Reaction Rate Data, No. 2, Nov. 1968, University of Leeds
43. S. Krongelb and M. W. P. Strandberg, *J. Chem. Phys.*, **31**, 1196 (1959)
44. A. A. Westenberg and N. de Haas, *J. Chem. Phys.*, **40**, 3087 (1964)
45. A. A. Westenberg and N. de Haas, *J. Chem. Phys.*, **51**, 5215 (1969)
46. J. P. Wittke and R. H. Dicke, *Phys. Rev.*, **103**, 620 (1956)
47. A. A. Westenberg and N. de Haas, *J. Chem. Phys.*, **48**, 4405 (1968)
48. A. A. Westenberg and N. de Haas, *J. Chem. Phys.*, **47**, 4241 (1967); ibid., *J. Chem. Phys.*, **50**, 2512 (1969)
49. J. M. Brown and B. A. Thrush, *Trans. Faraday Soc.*, **63**, 630 (1967); J. M. Brown, P. B. Coates and B. A. Thrush, *Chem. Commun.*, 843 (1966)
50. A. A. Westenberg and N. de Haas, *J. Chem. Phys.*, **46**, 490 (1967)
51. E. A. Albers, K. Hoyermann, H. G. Wagner and J. Wolfrum, 12th Symp. Combustion, 313 (1968)
52. P. B. Davies, B. A. Thrush and A. F. Tuck, *Trans. Faraday Soc.*, **66**, 886 (1970)
53. A. A. Westenberg and N. de Haas, 12th Symp. Combustion, 289 (1968)
54. K. Hoyermann, H. G. Wagner and J. Wolfrum, *Berichte physik. Chem.*, **71**, 599 (1967)
55. J. E. Bennett and D. R. Blackmore, *Proc. Roy. Soc.*, **A305**, 553 (1968)
56. (a) A. A. Westenberg, *J. Chem. Phys.*, **43**, 1544 (1965);
    (b) A. A. Westenberg and N. de Haas, *J. Chem. Phys.*, **43**, 1550 (1965);
    (c) G. Dixon-Lewis, W. E. Wilson and A. A. Westenberg, *J. Chem. Phys.*, **44**, 2877 (1966);
    (d) A. A. Westenberg and W. E. Wilson, *J. Chem. Phys.*, **45**, 338 (1966)
57. A. Carrington, *Proc. Roy. Soc.*, **A302**, 291 (1968)
58. J. C. Boden and B. A. Thrush, *Proc. Roy. Soc.*, **A305**, 107 (1968)
59. J. U. White, *J. Opt. Soc. Am.*, **32**, 285 (1942)
60. T. C. Clark and M. A. A. Clyne, *Trans. Faraday Soc.*, **65**, 2994 (1969); ibid., *Trans. Faraday Soc.*, **66**, 372 (1970)
61. M. A. A. Clyne and J. A. Coxon, *Proc. Roy. Soc.*, **A303**, 207 (1968)
62. M. A. A. Clyne, J. A. Coxon and A. R. Woon-Fat, *Trans. Faraday Soc.*, **67**, 3351 (1971)
63. J. R. McNesby and H. Okabe, *Advan. Photochem.*, **3**, 157 (1964); see also J. A. R. Samson, Techniques of Vacuum Ultraviolet Spectroscopy (Wiley, 1967), chapter 5
64. D. A. Parkes, L. F. Keyser and F. Kaufman, *Astrophys. J.*, **149**, 217 (1967)
65. W. Braun and T. Carrington, *J. Quant. Spectry. Radiative Transfer*, **9**, 1133 (1969)
66. M. A. A. Clyne, J. A. Coxon and D. J. McKenney, to be published

67. A. A. Passchier, J. D. Christian and N. W. Gregory, *J. Phys. Chem.*, **71**, 937 (1967)
68. R. G. W. Norrish and B. A. Thrush, *Quart. Rev.*, **10**, 149 (1956)
69. G. Burns and J. K. K. Ip, *J. Chem. Phys.*, **51**, 3414 (1969); J. A. Blake and G. Burns, *J. Chem. Phys.*, **54**, 1480 (1971)
70. S. N. Foner, *Advan. At. Mol. Phys.*, **2**, 385 (1966)
71. A. C. G. Mitchell and M. W. Zemansky, Resonance Radiation and Excited Atoms (Camb. Univ. Press 1934)
72. G. B. Kistiakowsky and G. G. Volpi, *J. Chem. Phys.*, **27**, 1114 (1957)
73. D. S. Jackson and H. I. Schiff, *J. Chem. Phys.*, **21**, 2233 (1953)
74. S. N. Foner and R. L. Hudson, *J. Chem. Phys.*, **21**, 1608 (1953)
75. L. F. Phillips and H. I. Schiff, *J. Chem. Phys.*, **37**, 1233 (1962)
76. F. P. Lossing, chapter in Mass Spectrometry (Editor, McDowell), (McGraw-Hill, 1963)
77. See, for example, a review by J. B. Anderson, R. P. Andres and J. B. Fenn, in *Advan. At. Mol. Phys.*, **1**, 345 (1965)
78. S. N. Foner and R. L. Hudson, *J. Chem. Phys.*, **36**, 2681 (1962)
79. M. Gehring, K. Hoyermann, H. G. Wagner and J. Wolfrum, *Berichte physik Chem.*, **73**, 956 (1969)
80. S. N. Foner, *J. Chem. Phys.*, **49**, 3724 (1968)
81. F. C. Fehsenfeld, K. M. Evenson and H. P. Broida, *Rev. Sci. Instr.*, **36**, 294 (1965)
82. J. M. Brown, B. A. Thrush and A. F. Tuck, *Proc. Roy. Soc.*, **A302**, 311 (1968)
83. (a) F. A. Morse and F. Kaufman, *J. Chem. Phys.*, **42**, 1785 (1965); (b) C-L. Lin and F. Kaufman, *J. Chem. Phys.*, **55**, 3760 (1971)
84. See, for example, I. M. Campbell and B. A. Thrush, *Trans. Faraday Soc.*, **64**, 1275 (1968)
85. R. P. Wayne, *Advan. Photochem.*, **7**, 311 (1969)
86. L. W. Bader and E. A. Ogryzlo, *Discussions Faraday Soc.*, **37**, 46 (1964); M. A. A. Clyne, B. A. Thrush and R. P. Wayne, *Nature*, **199**, 1057 (1963)
87. F. S. Klein and J. T. Herron, *J. Chem. Phys.*, **41**, 1285 (1964)
88. J. T. Herron, *J. Chem. Phys.*, **45**, 1854 (1966)
89. J. T. Herron and R. D. Penzhorn, *J. Phys. Chem.*, **73**, 191 (1969)
90. L. F. Phillips and H. I. Schiff, *J. Chem. Phys.*, **36**, 1509 (1962)
91. L. F. Phillips and H. I. Schiff, *J. Chem. Phys.*, **41**, 3171 (1965)
92. C. G. Freeman and L. F. Phillips, *J. Phys. Chem.*, **72**, 3025 (1968)
93. H. Niki, E. E. Daby and B. Weinstock, 12th Symp. Combustion, 277 (1968); H. Niki and E. E. Daby, *J. Chem. Phys.*, **51**, 1255 (1969)
94. See, for example, J. W. Linnett and D. G. H. Marsden, *Proc. Roy. Soc.*, **A234**, 489, 504 (1956)
95. W. V. Smith, *J. Chem. Phys.*, **11**, 110 (1943)
96. E. L. Tollefson and D. J. LeRoy, *J. Chem. Phys.*, **16**, 1057 (1948)
97. (a) F. S. Larkin and B. A. Thrush, *Discussions Faraday Soc.*, **37**, 112 (1964); (b) F. S. Larkin, *Can. J. Chem.*, **46**, 1005 (1968)
98. L. Elias, E. A. Ogryzlo and H. I. Schiff, *Can. J. Chem.*, **37**, 1680 (1959)
99. L. W. Bader and E. A. Ogryzlo, *Discussions Faraday Soc.*, **37**, 46 (1964)
100. D. B. Gibbs and E. A. Ogryzlo, *Can. J. Chem.*, **43**, 1905 (1965)
101. M. A. A. Clyne and J. A. Coxon, *J. Mol. Spectry*, **23**, 258 (1967)
102. R. A. Young and R. L. Sharpless, *Discussions Faraday Soc.*, **33**, 228 (1962)
103. S. Ticktin, G. Spindler and H. I. Schiff, *Discussions Faraday Soc.*, **44**, 218 (1967)
104. M. A. A. Clyne and J. A. Coxon, *Proc. Roy. Soc.*, **A298**, 424 (1967)
105. Y. Tanaka, A. Jursa and F. LeBlanc, The Threshold of Space (Pergamon Press, 1957) p. 89
106. I. M. Campbell and B A. Thrush, *Trans. Faraday Soc.*, **65**, 32 (1969)
107. S. W. Benson, *J. Chem. Phys.*, **48**, 1765 (1968)

108. H. P. Broida, H. I. Schiff and T. M. Sugden, *Trans. Faraday Soc.*, **57**, 259 (1961)
109. R. R. Reeves, P. Harteck and W. H. Chace, *J. Chem. Phys.*, **41**, 764 (1964); D. Applebaum, P. Harteck and R. R. Reeves, *Photochem. Photobiol.*, **4**, 1003 (1965)
110. A. McKenzie and B. A. Thrush, *Chem. Phys. Lett.*, **1**, 681 (1968)
111. D. B. Hartley and B. A. Thrush, *Proc. Roy. Soc.*, **A297**, 520 (1967)
112. A. Fontijn, C. B. Meyer and H. I. Schiff, *J. Chem. Phys.*, **40**, 64 (1964)
113. P. N. Clough and B. A. Thrush, *Trans. Faraday Soc.*, **63**, 915 (1967); M. A. A. Clyne, B. A. Thrush and R. P. Wayne, *Trans. Faraday Soc.*, **60**, 359 (1964)
114. C. J. Halstead and B. A. Thrush, *Proc. Roy. Soc.*, **A295**, 380 (1965); ibid., *Photochem. Photobiol.*, **4**, 1007 (1965)
115. J. C. Polanyi, *Appl. Opt. Suppl.*, **2**, 109 (1965)
116. G. C. Pimentel and J. H. Parker, *J. Chem. Phys.*, **51**, 10 (1969)
117. J. C. Polanyi, *Discussions Faraday Soc.*, **44**, 293 (1967)
118. See, for example, K. G. Anlauf, P. J. Kuntz, D. H. Maylotte, P. D. Pacey and J. C. Polanyi, *Discussions Faraday Soc.*, **44**, 183 (1967)
119. P. E. Charters and J. C. Polanyi, *Can. J. Chem.*, **38**, 1742 (1960)
120. A. M. Bass and D. Garvin, *J. Mol. Spectry*, **9**, 114 (1962)
121. D. Garvin, H. P. Broida and H. J. Kostkowski, *J. Chem. Phys.*, **32**, 880 (1960)
122. P. N. Clough and B. A. Thrush, *Trans. Faraday Soc.*, **65**. 23 (1969)
123. S. J. Arnold, M. Kubo and E. A. Ogryzlo, *Advan. Chem. Ser.*, **77**, 133 (1968)
124. (a) R. E. March, S. G. Furnival and H. I. Schiff, *Photochem. Photobiol.*, **4**, 971 (1965);
     (b) S. J. Arnold, R. J. Browne, E. A. Ogryzlo and H. Witzke, *J. Chem. Phys.*, **40**, 1769 (1964)
125. A. M. Falick, B. H. Mahan and R. J. Myers, *J. Chem. Phys.*, **42**, 1837 (1965)
126. M. Jeunehomme and A. B. F. Duncan, *J. Chem. Phys.*, **41**, 1692 (1964)
127. B. A. Thrush, *J. Chem. Phys.*, **47**, 3691 (1967); K. L. Wray, *J. Chem. Phys.*, **44**, 623 (1966)
128. I. D. Clark and R. P. Wayne, *Proc. Roy. Soc.*, **A316**, 539 (1970)
129. S. J. Arnold and E. A. Ogryzlo, *Can. J. Phys.*, **45**, 2053 (1967); S. H. Whitlow and F. D. Findlay, *Can. J. Chem.*, **45**, 2087 (1967)
130. R. J. Derwent, D. R. Kearns and B. A. Thrush, *Chem. Phys. Lett.*, **6**, 115 (1970); M. A. A. Clyne, J. A. Coxon and H. W. Cruse, *Chem. Phys. Lett.*, **6**, 57 (1970)
131. P. N. Clough and B. A. Thrush, *Proc. Roy. Soc.*, **A309**, 419 (1969)
132. A. M. Bass and K. G. Kessler, *J. Opt. Soc. Am.*, **49**, 1223 (1959)
133. See, for example, K. Schofield and H. P. Broida, *Photochem. Photobiol.*, **4**, 989 (1965)
134. M. A. A. Clyne and J. A. Coxon, *J. Phys.*, **B3**, L9 (1970)
135. J. F. Prince, C. B. Collins and W. W. Robertson, *J. Chem. Phys.*, **40**, 2619 (1964); C. B. Collins and W. W. Robertson, *J. Chem. Phys.*, **40**, 701 (1964)
136. D. W. Setser and D. H. Stedman, *J. Chem. Phys.*, **49**, 469 (1968); ibid., *Chem. Phys. Lett.*, **4**, 542 (1968); D. H. Stedman, J. A. Meyer and D. W. Setser, *J. Am. Chem. Soc.*, **90**, 6856 (1968); M. A. A. Clyne, J. A. Coxon, D. W. Setser and D. H. Stedman, *Trans. Faraday Soc.*, **65**, 1177 (1969)
137. D. W. Setser, D. H. Stedman and J. A. Coxon, *J. Chem. Phys.*, **53**, 1004 (1970)
138. See, for example, article by H. Wise and B. J. Wood in *Advan. At. Mol. Phys.*, **3** (1968)
139. D. W. Setser and D. H. Stedman, *J. Chem. Phys.*, **52**, 3957 (1970); D. H. Stedman, *J. Chem. Phys.*, **52**, 3966 (1970)
140. J. A. Meyer, D. W. Setser and D. H. Stedman, *J. Phys. Chem.*, **74**, 2238 (1970)
141. R. L. Johnson, M. J. Perona and D. W. Setser, *J. Chem. Phys.*, **52**, 6372, 6384 (1970)
142. See, for example, references 7b, 8, 9, 10, 143, 165

143. M. A. A. Clyne, *Ann. Rept. Chem. Soc.*, **A65**, 165 (1968)
144. K. M. Evenson and D. S. Burch, *J. Chem. Phys.*, **45**, 2450 (1966)
145. I. M. Campbell and B. A. Thrush, *Proc. Roy. Soc.*, **A296**, 201 (1967); ibid., *Trans. Faraday Soc.*, **64**, 1275 (1968)
146. M. A. A. Clyne and D. H. Stedman, *J. Phys. Chem.*, **71**, 3071 (1967)
147. H. S. Johnston, Gas Phase Reaction Rate Theory (Ronald Press, New York, 1966)
148. J. C. Keck, *J. Chem. Phys.*, **29**, 410 (1958)
149. J. C. Light, *J. Chem. Phys.*, **36**, 1016 (1962)
150. J. T. Herron, J. L. Franklin, P. Bradt and V. H. Dibeler, *J. Chem. Phys.*, **30**, 879 (1959)
151. C. Mavroyannis and C. A. Winkler, *Can. J. Chem.*, **39**, 1601 (1961)
152. C. B. Kretschmer and H. L. Petersen, *J. Chem. Phys.*, **39**, 1772 (1963)
153. L. W. Bader and E. A. Ogryzlo, *Nature*, **201**, 491 (1964)
154. R. J. Browne and E. A. Ogryzlo, *J. Chem. Phys.*, **52**, 5774 (1970)
155. H. B. Dunford, *J. Phys. Chem.*, **67**, 258 (1963)
156. T. C. Marshall, *Phys. Fluids*, **5**, 743 (1962)
157. F. Kaufman and J. R. Kelso, *Discussions Faraday Soc.*, **37**, 26 (1964)
158. M. F. R. Mulcahy and D. J. Williams, *Trans. Faraday Soc.*, **64**, 59 (1968)
159. M. A. A. Clyne, D. J. McKenney and B. A. Thrush, *Trans. Faraday Soc.*, **61**, 2701 (1965)
160. F. Kaufman and J. R. Kelso, Symp. Chemiluminescence, Durham, North Carolina (1965)
161. T. C. Clark, M. A. A. Clyne and D. H. Stedman, *Trans. Faraday Soc.*, **62**, 3354 (1966)
162. T. C. Clark and M. A. A. Clyne, *Chem. Commun.*, 287 (1966)
163. C. J. Halstead and B. A. Thrush, *Proc. Roy. Soc.*, **A295**, 363 (1966)
164. M. A. A. Clyne and J. A. Coxon, *Proc. Roy. Soc.*, **A303**, 207 (1968)
165. J. A. Kerr, *Ann. Rept. Chem. Soc.*, **A64**, 73 (1967)
166. K. Schofield, *Planetary Space Sci.*, **15**, 643 (1967)
167. W. R. Schulz and D. J. Le Roy, *J. Chem. Phys.*, **42**, 3879 (1965); ibid., *Can. J. Chem.*, **42**, 2480 (1964); B. A. Ridley, W. R. Schulz and D. J. LeRoy, *J. Chem.* 44 3344 (1966); D. J. LeRoy, B. A. Ridley and K. A. Quickert, *Discussions Faraday Soc.*, **44**, 92 (1967)
168. A. A. Westenberg and N. de Haas, *J. Chem. Phys.*, **47**, 1393 (1967)
169. R. E. Weston, Jr., *J. Chem. Phys.*, **31**, 892 (1959)
170. E. L. Wong and A. E. Potter, *J. Chem. Phys.*, **43**, 3371 (1965)
171. A. A. Westenberg and N. de Haas, *J. Chem. Phys.*, **50**, 2512 (1969)
172. H. Steiner and E. K. Rideal, *Proc. Roy. Soc.*, **A173**, 503 (1939); W. H. Rodebush and W. C. Klingelhoefer, *J. Am. Chem. Soc.* **55**, 130 (1933); P. G. Ashmore and J. Chanmugam, *Trans. Faraday Soc.*, **49**, 254 (1953)
173. S. W. Benson, F. R. Cruickshank and R. Shaw, *Int. J. Chem. Kinetics*, **1**, 29 (1969)
174. H. I. Schiff, *Ann. Geophys.*, **20**, 115 (1964)
175. O. R. Lundell, R. D. Ketcheson and H. I. Schiff, 12th Symp. Combustion, 307 (1968)
176. D. Garvin and H. P. Broida, 9th Symp. Combustion, 678 (1962)
177. See, for example, B. S. Rabinovitch and D. W. Setser, *Advan. Photochem.*, **3**, 1 (1964)
178. W. Braun and M. Lenzi, *Discussions Faraday Soc.*, **44**, 252 (1967)
179. C. J. Halstead and B. A. Thrush, *Proc. Roy. Soc.*, **A295**, 363 (1966)
180. See reference 7b for recalculated value originally published in reference 5
181. M. A. A. Clyne and J. A. Coxon, *Trans. Faraday Soc.*, **62**, 1175 (1966)
182. H. S. Johnston, E. D. Morris and J. Van den Bogaerde, *J. Am. Chem. Soc.*, **91**, 7712 (1969)
183. M. A. A. Clyne and R. F. Walker, to be published

184. A. P. Modica and D. F. Hornig, *J. Chem. Phys.*, **49**, 629 (1968)
185. R. E. Walker, *Phys. Fluids*, **4**, 1211 (1961)
186. R. V. Poirier and R. W. Carr, *J. Phys. Chem.*, **75**, 1593 (1971)
187. J. E. Breen and G. P. Glass, *J. Chem. Phys.*, **52**, 1082 (1970)
188. M. F. R. Mulcahy and R. H. Smith, *J. Chem. Phys.*, **54**, 5215 (1971)
189. M. A. A. Clyne and H. W. Cruse, *J.C.S. Faraday* II, **68**, 1281 (1972)
190. M. A. A. Clyne and H. W. Cruse, *J.C.S. Faraday* II, **68**, 1377 (1972)
191. E. E. Ferguson, F. C. Fehsenfeld and A. L. Schmeltekopf, *Adv. At. Mol. Phys.*, **5**, 1 (1969)
192. K. H. Becher, E. H. Fink, W. Groth, W. Jud and D. Kley, *Discuss. Faraday Div.*, **53** (1972)
193. M. A. A. Clyne, H. W. Cruse and R. T. Watson, *J.C.S. Faraday* II, **68**, 153 (1972)
194. M. A. A. Clyne and H. W. Cruse, *Trans. Faraday Soc.*, **67**, 2869 (1971)
195. P. P. Bemand and M. A. A. Clyne, to be published
196. D. D. Davis and W. Braun, *Appl. Optics*, **7**, 2071 (1968)
197. M. A. A. Clyne and A. R. Woon-Fat, *J.C.S. Faraday* II, submitted
198. M. A. A. Clyne and R. T. Watson, to be published
199. J. R. Kanofsky, D. Lucas, F. Pruss and D. Gutman, 14th Symp. Combustion (1972), in press
200. I. T. N. Jones and K. D. Bayes, 14th Symp. Combustion (1972), in press
201. H. Gg. Wagner, C. Zetzsch and J. Warnatz, *Ber. phys. Chem.*, **76**, 526 (1972)
202. D. O. Ham, D. W. Trainor and F. Kaufman, *J. Chem. Phys.*, **53**, 4395 (1970)
203. I. M. Campbell and S. B. Neal, *Discuss. Faraday Div.* **53**, 000 (1972)
204. K. H. Becker, W. Groth and D. Thran, 14th Symp. Combustion (1972), in press
205. N. Jonathan, A. Morris, K. J. Ross and D. J. Smith, *J. Chem. Phys.*, **54**, 4954 (1971)
206. M. A. A. Clyne, J. A. Coxon and L. W. Townsend, *J.C.S. Faraday* II, **69** (1973) in press.
207. M. J. Perona, *J. Chem. Phys.*, **54**, 4024 (1971)
208. H. W. Chang, D. W. Setser and M. J. Perona, *Chem. Phys. Letts.*, **9**, 587 (1971)
209. D. W. Setser and D. H. Stedman, *Prog. Reaction Kinetics*, **6**, (1971)
210. J. Troe and H. G. Wagner, this volume, chapter 1
211. A. A. Westenberg and N. de Haas, *J. Phys. Chem.*, **76**, 1586 (1972)
212. W. C. Richardson, D. W. Setser, D. L. Allbritton and A. L. Schmeltekopf, *Chem. Phys. Letts.*, **12**, 349 (1971)
213. M. A. A. Clyne and J. Conner, *J.C.S. Faraday* II, **68**, 1220 (1972)
214. R. E. Huie, private communication (H + O_2 + M, O + O_2 + M); H. Niki and F. Stuhl, *J. Chem. Phys.*, **55**, 3943 (1971)
215. T. Hikida, J. A. Eyre and L. M. Dorfman, *J. Chem. Phys.*, **54**, 3422 (1971)— (H + NO + M, H + O_2 + M)
216. I. Shavitt, *J. Chem. Phys.*, **49**, 4048 (1968); K. A. Quickert and D. J. Le Roy, *J. Chem. Phys.*, **53**, 1325 (1970)
217. Since this section was written (July 1970), a somewhat similar interpretation has been advanced by N. S. Snider, *J. Chem. Phys.*, **53**, 4116 (1970)
218. Including the recent value of W. Braun, A. M. Bass and D. D. Davis, *Int. J. Chem. Kin.*, **2**, 101 (1970)
219. Ar $^3P_2$ has been observed using e.p.r. spectrometry, by W. H. Breckenridge and T. A. Miller, *Chem. Phys. Letts.*, **12**, 437 (1972)
220. M. A. A. Clyne and H. W. Cruse, to be published
221. H. G. Wagner, C. Zetzsch and J. Warnatz, *Ber. phys. Chem.*, **76**, 526 (1972)
222. D. Beck and H. J. Loesch, *Ber. phys. Chem.*, **75**, 736 (1971); Y. T. Lee, J. D. McDonald, P. R. Le Breton and D. R. Herschbach, *J. Chem. Phys.*, **49**, 2447 (1968); **51**, 455 (1969); N. C. Blais and J. B. Cross, *J. Chem. Phys.*, **50**, 4108 (1969); **52**, 3580 (1970)

223. M. 1. Christie, R. S. Roy and B. A. Thrush, *Trans. Faraday Soc.*, **55,** 1139 (1959); **55,** 1149 (1959); W. G. Burns and F. S. Dainton, *Trans. Faraday Soc.*, **48,** 52 (1952)
224. M. J. Kurylo, N. C. Peterson and W. Braun, *J. Chem. Phys.*, **53,** 2776 (1970)
225. (a) A. A. Westenberg, private communication; (b) F. Kaufman, private communication; (c) M. A. A. Clyne and S. Down, to be published
226. However, N. Basco and Dogra, *Proc. Roy. Soc.* A, **323,** 401 (1971), using flash photoysis and photographic absorption spectrometry to follow [ClO], found no pressure effect on the rate constant k_5 in the same range of pressures as ref. 182
227. M. A. A. Clyne and R. T. Watson, *Chem. Phys. Letts.*, **12,** 344 (1971)
228. H. Gg. Wagner, C. Zetzsch and J. Warnatz, *Angew. Chemie*, **10,** 564 (1971)
229. M. F. Golde and B. A. Thrush, *Discuss. Faraday Div.*, **53** (1972)

Subject Index